Analytical Techniques for Biopharmaceutical Development

Analytical Techniques for Biopharmaceutical Development

Edited by

Roberto Rodriguez-Diaz
Dynavax Technologies, Berkeley, California

Tim Wehr
Bio-Rad Laboratories, Hercules, California

Stephen Tuck
Dynavax Technologies, Berkeley, California

informa
healthcare

New York London

FIRST INDIAN REPRINT 2008

Although great care has been taken to provide accurate and current information, neither the author(s) nor the publisher, nor anyone else associated with this publication, shall be liable for any loss, damage, or liability directly or indirectly caused or alleged to be caused by this book. The material contained herein is not intended to provide specific advice or recommendations for any specific situation.

Trademark notice: Product or corporate names may be trademarks or registered trademarks and are used only for identification and explanation without intent to infringe.

Library of Congress Cataloging-in-Publication Data
Analytical techniques for biopharmaceutical development / Tim Wehr, Roberto rodriguez-Diaz, Stephen Tuck, editors.
    p. ; cm.
    Includes bibliographical references and index.
    ISBN 0-8247-2667-7 (alk. paper)
  1. Protein drugs--Analysis--Laboratory manuals.
    [DNLM: 1. Pharmaceutical Preparations--analysis--Laboratory Manuals. 2. Biopharmaceutics--methods--Laboratory Manuals. 3. Chromatography--methods--Laboratory Manuals. 4. Electrophoresis--methods--Laboratory Manuals. 5. Spectrum Analysis--methods--Laboratory Manuals. QV 25 A5338 2005] I. Wehr, Tim. II. Rodríguez-Díaz, Roberto. III. Tuck, Stephen (Stephen F.)
    RS431.P75A536 2005
    615′.7--dc22   2004058292

ISBN: 0-8247-0706-0

Headquarters
Marcel Dekker, Inc., 270 Madison Avenue, New York, NY 10016, U.S.A.
tel: 212-696-9000; fax: 212-685-4540

World Wide Web
http://www.dekker.com

Printed in India by Saurabh Printers Pvt. Ltd.

FOR SALE IN SOUTH ASIA ONLY

Contents

About the Editors

Roberto Rodriguez-Diaz is Senior Scientist at Dynavax Technologies in Berkeley, California. He has extensive experience in product development in the biopharmaceutical industry, and his research has focused on development of analytical methodology ranging from determination of low molecular weight reactants to analysis of protein-oligonucleotide conjugates. He holds a B.S. degree from the University of Michoacan, Morelia, Mexico.

Tim Wehr is Staff Scientist at Bio-Rad Laboratories in Hercules, California. He has more than 20 years of experience in biomolecule separations, including development of HPLC and capillary electrophoresis methods and instrumentation for separation of proteins, peptides, amino acids, and nucleic acids. He has also worked on development and validation of LC-MS methods for small molecules and biopharmaceuticals. He holds a B.S. degree from Whitman College, Walla Walla, Washington, and earned his Ph.D. from Oregon State University in Corvallis.

Stephen Tuck is Vice President of Biopharmaceutical Development at Dynavax Technologies in Berkeley, California. He has over 14 years of experience in pharmaceutical chemistry. He was involved in the development of Fluad™ adjuvated flu vaccine as well as various subunit vaccines, adjuvants, vaccine conjugates, and protein therapeutics. He earned his B.Sc. and Ph.D. degrees from Imperial College, University of London, United Kingdom.

Contributors

Alain Balland   Analytical Sciences, Amgen, Seattle, Washington

Emil W. Ciurczak   Integrated Technical Solutions, Golden's Bridge, New York

Pete Gagnon   Validated Biosystems, Inc., Tucson, Arizona

David E. Garfin   Garfin Consulting, Kensington, California

Fiona Haycock   Dynavax Technologies, Berkeley, California

Claudia Jochheim   Analytical Biochemistry, Corixa, Seattle, Washington

Yung-Hsiang Kao   Genentech, Inc., South San Francisco, California

Joanne Rose Layshock   Chiron Corporation, Emeryville, California

Rowena Ng   Dynavax Technologies, Berkeley, California

Richard L. Remmele, Jr.   Pharmaceutics, Amgen, Inc., Thousand Oaks, California

Roberto Rodriguez-Diaz   Dynavax Technologies, Berkeley, California

Stephen Tuck   Dynavax Technologies, Berkeley, California

Martin Vanderlaan   Genentech, Inc., South San Francisco, California

Tim Wehr   Bio-Rad Laboratories, Hercules, California

Ping Wong   Genentech, Inc., South San Francisco, California

Mingde Zhu   Bio-Rad Laboratories, Hercules, California

1

Analytical Techniques for Biopharmaceutical Development

Stephen Tuck

INTRODUCTION

Before the days of mass literacy, medicine was more art than science, and people recognized a pharmacy by the four traditional colored bottles that represent earth, fire, air, and water. Medicine has come a long way since the days of the apothecary with its impressive collection of powders and bottles; drugs today are highly regulated and must comply with standards set by the U.S., Europe, Japan, and other countries. A drug must be shown to be efficacious and meet rigorous standards of purity, composition, and potency before being approved for use in the patient population. These regulations provide confidence to the patient that a prescribed medicine will achieve its therapeutic goal. Whether in the form of a pill, a capsule, an injection, a tablespoon of syrup, or an inhaler, analytical methods ensure the identity, purity, potency, and ultimately the performance of these drugs.

Analytical methods are important not only in the development and manufacture of commercial biopharmaceutical drugs, they also play a vital role in the whole drug development life cycle. Drug discovery and preclinical research require development and application of analytical methodologies to support identification, quantitation, and characterization of lead molecules. It is difficult to perform a comparative potency assay on lead molecules if one does not know how much of each is going into the assay or how pure the molecule is. Analytical methods are typically developed, qualified, and validated in step with the clinical phase of the molecule. Techniques used during discovery and preclinical development will be qualified for basic performance. When the drug is approaching early human clinical trials, and compliance to regulations becomes the order of the day, the analytical scientist begins developing assays that International Conference on Harmonization (ICH) guidelines define as "appropriate for their intended applications." Analytical methods will be required for characterizing the protein's physical-chemical and biological properties, developing stable formulations, evaluating real-time and accelerated stability, process development, process validation, manufacturing, and quality control.

The objective of this book is to provide both an overview and practical uses of the techniques available to analytical scientists involved in the development and application of methods for protein-based biopharmaceutical drugs. The emphasis is on considering the analytical method in terms of the stage of the development process and its appropriateness for the intended application. The availability of techniques will reveal whether or not the analytical problem has a potential solution. Then will come the question of whether or not the technique is a truly appropriate solution. The theoretical considerations behind choosing the technique may be solid. However, the practicality of the method may not hold up to inspection.

Consider this question: "Can one develop a stable 2 to 8°C formulation of a protein that has a propensity to aggregate and lose activity?" Several challenges face the analytical scientist. Activity is obviously a key stability-indicating assay, but is best used as a confirmatory assay because it is usually an expensive *in vitro* or *in vivo* assay that is time consuming and may not be sensitive enough to differentiate between formulations. A highly automatable or high-throughput surrogate assay would be more appropriate if it can be demonstrated to be stability indicating and to correlate with activity. If one simply wants to detect a conformational change, then there are techniques; one might consider nuclear magnetic resonance (NMR) as a technique that has the potential for detecting such a structural change. On further inspection, NMR is a technique that requires expensive equipment, highly trained operators, and significant quantities of protein. In addition, sample throughput time is slow, so all of these factors suggest that it is probably not a good screening assay for sample-intensive formulation-screening studies. However, NMR could be a good assay for characterizing the structure of the molecule and confirming that its conformation has changed. NMR is an assay requiring serious consideration prior to development, whereas gel electrophoresis is a workhorse method that is used throughout the development process and across many areas.

Each chapter of this book describes an analytical technique and discusses its basic theory, applications, weaknesses and strengths, and advantages and disadvantages, and, where possible, compares it to alternate methods. The aim is not to go into significant theoretical considerations of the technique, but rather to provide information on how and when to apply the method with examples.

The basic theory allows the reader to discern what considerations need to be addressed in order to evaluate the technique for the application at hand.

The chapters are organized to follow the order in which one might need to employ the methods during the biopharmaceutical development cycle. It is difficult to do much of anything analytical with a protein if one does not know its concentration. How much is being loaded on the gel, or how much protein is being expressed in cell culture? Purity is also an early key consideration. Developing an impurity in your protein is a simple recipe for disaster. Column chromatography and electrophoresis are the most common techniques for assessing purity and can be used orthogonally. A protein with an assigned concentration and known purity becomes much easier to develop. Finally, the "fine-tuning" assays that are used to characterize the properties of the protein, which affect such attributes as stability and activity, are described.

Analytical scientists will provide support for many of the activities in a biopharmaceutical company. They are responsible for characterizing the molecules in development, establishing and performing assays that aid in optimization and reproducibility of the purification schemes, and optimizing conditions for fermentation or cell culture to include product yields. Some of the characterization techniques will eventually be used in quality control to establish purity, potency, and identity of the final formulation. The techniques described here should provide the beginning of a palette from which to develop analytical solutions.

2

Introduction to the Development of Biopharmaceuticals

Roberto Rodriguez-Diaz

INTRODUCTION

Although the purpose of this book is not to serve as a guideline for all aspects of biopharmaceutical development, and even less as a guideline to regulatory compliance, acquiring a general idea of these subjects is of increasing importance to the understanding of the development of biological drugs and ultimately to the role of analysts in the process. The following introduction is meant as a bird's-eye view of the landscape for scientists who are new to the field or are removed from big-picture considerations of their particular projects.

Biopharmaceutical development involves a complex interaction of multiple entities. At the core of the interaction are the pharmaceutical company and the regulatory agencies. Often, companies designate a specialized group of individuals to serve as an interface with the regulatory agencies. The function of this group is to stay up-to-date with continuously evolving regulations and to assess the impact those changes have in the company's programs.

The company is composed of multiple groups of people with experience in one or more of several disciplines working together to transform a drug candidate into a product designed to improve the quality of people's lives and ensure commercial success.

The journey from end of discovery to commercialization is the development process of biopharmaceuticals. Scientists are an integral part of the company assessing the advantages and drawbacks of a candidate molecule from the discovery laboratory to the market. These individuals must become proficient not only in their particular discipline (e.g., laboratory techniques) but must be aware of the regulations that affect their particular work and the project as a whole. Because regulations affect the analyst's role, a common impression is that pharmaceutical development is a rigid discipline. But in reality, regulations are seldom used as unmodifiable templates that ultimately lead to development success. This is because pharmaceuticals, and especially biological products, are fairly unique, and thus require a tailored developmental scheme. In other words, what works for the development of one drug does not necessarily work for the development of another one. For example, highly specific issues about drugs often necessitate a "case-by-case" approach by regulatory agencies. What is most important for a company is to demonstrate the reasoning, safety profile, and impact of the issue on the decision-making process. This leads to a final regulatory package that is based on the characteristics of the drug itself. This is often frustrating to newcomers because there is an impression that the regulatory guidelines are marred with vagueness. However, in the production of pharmaceutical molecules the industry (the science) and the Federal Drug Administration (FDA) (the regulations) establish a close interaction in which the participants mold (or at least influence) each other and issues get clarified as the process progresses (the disagreements and the need for agreements usually prevent the drug development process from being smooth). To be effective, regulations must be based in good science. To provide safety and evaluate benefits, regulations must force the industry into performing good science. Nobody debates these two points. The major roadblock to this philosophy is to reach an agreement as to what (or who) defines good science.

Bioanalytical laboratories provide support for most of the activities at the biopharmaceutical company. They are responsible for characterizing the molecules in development, establishing and performing assays that aid in the optimization and reproducibility of the purification schemes, and optimizing the conditions for fermentation or cell culture, including product yields. Some of the characterization techniques will eventually be used in quality control to establish the purity, potency, and identity of the final formulation.

Because a great deal of the characterization knowledge resides in the analytical laboratory, this is where most stability and formulation work occurs. It is not unusual for the bioanalytical laboratory to be involved in the support of clinical studies (i.e., patient sample analysis).

Biopharmaceutical companies are highly diverse not only in their products but also in their size, capabilities, and approaches to development. Some biopharmaceutical companies (especially small companies) outsource part or most of the analytical work, some outsource manufacture and filling, and some invest and develop expertise to do everything in their own facilities. Most biotechnology companies are small, and sometimes it is faster and more cost-effective to outsource a task that requires expertise or expensive pieces of laboratory equipment not available in the company. Nevertheless, the analysts in the company will interface with the contract laboratory to ensure that proper assays are performed, and most likely will participate in the decision-making process derived from the data obtained.

Because biopharmaceutical development is a lengthy, expensive process, the odds of commercialization for a drug are maximized by developmental groups screening large numbers of compounds, each typically produced at the bench scale necessary for activity testing. Once a molecule is chosen, product development is initiated. During development there are a number of overlapping activities that include process development (production of the drug), analytical testing, formulation (conditions that preserve the activity of the molecule), physical-chemical characterization (e.g., molecular size and shape), preclinical and clinical studies (to define or elucidate toxicity, bioactivity, and bioavailability), and regulatory activities (including process validation, equipment qualification, quality assurance and quality control, and documentation). The correct performance of all these activities is vital to the successful development of new pharmaceuticals.

Process development of biopharmaceuticals is particularly challenging because biomolecules are too complex to be manufactured by traditional chemical synthesis. Biopharmaceuticals produced by living cells or cultures can be heterogeneous and exhibit characteristics that can change over time even if the same system is used to generate the product. For small molecules, analytical techniques can be used at the end of the production process to characterize (define) the product. Because of the complexity of biopharmaceuticals, this approach is difficult to implement. Instead, it is hypothesized that a consistent manufacturing process will yield a consistent product. So, for the production of biologics, more emphasis is placed on the manufacturing details (which encompass the chemistry, manufacturing, and controls section of regulatory applications). The more robust a manufacturing process, the less need for characterization of the end product.

Other concepts of high importance in biopharmaceutical development are formulation, stability, and delivery. This is because proteins are highly complex biomolecules that are sensitive to their environment (defined by the drug's formulation and storage). A formulation is developed in the preclinical stage and evaluated continuously until final approval of the product. The key aspects of formulation are based on determination of the stability of the drug in the presence of particular conditions or excipients or both. Usually accelerated stability and intended storage conditions studies are performed. In these assays, the effect of exposure to physical and chemical agents (such as heat and light) on the drug is evaluated. These studies require techniques capable of resolving impurities generated during exposure of the sample to harsh conditions. Such methods are said to be stability-indicating. These methods may be the same or different from those used to resolve and detect impurities generated during production of the drug.

It is important for development scientists to familiarize themselves with the regulatory process, which defines the development stages of a biopharmaceutical. Along this path there are several checkpoints that must be passed before reaching the next plateau. These checkpoints (or phases) affect all groups within a company. For example, for methods development and characterization scientists, the required level of knowledge on the behavior of methods in establishing the structure, purity, potency, and stability of the drug increases as the process advances.

Because bioanalytical methods constantly improve, development scientists' ability to find impurities increases. The cycle could go on forever, and a drug may never be considered truly pure. Developers must strike a balance between creating a process that is far too complex and expensive (both for the manufacturer and ultimately for the patients) and one that will produce a safe drug.

STAGES OF THE DRUG DEVELOPMENT PROCESS

The following paragraphs provide a brief and simplified description of the different stages of the development process. The concepts are presented here to position everyday work within the greater perspective of drug development.

Drug discovery — Although some drugs are developed by fairly large biotechnology companies, most of the promising drug candidates in development were discovered in academic research laboratories, often as a result of disease investigation, and not because of active research pursuing particular drugs for their activity. It is common to identify the individuals who discover the drug as part of the founders of a company. At this stage, drugs are usually produced in small quantities used for activity studies. Some characterization and analytical drug testing is necessary to ensure that the observed results are due to the drug itself and that the observed activity is reproducible.

Preclinical development — Preclinical development is charged with defining the initial safety and activity profiles of promising new drugs. The industry is hard at work developing alternative systems to evaluate drugs, but at present the bioactivity and efficacy of a protein therapeutic can only be determined through testing in biological systems (animal studies). One of the first characteristics to be evaluated after activity is the toxicity profile and pharmacokinetics of the drug. Toxicity studies are used to determine the safe range of dosing for initial (phase I) clinical trials in humans. Pharmacokinetics studies provide data on absorption, distribution, metabolism, and excretion (ADME) of the drug. At least two different species of animals (typically, the early studies are performed in rodents, and the late, more expensive studies in nonhuman primates) are used in toxicity testing of biopharmaceuticals. At the preclinical stage, the production group actively evaluates processes that are potentially suitable for the generation of the lead molecule. Communication between the preclinical and process development groups is crucial because production modifications may result in activity changes. The portfolio of techniques, which is a work in progress at this stage, is used to continue product characterization and often to evaluate discrepancies in activity (e.g., when there are no physical-chemical changes detected, yet the biological assays indicate large differences between two preparations of the same drug).

Investigational new drug application — There are two major regulatory documents in the life of a pharmaceutical biological product. One is the Investigational New Drug (IND) application, and the other is the Biologic License Application (BLA). The IND document is required because companies cannot administer drugs to humans without FDA authorization; thus, the IND application is a company's request to regulatory bodies to allow the exposure of volunteers and patients to the drug under study. The IND designation is a "living document" in the sense that there is flow of information during its different stages of development. It is in effect until the approval of the drug for commercialization (at which point a BLA is filed) or until the company decides to stop clinical trials for the drug.

The drug development phases are aimed at determining the safety profile, dosage range, clinical end points, ADME, and effectiveness (efficacy) of the drug candidate. Drug development is a *Process*, and therefore information, data, and knowledge are accumulated over time. Thus, it should be anticipated that many things will be reevaluated as the drug progresses from one phase to another and that unforeseen issues may result that require resolution before continuation of the studies.

Phase I clinical development — Phase I clinical development is carried out in a relatively small group of volunteers and patients (usually 15 to 100), where the main goal of the trial is to establish the safety characteristics of the molecule. Because the number of volunteers is low, only frequent adverse effects are observed. It is also common to explore dose range and dose scheduling during phase I. Often doses below the expected treatment level are used first for safety reasons and the amount of drug is increased over time. This phase is initiated 30 days after submission of the IND to the FDA, unless the regulatory agency has concerns about toxicity or the design of the study. If this occurs, the clinical trial is put on hold until the issue is resolved. Clinical trials are usually double-blinded (the clinicians and patients do not know if the substance administered contains drug or placebo). The trials are blinded to minimize the so called placebo effect, in which patients respond to the treatment even in the absence of true drug effect. This response can be positive (the patient feels better) or negative (the patient has adverse effects), and thus can blur the true benefits and risks of the biopharmaceutical.

During phase I the analytical laboratory continues characterization of the drug molecule and optimization and refinement of the methodology. Production is also refining the process to increase purity and yield and make it amenable to scaling up. Formulation studies usually consist of excipient screening during this phase.

Phase II clinical development — Phase II involves a greater number of patients (usually 100 to 300) than phase I clinical studies. At this stage more emphasis is placed on activity, dosing, and efficacy than in phase I, and thus, only patients are used for phase II studies and beyond. Sometimes, reevaluation of the safety profile may be necessary when initially testing on a new population (i.e., if only healthy volunteers were used in phase I, and phase II is performed only on the target population). Just as in phase I, phase II studies are usually blinded, but at this stage a control group and multiple centers of study may be added depending on the complexity of the trial, which in turn depends on the observations achieved during phase I. Phase II studies are also useful in identifying populations that will be more likely to benefit from the treatment. Because the number of subjects is higher, phase II studies can reveal adverse effects that are less common but not large enough to gather unambiguous statistical information to prove efficacy and safety.

The production and purification groups continue to evaluate raw materials and purification processes and to perform lot-release tests. More emphasis is placed on manufacturing scale-up as phase II studies progress. Quality assurance, quality control, compliance and regulatory affairs, and clinical development are actively preparing to organize the package of information describing drug characteristics and activity data, in preparation for the post-phase II meeting with the regulatory agencies.

Phase III clinical development — Phase III clinical studies are conducted in fairly large groups of individuals. The trial can be designed to provide data that support the licensure to market the drug (pivotal trial), or it can be used to further define the characteristics of the molecule in a clinical setting (e.g., when a larger group of individuals is needed to establish the efficacy or dosing of the drug). The number of patient volunteers needed for phase III trials depends on many factors, but most studies enroll 1000 to 3000 individuals. The goal of phase III studies is to gather enough evidence on the risk–benefit relationship in the target population. In this phase, long-term effects are analyzed for drugs that are intended for multiple- or extended-time usage. Phase III trials are expensive, and therefore only drugs with a very high potential for commercialization are evaluated.

During phase III the analytical laboratory performs systematic methods validation and continues with product characterization. A suitable formulation or a formulation candidate is in place and testing for stability continues. Production evaluates the consistency of the manufacturing process, which should be at a scale capable of delivering commercial quantities. Advanced studies are continued or initiated to evaluate chronic toxicology and reproductive side effects in animal models. Parallel to phase III studies, preparations are made for the submission of the BLA.

Biologics license application (BLA) — In this document, nonclinical, clinical, chemical, biological, manufacturing, and related information is included. The goal of the manufacturer is to demonstrate that the drug is safe and effective, and the manufacturing and quality control are appropriate to ensure identity, strength, potency and purity, consistency of the process, and adequate labeling. The BLA is supported by all the data collected during the clinical trials, but because phase III studies are larger and thus statistically more significant, more emphasis is placed on them. The BLA is a request to market a new biologic product, and it contains data which demonstrate that the benefits of the drug outweigh any adverse effects. Because a large amount of information needs to be reviewed (and therefore presented in a clear manner), a pre-BLA meeting is scheduled with regulatory agencies.

Phase IV clinical surveillance — When the drug has reached the market, further studies are conducted to create profiles on adverse effects, evaluate the drug's long-term effects, and further tune dosage for maximum efficacy. Potential interactions with other therapies are monitored closely. By observing the behavior of the drug after introduction to the general public or by extending the use of the product to populations not included in the trials, sometimes additional indications are uncovered or confirmed. Safety and efficacy comparisons to existing therapies may also be performed.

RECOMMENDED READING

1. *Biologics Development: a Regulatory Overview*, Mark Mathieu, Ed., Paraxel International Corporation, Waltham, MA, 1997.
2. *Validation and Qualification in Analytical Laboratories*, Ludwig Huber, Interpharm Press, Buffalo Grove, IL, 1999.
3. *The Biopharm Guide to Biopharmaceutical Development,* A supplement to Bio-Pharm, Patrick Clinton, editor-in-chief, 2002.
4. *The Biopharm Guide to GMP History*, 2nd ed., A supplement to BioPharm, S. Anne Montgomery, editor-in-chief, 2002.
5. *Analytical Chemistry and Testing, a Technology Primer,* supplement to Pharmaceutical Technology, John S. Haystead, editor-in-chief, Advanstar.

3

Protein Assay

Stephen Tuck and Rowena Ng

INTRODUCTION

If analytical methods are at the heart of biopharmaceutical development and manufacturing, then protein concentration methods are the workhorse assays. A time and motion study of the discovery, development, and manufacture of a protein-based product would probably confirm the most frequently performed assay to be protein concentration. In the 1940s Oliver H. Lowry developed the Lowry method while attempting to detect miniscule amounts of substances in blood. In 1951 his method was published in the *Journal of Biological Chemistry*. In 1996 the Institute for Scientific Information (ISI) reported that this article had been cited almost a quarter of a million times, making it the most cited research article in history. This statistic reveals the ubiquity of protein measurement assays and the resilience of an assay developed over 60 years ago. The Lowry method remains one of the most popular colorimetric protein assays in biopharmaceutical development, although many alternative assays now exist.

As described in the following chapter, there are many biopharmaceutical applications of protein assays. Assigning the protein concentration for the drug substance, drug product, or in-process sample is often the first task for subsequent analytical procedures because assays for purity, potency, or identity require that the protein concentration be known. Hence it is typical for several different methods to be employed under the umbrella of protein concentration measurement, depending on the requirements of speed, selectivity, or throughput. The protein concentration is valuable as a stand-alone measurement for QC and stability of a protein. However, protein concentration methods provide no valuable

13

information with respect to conformation or structure beyond the different affinities of proteins for the various dyes used.

Fortunately, protein concentration methods are relatively simple (low-tech) and inexpensive. The simplest assays require only a spectrophotometer calibrated for wavelength and absorbance accuracy, basic laboratory supplies, and good pipetting techniques. Protein concentration assays are quite sensitive, especially given the typical detection limits required for most biopharmaceuticals.

What follows is not an exhaustive or up-to-the-minute survey of the methods available for protein quantitation, but a practical guide to selecting the appropriate assay for each stage of drug development. A case study further illustrates the application of the standard protein methods to the drug development process. The reader is referred to reviews on the topic for further details.[1,2]

PROTEIN ASSAY METHODS

There are five categories of protein assay: colorimetric assays, direct absorbance methods, fluorescence methods, amino acid analysis, and custom quantitation methods. A brief summary of the principles, advantages, and limitations of these methods follows.

Colorimetric Assays

The colorimetric methods depend on a chemical reaction or interaction between the protein and the colorimetric reagent. The resulting generation of a chromophore, whose intensity is protein-concentration dependent, can be quantified using a spectrophotometer. Beer's Law is employed to derive the protein concentration from a standard curve of absorbances. Direct interaction of the protein with a chromogenic molecule (dye) or protein-mediated oxidation of the reporter molecule generates a new chromophore that can be readily measured in the presence of excess reagent dye.

If the concentration of the test protein is in the 100- to 2000-μg/ml range and >50 μg of sample are available, sensitivity is not a problem for colorimetric methods and a few samples can be accurately measured in 2 to 3 h by any of the commercially available assays described in the following subsections. Protein concentrations lower than 100 μg/ml require the use of microassays which, when compared with their regular counterparts, may require more sample volume, longer reaction times, and higher incubation temperatures. A common drawback of the larger sample volume is a greater potential for interference by the increased amounts of excipients present in the final reaction volume. Alternatively, techniques can be employed to increase the concentration of samples prior to analysis, usually with the added advantage that interfering excipients are removed in the process. One such example is protein precipitation with acetone or trichloroacetic acid. However, the additional sample handling will probably decrease the accuracy and precision of the final result, and protein recovery studies should be performed.

With respect to accuracy, these colorimetric methods require calibration of the absorbance of the chromophore that is created by the protein–reagent interaction. This is typically achieved by preparing a standard curve with either a readily available standard protein or the target protein itself. If a standard protein such as bovine serum albumin (BSA) is to be used, a correction factor will need to be determined to generate an accurate value for the target protein. This can be achieved by using amino acid analysis to establish a true value for the target protein, comparing it with the value obtained for the target protein from the BSA standard curve, and then generating the correction factor for the BSA-derived value. Clearly, sufficient replicates of both assays are necessary to generate an accurate correction factor. Once this has been done for a given colorimetric technique, target protein, standard protein, and a given set of assay conditions, an accurate target protein concentration can be obtained. It will be necessary to empirically calibrate the response of each new target protein to these reagents. Changes in the buffers and excipients will also require recalibration of the assay because the method may be sensitive to the buffer components.

Lowry Method

The Lowry method is probably the most widely used method for protein concentration. The chemistry behind the method involves redox reactions. The target protein is treated with alkaline cupric sulfate in the presence of tartrate, which results in the reduction of the cupric ion by the protein and complexation of the resulting cuprous ion by the tartrate. This tetradentate cuprous ion complex is then reacted with Folin–Ciocalteau phenol reagent. Reduction of this reagent by the cuprous complex yields a water-soluble blue product that has an absorbance at 750 nm. This method only requires approximately 1 h of total incubation, but has the disadvantage of two incubations with exact incubation times. The practical limit for the number of samples per assay is approximately 20. Interfering substances include detergents and reductants (thiols, disulfides, copper chelators, carbohydrates, tris, tricine, and potassium ions). The Lowry assay has a working range of 1 to 1500 μg.

DC Method

The DC (detergent-compatible) method is based on the Lowry assay. The Bio-Rad DC protein assay requires only a single 15-min incubation, and absorbance is stable for at least 2 h. The standard assay has a working range of 100 to 2000 μg/ml, whereas the microassay is suitable for use in the 5- to 250-μg/ml range. The microtiter plate assay procedures available for both protein concentration ranges provide automation for high throughput.

BCA Method

The BCA method is simpler than the Lowry method and relies on the same redox reaction. The target protein is treated with alkaline cupric sulfate in the presence of tartrate, which results in the reduction of the cupric ion to cuprous by the protein. The cuprous ion is then treated with bicichoninic acid (BCA) and two

BCA molecules complex with the cuprous ion to yield a water-soluble purple product that has an absorbance at 562 nm. This method only requires 30 min of incubation at 37°C but has the disadvantage of not being a true end-point assay because the color will keep developing with time. In reality, the rate of color development is slowed sufficiently following incubation to permit large numbers of samples to be assayed in a single run.

The structure of the protein, the number of peptide bonds, and the presence of cysteine, cystine, tryptophan, and tyrosine have all been reported to be responsible for color formation.[3] However, studies with model peptides suggest that color formation is not simply due to the sum of the contributions of the individual functional groups; hence it is not possible to predict the response of the target protein in this assay. Interfering substances include reductants and copper chelators in addition to reducing sugars, ascorbic and uric acids, tyrosine, tryptophan, cysteine, imidazole, tris, and glycine. Increasing the amount of copper in the working reagent can eliminate interfering copper-chelating agents.

The BCA assay has a working range of 20 to 2000 µg/ml. If the target protein is in a dilute aqueous formulation, the concentration can be determined with the micro BCA assay. The BCA protocol is modified by increasing the BCA concentration, incubation time, and incubation temperature. These modifications permit the detection of BSA at 0.5 µg/ml. The major disadvantage of these modifications is that the presence of interfering substances decreases the signal-to-noise ratio and thus the sensitivity of the assay.

Bradford Method

The Bradford method is probably the simplest colorimetric method, relying on only the immediate binding of the target protein to Coomassie® Brilliant Blue G-250 in acidic solution. The water-soluble blue product has an absorbance at 595 nm. Mechanistic studies suggest that the sulfonic acid form of the dye is the species that binds with the protein.[4] The binding of the target protein's basic and aromatic side chains (arginine, lysine, histidine, tyrosine, tryptophan, and phenylalanine) to the anionic form of the Coomassie® dye results in an absorbance change from red to blue. Van der Waals and hydrophobic interactions are also believed to have a role in the binding. This method consists only of mixing with no requirement for incubation time or elevated temperatures. Detergents are the major interfering substances for this assay. The Bradford assay has a working range of 100 to 1000 µg/ml. A micro version of this assay exists with sensitivity down to 1 µg/ml. Despite these advantages, the Bradford assay exhibits high interassay variability, which limits its use in situations where high precision is required.

Direct Absorbance Methods

The direct absorbance methods require only a protein-specific extinction coefficient to deliver an accurate protein concentration. These methods typically require minutes to perform and require only a spectrophotometer and a good quantitative sample preparation technique. In addition, these methods are amenable to auto-mation. They do not require a standard curve for quantitation but are protein composition and structure dependent. Absorbance methods typically rely upon the intrinsic absorbance of a polypeptide or protein at 280 nm. The aromatic amino acids that absorb at this wavelength are tyrosine, tryptophan, and phenyl-alanine. Because these residues remain constant for a given protein, the absolute absorption remains constant. An extinction coefficient needs only to be deter-mined once and is then absolute for the target protein in that buffer system. Thus, protein concentration may be determined by Beer's Law, $A = \varepsilon \, l \, c$, where A is the absorbance, ε the molar extinction coefficient, l the detection cell path length in centimeters, and c the sample concentration in mol/l.

Determination of the extinction coefficient is a relatively straightforward task. The target protein is diluted to give five different concentrations. These samples are then divided into two aliquots. Amino acid analysis (AAA) accurately determines the protein concentration of one set of samples at the five concentrations, and the absorbance at 280 nm (A_{280}) is measured for the other set of samples. The slope of a plot of A_{280} vs. protein concentration by AAA yields the extinction coefficient.

Fluorescence Methods

Molecules with intrinsic fluorescence absorb energy at a specific excitation wave-length (λ_{ex}) and rise to an excited state. The energy is released at a longer emission wavelength (λ_{em}) as the molecules return to ground state. Fluorescence at distinct wavelengths where there is little interference from other sample components provides high selectivity for the fluorescent molecules. In addition, sensitivity with these methods is high because there is little interference from background light at the emission wavelengths.

Native Fluorescence

Native fluorescence of a protein is due largely to the presence of the aromatic amino acids tryptophan and tyrosine. Tryptophan has an excitation maximum at 280 nm and emits at 340 to 350 nm. The amino acid composition of the target protein is one factor that determines if the direct measurement of a protein's native fluorescence is feasible. Another consideration is the protein's conforma-tion, which directly affects its fluorescence spectrum. As the protein changes conformation, the emission maximum shifts to another wavelength. Thus, native fluorescence may be used to monitor protein unfolding or interactions. The conformation-dependent nature of native fluorescence results in measurements specific for the protein in a buffer system or pH. Consequently, protein denatur-ation may be used to generate more reproducible fluorescence measurements.

Derivatization with Fluorescent Probes

Proteins that do not contain tryptophan or tyrosine must be derivatized prior to fluorescence detection. A common derivatization chemistry involves the reaction of amines with fluorescamine or *o*-phthalaldehyde (OPA). The selectivity provided by the derivatization of the amines can be further enhanced by separation of the fluorescent probes and derivatized sample components using an analytical method such as high-performance liquid chromatography (HPLC). Alternatively, postcolumn derivatization can occur following separation of the target protein from other sample components. Fluorescent probes that react with other functional groups offer different selectivities. Although derivatization with a fluorescent probe may provide selectivity and sensitivity within a complex sample matrix, this labor-intensive method is less precise than direct measurement methods or even colorimetric assays that require less extensive sample preparation. Tris interferes with amine derivatization, and care should be taken to determine if other buffer components affect the derivatization chemistry of choice.

Amino Acid Analysis

The fourth category of protein assay is amino acid analysis. This method is the most accurate and robust method for determination of protein concentration, but is appropriate only for pure proteins. In addition, it is relatively slow and requires specialized instrumentation and knowledge of the target protein's theoretical amino acid composition.

AAA usually involves hydrolysis of the protein into its constituent amino acids, which are then derivatized with a UV or fluorescent label and quantified by HPLC against known amino acid standards. Hydrolysis occurs with strong acid at high temperatures. Hence some amino acids are modified (e.g., glutamine to glutamic acid), whereas others, such as tryptophan, are completely destroyed. Peptide bonds between hydrophobic residues such as leucine, phenylalanine, or valine are hard to break and may require extended hydrolysis detrimental to the recovery of other amino acids. Although special hydrolysis conditions exist for the recovery of labile residues such as threonine, serine, tyrosine, and tryptophan, no one set of hydrolysis conditions quantitatively yields all amino acids. After hydrolysis, the liberated amino acids are typically derivatized with phenylisothiocyanate (PITC). The resulting phenylthiocarbamyl (PTC) amino acids are then separated and quantified by HPLC. Alanine, leucine, valine, and phenylalanine are among the most stable residues and are typically used for protein quantitation.[5,6]

Unlike the previous techniques, sensitivity is not an issue for AAA. There are few interfering substances because the method involves hydrolysis, derivatization, and chromatography with detection at a unique wavelength. Most excipients will not affect the hydrolysis step, but one has to be careful to ensure that the amino acids used to quantitate the protein are not destroyed. In addition, it must be determined if the excipients interfere with the derivatization chemistry or the chromatography. A BSA standard in the same buffer formulation is routinely run in parallel to the target protein to ensure the accuracy of the method.

Custom Quantitation Methods

Finally, there are custom two-step quantitation methods such as chromatography or ELISA that require a capture step for isolating the protein and then a quantitation step based on a standard curve of the purified target protein. The preliminary capture step may also concentrate the protein for increased sensitivity. These techniques are typically not available in a commercial kit form and may require extensive method development. They are more labor intensive and complex than the colorimetric or absorbance-based assays. In addition, recovery of the protein from and reproducibility of the capture step complicate validation. Despite these disadvantages, the custom two-step quantitation methods are essential in situations requiring protein specificity.

APPLICATIONS FOR PROTEIN ASSAYS

Drugs produced by the biotechnology and pharmaceutical industries are highly regulated by the Food and Drug Administration (FDA) in the U.S., the European Medicines Agency (EMEA) in Europe, and the Ministry of Health, Labour, and Welfare (MHLW) in Japan. These three regulatory regions have combined to produce International Conference on Harmonization (ICH) guidelines on many common technical regulatory issues such as analytical assay validation, test procedures, and specifications. ICH, as well as common sense, dictates that an analytical method is suitable for its intended application. Accordingly, all or a combination of the described protein assay methods may be required during the development of a protein biopharmaceutical depending on the particular requirement, be it speed, accuracy, or throughput.

If the drug development process starts with the discovery of a target protein, protein assays will be required from cell culture/fermentation and purification to determining the concentration of the purified target. The latter value is probably more important because a well-characterized production process is of low priority at this early stage of development. The activity of the target will be the yardstick by which its suitability for further development will be determined. However, the protein assay precision will be superior to a bioassay by a log, hence activity differences do not result from dosing markedly different quantities of the target protein.

Once a target protein has been identified and becomes a clinical candidate, drug development begins in earnest. The requirements for protein assay change during the production process from cell culture/fermentation to harvesting, purification, or formulation. Any combination of speed, throughput, limit of quantitation, or selectivity could be critical for protein assay at a particular process step. As the purification process progresses, protein purity increases. The difference between total-protein and target-protein concentrations is greatest during cell culture/fermentation, where the assay must be capable of selectively detecting and quantitating a protein that is a minor component amid media and host cell proteins. Media containing 10% serum have protein concentrations ranging from 6,200 to 10,000 mg/l. In contrast, recombinant protein expression in mammalian cells ranges from a few mg/l to 1000 mg/l. Clearly, specificity and limit of detection of the assay are critical for this application. Accuracy and speed are of relatively little importance because the primary concern is relative expression levels and not absolute quantitation. One is looking for order of magnitude differences in expression at the early development stage rather than small percentage optimizations.

At the cell culture/fermentation process step, a highly specific assay such as ELISA is required because the target protein may represent only 1% of the total protein. The ELISA is an example of a two-step method capable of separating the target protein from the milieu by binding the protein to a target-specific antibody and quantitating it via a secondary labeled antibody. A chromatographic method could also be employed here to separate the target protein from the milieu, followed by spectrophotometric detection and quantitation. Clearly, for applications in which the protein to be quantitated is a minor component of a protein mixture, a custom two-step quantitation method is essential.

After cell culture/fermentation, the purification process will require application of the complete portfolio of protein assays: a two-step method for the early process steps to counter low purity, followed by less selective methods as the protein becomes >90% pure. In contrast to the two-step methods, direct absorbance methods offer speed, simplicity, and accuracy. For these reasons, they are favored in the production area. A 10-min assay is virtually invisible in the manufacturing process; a 4-h assay consumes a workday. In addition to the technical considerations, there are economic aspects to choosing protein assays. Simply diluting the sample into a quartz cuvette, reading an optical density, and dividing by a dilution-adjusted extinction coefficient is very attractive when the manufacturing operator overhead is $2000/h. The major weakness of direct absorbance methods is that they require the presence of aromatic residues in the protein. Thus other biomolecules, such as nucleic acids, that strongly absorb in the UV region can generate erroneous values.

In addition to the direct absorbance methods, colorimetric methods are suited for relatively pure proteins as purification progresses. They are accurate if calibrated from a standard curve of the test protein reference sample and fast if automated. However, they are not as simple to perform as direct absorbance methods. Hence they are not as suitable for production as direct absorbance methods. The relative simplicity of colorimetric methods makes them more suited to automated formulation and stability studies and total-protein assays of complex mixtures. Microtiter plate versions of colorimetric assays allow for automation and consumption of relatively small sample sizes while requiring little specialized equipment or training.

Once the protein is purified, it will be formulated to produce the drug product. This could be as simple as diluting the protein in phosphate-buffered saline, or as complex as addition of excipients and lyophilization. The mark of a successfully formulated protein is that it is stable for its intended shelf life and the full dose can be recovered from the vial. Hence protein assay is the true workhorse for formulation. Loss of recoverable protein is the first clue to the occurrence of instability such as denaturation, aggregation, precipitation, or surface adsorption. As described earlier, the highly automatable colorimetric and direct absorbance methods are well suited for use at this stage because many buffers, excipients, and protein contents will be screened during accelerated stability testing to arrive at the drug product formulation. One commonly occurring problem with formulation studies is interference of excipients with the concentration method. Several different options exist to overcome interference, examples of which are protein precipitation, the use of dyes that are unaffected by the interfering agent, and the use of dyes that form protein complexes with absorbance maxima that are different from the absorbance of the interfering agent. All of the commercially available colorimetric methods describe interfering agents in their literature.

Next, the formulated protein, or drug product, needs to be tested for protein content. The requirement here is to have a product-specific method that accurately determines the dose of drug in the container, is capable of supporting the product through pivotal phase III human clinical trials, and can be validated. Ideally, the method used here will be the method anticipated for commercial production. As the drug goes through development, it is usual for formulations, dosage strengths, and even delivery vehicles to change. Hence a major challenge for the protein assay at this juncture is for it to remain suitable through the development life cycle. It is worthwhile to develop a drug product protein assay to be as robust and rugged as soon as possible to minimize problems with dose strength later. For example, if one employs an absorbance method during the early phases of product development and fails to identify that the drug is highly aggregated, then light scattering will result in an overestimation of the protein concentration and an unknown true dose. Clearly, this inaccuracy can be corrected at a later point in the development cycle, but one will probably not be able to accurately assign the doses given in earlier studies because the aggregation state of the protein may well have increased over time. The release of a very expensive product relies on the "suitability of the method for its application," so the protein assay needs to give predictably precise and accurate results. Predictability is achieved through validation as described by the ICH guidelines.

Finally, the protein assay for the drug product will also be used for real-time and accelerated stability testing if it has been validated to be stability indicating. A stability-indicating protein concentration method usually translates to a method that can reveal how much protein can be recovered from the dosage form. Many protein instabilities result in precipitation of the protein and adsorption to the container. An instability that results in only a modification of the protein structure but not in loss of protein from solution will not be detected by a sequence-independent protein assay such as a colorimetric assay.

CASE STUDY

AIC is a protein–oligonucleotide conjugate being developed for ragweed immunotherapy by Dynavax Technologies (Berkeley, CA). Specifically, it is Amb a 1, the major allergen of short ragweed pollen, linked to multiple immunostimulatory oligonucleotides (1018 ISS). Amb a 1, which is purified from ragweed pollen, is activated with a heterobifunctional cross-linker and linked to 1018 ISS to produce AIC. Hence, protein concentration assays are employed from the Amb a 1 extraction step through to determining the strength of the AIC drug product. The range of purities and environments in which these protein assays are performed during the production process requires several different protein concentration methods to be employed.

The first assay to be employed for protein concentration is the Bradford assay, a commercially available colorimetric assay used to quantitate the total extracted protein. Amb a 1 is approximately 1% of the total protein extracted from ragweed pollen; hence the Bradford assay does not reflect Amb a 1 concentration. However, at this step of the production process, the protein concentration is used to calculate final yields and not to make time-dependent or expensive decisions. Hence the nonspecific Bradford assay is ideal. A simpler direct absorbance method is not suitable due to the presence of a nonprotein chromophore in the ragweed extract.

The actual Amb a 1 concentration of the extract can be quantitated using a reversed-phase HPLC method developed at Dynavax. This is a custom two-step method that employs chromatography to separate the Amb a 1 from the other extracted proteins. The Amb a 1 concentration is then determined from the resolved Amb a 1 peak area and a standard curve of purified Amb a 1. This is the only step at which the Amb a 1 concentration of the process material is measured by a two-step process. Following the extraction step, the Amb a 1 rapidly becomes enriched over two purification steps, and the Bradford assay adequately reflects Amb a 1 concentration through the remainder of the process.

The Amb a 1 concentration of the final purified intermediate bulk is determined by an absorbance method chosen for its precision, accuracy, and simplicity. Because Amb a 1 bulk intermediate will now be conjugated to 1018 ISS (and the number of linked 1018 ISS affects the activity of the resulting AIC), it is essential to quantitate the Amb a 1 concentration accurately and precisely. A significant over- or underestimation of protein concentration will result in an over- or underestimation of the heterobifunctional linker required to activate the protein for coupling to 1018 ISS. The absorbance method, more dependent on well-calibrated instrumentation than lab technique, was chosen because it is an easy procedure to transfer to the production site. Dilution skills are the only requirement for robust performance of a well-developed and validated absorbance method. Hence a contract manufacturing site could readily quantitate Amb a 1 without the complications of preparing standard curves and assigning protein concentrations to reference standards. In this case, Amb a 1 has been assigned an experimentally determined extinction coefficient. The accuracy of the direct absorbance method would be improved if the Amb a 1 extinction coefficient is determined from the slope of the A_{280} vs. AAA-quantitated protein concentration plot.

As soon as the protein is activated with the heterobifunctional crosslinker, the extinction coefficient determined for pure Amb a 1 no longer applies because the heterobifunctional crosslinker absorbs at 280 nm. At this step in the production of AIC, the manufacturing overhead cost requires the use of a fast protein assay, whereas the exact stoichiometry of the subsequent reaction dictates the use of an accurate and precise method. Hence we developed a new extinction coefficient for the activated protein based on experimental data and demonstrated that within the normal activation range of 9 to 12 crosslinkers per Amb a 1, the new extinction coefficient remained constant. The concentration of the purified activated Amb a 1 determined by this direct absorbance A_{280} method is more precise and accurate than could be assigned by a colorimetric assay. Consequently, the activated Amb a 1 concentration allows for the accurate addition of 1018 ISS required to consistently produce AIC with optimal activity.

The purified AIC drug substance has an approximate molar ratio of four 1018 ISS to one Amb a 1. Because oligonucleotides absorb strongly in the UV region and contribute much more to absorbance in this range than the protein component, a direct absorbance method cannot be used to quantitate the protein directly. Among the colorimetric methods, the BCA assay is used to determine AIC protein concentration based on the availability of unmodified amino acids involved in the generation of the chromophore. Once the protein concentration has been assigned, the absorbance of AIC at 260 nm can be calculated to an oligonucleotide concentration and subsequently correlated to the protein concentration based on a fixed oligonucleotide-to-protein ratio. Thus, a direct absorbance method can now be used to indirectly determine the protein concentration. As the AIC drug substance is formulated and filled into vials, the A_{260} direct absorbance method provides a precise and accurate method to ensure correct dilution of the AIC drug substance. This simple and fast protein assay is invaluable for in-process control in the manufacturing environment.

Finally, the drug product strength is determined by the micro BCA assay, and the AIC drug product is released based on this value. In the deliberate and unhurried release testing environment, AAA can also be used to confirm protein concentration. When the AIC drug product is used for real-time and accelerated stability testing, the direct absorbance method once again offers a simple, precise, and accurate method to indirectly determine protein concentration based on the fixed oligonucleotide-to-protein ratio of the AIC drug substance. Instabilities resulting in AIC precipitation or adsorption to the container will be detected by a decrease in the measured absorbance.

SUMMARY

This list of applications for protein concentration assays and the different requirements for each application reveal that no one method is superior to another. Each
protein assay is chosen for a particular application based on its strengths in
meeting specific requirements outweighing its weaknesses. Choosing the most
suitable protein assay for the application is based on consideration of the following relevant questions and issues.

Speed and Simplicity

How many samples are there and how urgently is the result required? If protein
concentration of a stable drug substance is being determined as part of a QC lot
release program, then time is probably not an issue. Virtually all of the other
assays on the certificate of analysis will take longer to perform than protein
concentration. However, if fractions are being collected from a column during a
production process, then time is of essence and a fast and accurate method that
can be performed at the production site with minimal effort is required. The
production process stops until the appropriate fractions are evaluated, so a direct
absorbance method is more suitable for this application than more time-consuming protein concentration methods. If protein concentration is an element of the
stability and formulation program, one may well have hundreds of samples to
process. In this scenario, methods amenable to automation such as absorbance
or colorimetric methods are desirable.

Limit of Detection (LOD) and Concentration

What is the expected concentration of the target protein and how much sample
will be available? The linear ranges and sensitivities of methods often vary
depending on the protein being analyzed. Increased sensitivity formats of many
colorimetric assays are available, but these modifications usually come with
drawbacks such as increased sample requirement, incubation times, and temperatures. Again, the application will determine the pertinent issues to consider. It
is not desirable that the release or stability assay for protein concentration be
performed near the LOD as this is typically the range with the greatest potential
for inaccuracy and imprecision. The consequence of erroneous results in this
application is either failure to meet the release specifications or an uninterpretable
stability profile. Alternatively, if a protein concentration method for a cleaning
assay is required, the lowest possible LOD may be needed.

Accuracy and Specificity

Each protein will have a unique response to the technique of choice. For example,
if a colorimetric assay is to be used, it will be necessary to calibrate the method
with the test protein to ensure the accuracy of the response. This can be easily

achieved by accurately quantitating a sample of the test protein using AAA and generating a correlation factor for the protein (BSA) that is used to create the standard curve. Colorimetric methods can accurately measure the concentration of a single protein based on a correctly constructed standard curve, but only if the sample exclusively contains the target protein. Orthogonal or multistep processes are required for complex mixtures of proteins in which the quantitation of only one component is needed. The custom-designed methods involving chromatography, ELISA, or another protein-specific capture technique are discussed elsewhere in this book.

Variability and Reproducibility

As mentioned earlier, the response of each protein will vary. This is especially apparent with colorimetric assays or derivatization methods requiring a chemical reaction. These protein-to-protein reactivity differences mean that a protein assay suitable for one protein may not be suitable for another. Even for a given protein and a specific protein determination method, results may still vary based on limitations of the assay. Methods requiring extensive sample preparation including protein concentration, buffer exchange, and time-sensitive reactions are liable to be less reproducible than direct measurement techniques, which have fewer variable parameters. The application will determine the suitability of the method.

Effects of Excipients

Is there a component in your protein solution that will interfere with your chosen protein quantitation method? Many methods will provide misleading results in the presence of standard biopharmaceutical reagents such as detergents, chaotropes, bases, or reductants. It is imperative to determine if excipients present in the sample interfere with the chemistry of the protein quantitation method.

References

1. Stoscheck, C.M., *Methods in Enzymology, 182:* 50 (1990).
2. Ritter, N. and J. McEntire, *BioPharm, 15(4):* 12 (2002).
3. Wiechelman, K.J., R.D. Braun, and J.D. Fitzpatrick, *Anal. Biochem., 175:* 231 (1988).
4. Compton, S.J. and C.G. Jones, *Anal. Biochem., 151:* 369 (1985).
5. Anders, J.C., *BioPharm, 15(4):* 32 (2002).
6. Ozols, J., *Methods in Enzymology, 182:* 587 (1990).

4

Use of Reversed-Phase Liquid Chromatography in Biopharmaceutical Development

Tim Wehr

INTRODUCTION

Reversed-phase liquid chromatography (RPLC) is an essential tool for analytical and preparative separations of biopharmaceuticals. Its ubiquitous application in the discovery and development of protein-based drugs arises from its high resolving power, the robustness of the separation technique, and its success in handling a wide range of separation problems. The favorable kinetics provided by hydrocarbonaceous ligands bonded to microparticulate packings typically generate column efficiencies in excess of 25,000 plates/m. Silica- and polymer-based supports are stable over a wide range of solvent chemistries and operating pressures, and the hydrophobic stationary phases are quite robust compared to other chromatographic ligands. Reversed-phase columns are available in a wide range of dimensions for applications ranging from nanoscale analytical separations to process-scale chromatography, allowing RPLC methods to be scaled during product development. The technique usually employs volatile solvents, enabling easy solvent removal in preparative applications and allowing direct coupling to mass spectrometric detectors for analytical LC-MS. Most important, a vast array of mobile-phase chemistries and commercially available RPLC columns can be brought to bear on a separation problem. However, in practice, the success of a few generic methods for proteins and peptides usually simplifies the method-development task.

Together, these characteristics of RPLC make it the preferred analytical technique for assessing protein purity and for elucidating protein structure. Reversed-phase high-performance liquid chromatography (HPLC) can also be used alone or in concert with other chromatographic techniques for purification of proteins. For the separation of synthetic polypeptides or peptides derived by proteolysis of proteins, RPLC has no equal. Reversed-phase peptide mapping is the technique of choice for confirming protein structure and identity, for determining the sites and nature of posttranslational modifications, and for characterizing protein modifications and degradation products. More recently, it has played a central role in proteomic studies that are anticipated to generate targets and leads in the drug discovery environment. Here, reversed-phase chromatography is used as a complement to or replacement for two-dimensional (2-D) gel electrophoresis as the front-end separation for LC-MS-MS identification of proteins.

A fundamental limitation of RPLC is the denaturing properties of the hydrophobic stationary phases and eluting solvents. In preparative applications, this can limit the recovery of bioactive species. In analytical applications, reversed-phase separation conditions can perturb the conformational state of the protein, causing analytes to appear as broad or asymmetric peaks, and in extreme cases to be resolved as individual conformers. In such cases, the analyst must find conditions to minimize this behavior or resort to another separation technique. For preparative work, hydrophobic interaction chromatography (HIC) may be preferred over reversed-phase chromatography. The salt gradients used for HIC elution favor maintenance of native conformations and biological activity. Because HIC is based on hydrophobic interactions, it can provide a selectivity similar to RPLC without the strongly denaturing conditions.

THE MECHANISM OF REVERSED-PHASE CHROMATOGRAPHY

Reversed-phase chromatography employs a nonpolar stationary phase and a polar aqueous–organic mobile phase. The stationary phase may be a nonpolar ligand, such as an alkyl hydrocarbon, bonded to a support matrix such as microparticulate silica, or it may be a microparticulate polymeric resin such as cross-linked polystyrene-divinylbenzene. The mobile phase is typically a binary mixture of a weak solvent, such as water or an aqueous buffer, and a strong solvent such as acetonitrile or a short-chain alcohol. Retention is modulated by changing the relative proportion of the weak and strong solvents. Additives may be incorporated into the mobile phase to modulate chromatographic selectivity, to suppress undesirable interactions of the analyte with the matrix, or to promote analyte solubility or stability.

The mechanism of reversed-phase chromatography arises from the tendency of water molecules in the aqueous–organic mobile phase to self-associate by hydrogen bonding. This ordering is perturbed by the presence of nonpolar solute molecules. As a result, solute molecules tend to be excluded from the mobile phase and are bound by the hydrophobic stationary phase. This solvophobic exclusion provides the thermodynamic driving force for retention in RPLC. In contrast to other modes of chromatography, in which retention relies upon affinity of the solute for the stationary phase, it is the increased entropy of water in the mobile phase accompanying the transfer of a solute molecule from the mobile phase to the nonpolar stationary phase that promotes retention. Thus, retention is favored by the hydrophobic contact area of the solute and reduced by dipolar or hydrogen-bonding interaction of the solute with the mobile phase. Solvophobic theory[1] predicts and experimental observations confirm that retention (as expressed by the capacity factor, k) decreases with a decrease in surface tension and that a linear relationship exists between the logarithm of k and the volume percent of organic modifier in the mobile phase. Protein retention is thought to occur by adsorption to the hydrophobic stationary phase according to the solvophobic effect or to a sorbed layer of the nonpolar solvent component extracted from the bulk mobile phase.

An understanding of the phenomenology of protein retention in reversed-phase chromatography requires consideration of the forces involved in defining the three-dimensional (3-D) conformation of the protein. Small polypeptides (comprising a dozen amino acid residues or so) exist in solution as random coils, and their chromatographic retention in reversed-phase systems can be predicted from the summed hydrophobic properties of the individual amino acids. These can be derived from hydrophobic indices determined by water: octanol partition measurements[2] or chromatographic retention coefficients obtained from model peptides.[3-5] An example of the latter approach is represented in the retention studies of Guo et al.[6,7] These workers synthesized a family of model octapeptides with each of the 20 protein amino acids represented in tandem within the peptide sequence. Retention coefficients for each amino acid were obtained from the relative retention of each model peptide in a defined reversed-phase chromatographic system. The experimentally derived retention coefficients were used to predict the retention of 58 peptides ranging from 2 to 16 residues in length. The observed retention times in this study correlated well with the predicted values (Figure 4.1).

However, in a later study employing polymeric peptides of 5 to 50 residues assembled from block sequences, significant deviations of observed retention values from the predicted values were found for polypeptides larger than about 10 residues. These deviations for larger polypeptides were correctly interpreted as conformational effects.

Peptides larger than 10 to 20 residues adopt conformations in solution through the interplay of hydrogen bonding, electrostatic and hydrophobic interactions, positioning of polar residues on the solvated surface of the polypeptide, and sequestering of hydrophobic residues in the nonpolar interior. Protein shape is dynamic, changing continuously in response to the solvent environment. The retention process in RPLC is initiated as the protein approaches the stationary-phase surface. Structured water associated at the phase surface and adjacent to hydrophobic contact surfaces on the polypeptide is released into the bulk mobile

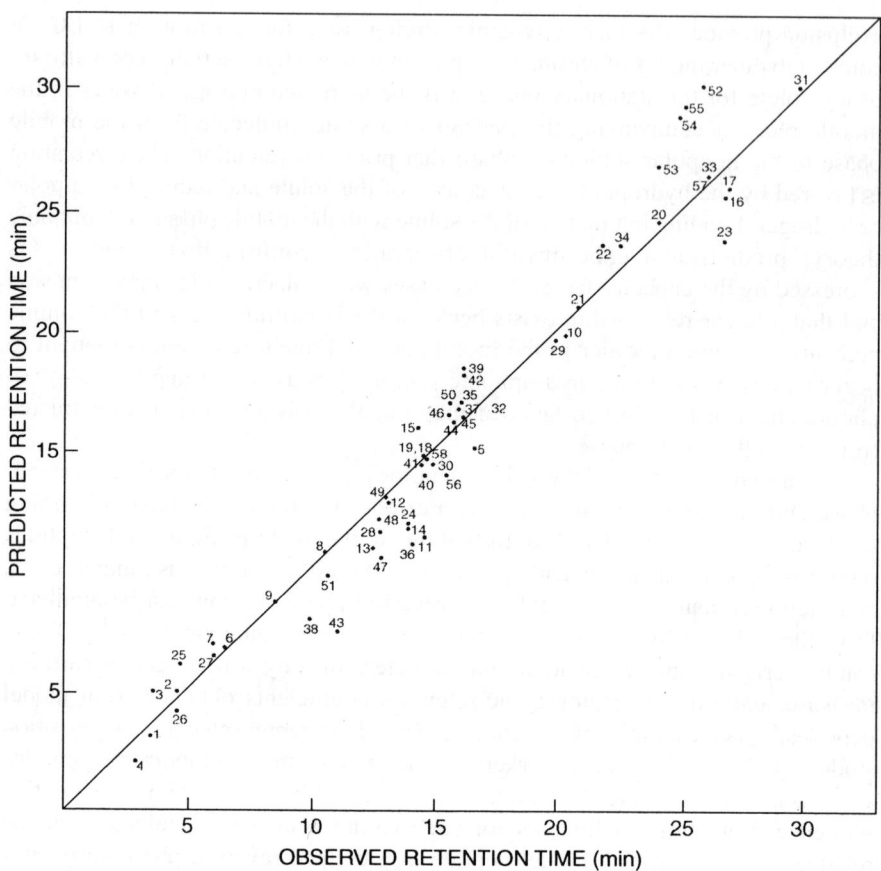

Figure 4.1 Correlation of predicted and observed retention times in reversed-phase chromatography. The predicted retention times for 58 peptides of 2 to 16 residues in length were obtained by summation of retention coefficients for each residue in the peptide. Retention coefficients were determined from the retention of model synthetic peptides with the structure Ac-Gly-XX-(Leu)$_3$-(Lys)$_2$-amide, where X was substituted by the 20 protein amino acids. (Reproduced from D. Guo, C.T. Mant, A.K. Taneja, and R.S. Hodges, *J. Chromatogr., 359:* 519 [1986]. With permission from Elsevier Science.)

phase, and this process of solvent exclusion is the driving force for protein binding to the hydrocarbonaceous phase. A protein in equilibrium between multiple conformational states may present different hydrophobic contact areas for binding (Figure 4.2). If the interconversion rate is fast (e.g., in the millisecond domain), the observed peak represents the average of the rapidly interconverting species. If the interconversion rate is slow (seconds to minutes), varying contact areas of the different conformational states may be manifested by broadened peaks or chromatographic resolution of discrete peaks.

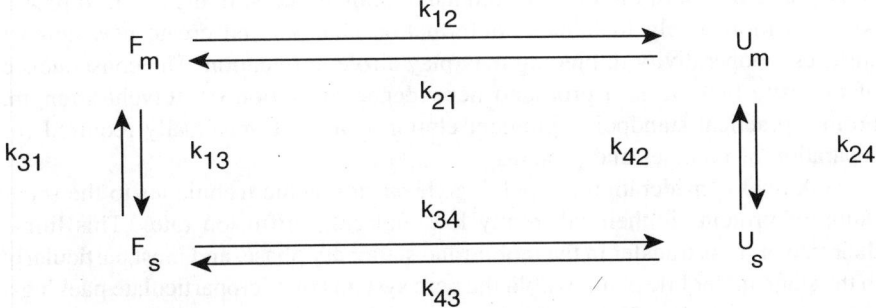

Figure 4.2 Protein transformations in reversed-phase chromatography for a two-state model. The native folded state can exist in either the mobile phase (F_m) or the stationary phase (F_s), as can the unfolded state (U_m, U_s). The equilibrium constants (k) for interconversions of the four species are indicated. (Reproduced from X.M. Lu, K. Benedek, and B.L. Karger, *J. Chromatogr., 359:* 19 [1986]. With permission from Elsevier Science.)

The molecular forces that are involved in maintaining the folded state of a protein are substantially the same as those involved in retention and elution. When the magnitude of the interactions involved in protein binding exceeds that of folding, the chromatographic process can induce conformation changes that expose hydrophobic groups and create new contact areas. As a result, the fully folded protein may undergo transitions to partially or fully denatured conformations. A globular protein may display excessive retention, conversion to chromatographically resolved folding intermediates, and loss of biological activity. For this reason, reversed-phase conditions are often considered sufficiently harsh to disallow the technique from being used as a preparative tool. In point of fact, RPLC has often proven successful for recovery of functional proteins. Small to medium mass proteins may be refractory to the hydrophobic environment, or can easily refold when returned to benign environments. Retention of biological activity will be favored by chromatographic conditions that minimize protein unfolding, e.g., more polar organic modifiers, less hydrophobic stationary-phase ligands, higher mobile-phase pH, reduced temperature, lower-capacity columns, and short column-residency times.

The characteristics that discourage the use of RPLC for preparative isolation of bioactive proteins favor its use as an analytical tool for studying protein conformation. Chromatographic profiles can provide information on conformational stability of a protein and the kinetics of folding and unfolding processes. Information about solvent exposure of certain amino acid residues (e.g., tryptophan) as a function of the folding state can be obtained by on-line spectral analysis using diode array UV-vis detection or fluorescence detection.

A common feature of protein retention in reversed-phase and other interactive chromatographic modes such as ion exchange and hydrophobic interaction is the participation of multiple sites in the binding process. In the reversed phase, which is more likely to induce conformation changes and create new contact surfaces, cooperativity in binding may play a role in retention. The consequence of multisite binding is a profound dependence of elution on solvent strength. From a practical standpoint, gradient elution is almost universally required for separation of peptides and proteins.

A final consideration in applying chromatographic techniques to the separation of proteins is their inherently low molecular diffusion rates. This limits their rate of mass transfer in the mobile and stationary phase, and most particularly in the stagnant mobile phase within the pore systems of microparticulate packings. When conventional HPLC packings are used for protein separations, significant peak broadening will be a consequence of unfavorable mass-transfer rates. Strategies for minimizing this problem will be discussed in the context of column selection and operation.

REVERSED-PHASE SEPARATION CONDITIONS

Support Matrix Composition

The ideal chromatographic support matrix should be mechanically stable under several hundred atmospheres of pressure, chemically stable in the presence of typical reversed-phase solvents, should possess a high surface area for good chromatographic capacity, be available in a range of particle sizes for analytical and preparative applications, be able to serve as an anchor for attaching chromatographic ligands, and should possess an inert surface with no potential for interactions with peptides. Silica has all of these qualities save two, and therefore it is the most widely used matrix for reversed-phase packings. Porous microparticulate silicas have large surface areas within their pore systems and can be manufactured in particle diameters of 2 to 10 µm for analytical applications and 15 to 30 µm for preparative work. The surface silanols on silica can serve as sites for covalent attachment of ligands.

A major limitation of silica as a matrix for chromatography of peptides and proteins is the potential for interaction of basic amino acid side chains with residual silanol groups on the silica surface. These can participate in secondary retention mechanisms through hydrogen bonding or ion exchange. The presence of highly active silanol sites can cause peak tailing. If the characteristics of the underlying silica change from one batch of column packing to another, columns may exhibit variations in peak symmetry and selectivity. It is advisable to use columns with high surface coverage of the reversed-phase ligand and columns that have been prepared with high-quality silica. The presence of metal ion contaminants in the silica can increase the activity of surface silanols, promoting their ability to participate in secondary interactions. Sol-gel techniques for preparing silicas reduce the level of metal contamination, and these high-purity silicas (sometimes termed "third-generation" or type B silicas) are the preferred matrix for columns to be used for polypeptide separations. Type B silicas also exhibit lower levels of isolated silanols (which are more likely to participate in unwanted interactions than vicinal or geminal silanols), producing a more inert surface.

A second limitation of silica is its solubility under alkaline conditions (pH >7). This problem is exacerbated under conditions of elevated temperature. This will be a concern if high-pH mobile phases are required to achieve the desired separation, or if exposure to alkaline conditions is required to clean the column after preparative applications. In the former situation, hybrid silicas are available that have extended lifetimes under conditions of elevated pH and temperature. Hybrid silicas are composed of a mixture of inorganic silica and alkyl silica. A caution in their use is the potential for selectivity changes relative to packings prepared with conventional silica. For preparative and process-scale applications in which alkaline cleaning regimes are anticipated, the use of a polymeric matrix will be preferred. Polymeric resins composed of polystyrene-divinylbenzene, polymethacrylate, or polyvinylalcohol are chemically stable at pH extremes (pH <2 and >10) and at elevated temperatures. Porous microparticulate resins of all three types are commercially available. The disadvantage of polymeric resins is their lower efficiency compared to silica-based packings.

Particle Size and Shape

Band spreading, a necessary evil in the chromatographic process, arises from the presence of multiple flow paths between particles and from resistance to analyte mass transfer in the flowing mobile phase and in the stagnant mobile phase within the particle pore system. These effects can be reduced by using particles that are spherical in shape, have a narrow distribution of particle sizes, and have reduced particle diameters. Spherical particles can be packed to form a homogeneous bed with reduced band broadening and higher porosity (and hence lower operating pressure). The advantage of small particle diameters for achieving high efficiency and resolution has long been appreciated, and HPLC packings have evolved from 10 μm in the 1970s to the 5-μm particles that are the workhorse materials in current usage. Particles of smaller diameters (e.g., 2.5 to 4 μm) can be used to obtain greater efficiency, but at the cost of increased operating pressure (note that the column pressure varies as the inverse square of the particle diameter). A useful characteristic of small particles is their reduced degradation of performance with increasing mobile-phase flow velocities. This is apparent in their shallower slopes in plots of plate height vs. flow (Figure 4.3). Because of this characteristic, short columns packed with sub-5-μm particles can be operated at high flow rates to achieve satisfactory resolution and still be within the pressure limits of the system. This strategy is being used in discovery environments in which heavy sample loads require high-throughput analytical techniques.

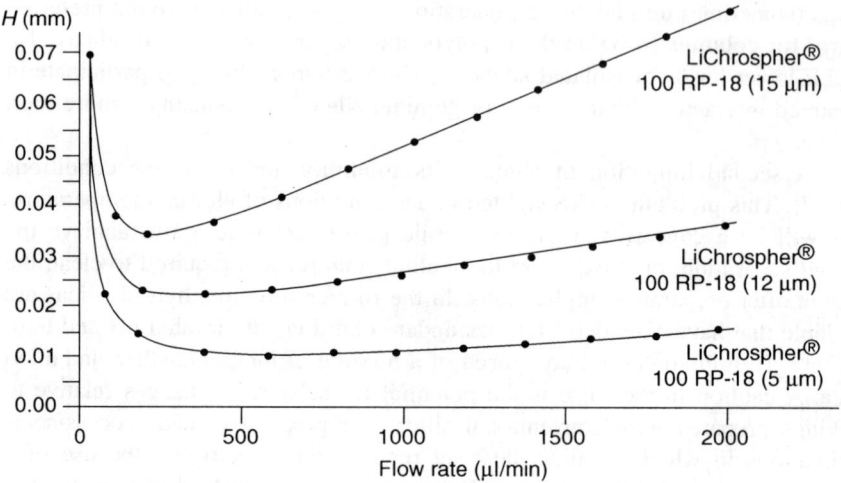

Figure 4.3 Effect of particle diameter on plate height. (Reproduced from Lichrospher & Lichroprep Sorbents Tailored for Cost Effective Chromatography, EM Separations, Gibbstown. With permission from Merck kGaA, Darmstadt, Germany, and EMD Chemicals, Inc.)

Table 4.1 Relationship Between Molecular Weight (M) and Stokes Radius (r)

Protein	M	r(Å)
Serum albumin (globular, solid spheres)	66,000	29.8
Catalase (globular)	225,000	39.8
Myosin (rodlike)	403,000	486
DNA (rodlike)	4×10^6	1170
Bushy stunt virus (globular)	10.6×10^6	120
Tobacco mosaic virus (rigid rod)	39×10^6	924

Pore Diameter

Microparticulate HPLC packings manufactured for chromatography of small molecules have pore diameters of 5 to 10 nm and surface areas greater than 200 m^2/g. These materials may be satisfactory for chromatography of peptides, but globular proteins above 10 kDa will have difficulty in accessing the internal surface of the pore system. Hence, the effective capacity of the packing will be limited to the external surface of the packing, amounting to a few m^2/g. Ideally, the pore should be at least threefold larger than the hydrodynamic diameter or Stoke's radius of the protein to allow permeation into the pore system (Table 4.1). A variety

Table 4.2 Relationship Between Pore
Diameter and Surface Area

Pore Diameter (nm)	Surface Area (m²/g)
10	250
30	100
100	20
400	5–10

of large-pore packings are commercially available for chromatography of proteins with pore diameters from 25 to 400 nm. Of these, particles with 30-nm pores have become the most popular and permit permeation of globular proteins up to 1000 kDa and random coil molecules up to 100 kDa. For larger proteins, macroporous packings may be preferred. However, increasing the particle porosity will compromise surface area (Table 4.2) and the mechanical stability of the particle. Operation of macroporous packings at elevated flow rates or pressures may not be recommended.

Alternatives to Porous Microparticulate Silicas

Restricted diffusion of a protein within a pore system can have two negative consequences for chromatography. The first is a greater dependence of efficiency on flow velocity such that resolution is compromised at high flow rates. This is particularly a limitation in preparative applications where high flow rates are desirable to achieve adequate throughput. This limitation of porous supports is shared by all modes of protein chromatography. A second problem encountered with porous materials is more characteristic of reversed-phase chromatography. Because reversed-phase conditions tend to be denaturing, there is the possibility that a folded species with facile permeation into the pore system can unfold to a partial or fully random coil configuration that will exhibit restricted diffusion within the pore. This can cause band broadening or, in worst case, can result in entrapment of the unfolded protein within the pore. This occurrence may account for the phenomenon of "ghosting" in reversed-phase gradient elution, that is, the appearance of peaks in a blank gradient following an analytical injection.

Three strategies have been employed to circumvent the problems of porous particles in protein chromatography. One strategy is to expand the pore diameter sufficiently to eliminate restricted diffusion. Of course, the consequences of this strategy are loss of mechanical stability and reduction in interactive surface area. A variation on this strategy that attempts to regain capacity through a bimodal pore system is represented by perfusion chromatography.[10] A perfusion particle contains a primary pore system of large "throughpores" that are wide enough to allow protein transport with little restriction. A secondary system of "diffusive" pores provides added surface area for capacity, but these are sufficiently shallow to minimize stagnant mobile-phase effects.

The simplest way to eliminate the problems of porous particles is to eliminate the pores. Nonporous particles of 10 μm diameter with a thin layer of stationary phase on their surfaces were in fact introduced early in the evolution of HPLC packings as a solution to the pore problem, but achieved little popularity because of their low capacity. Recently, nonporous materials have returned to the marketplace in the form of very small particles with diameters of 1.5 to 2.5 μm. The use of small particles compensates to some extent for the loss in capacity.[11] However, because of the high flow resistance of microparticulate nonporous packings, they are generally packed in short lengths and often operated at elevated temperatures.

The third strategy is the replacement of the microparticulate-packed bed with a monolith. Monolith columns contain a continuous interconnected skeleton with throughpores for transport of mobile phase and analytes. The large throughpores provide high permeability, thereby lowering the operating pressure. A secondary pore structure of shallow diffusive mesopores provides additional surface area for chromatographic capacity. Both polymer-based and silica monoliths are commercially available. Silica monolith rods[12] contain a skeleton with mesopores of approximately 13 nm diameter and macropores of about 2 μm. To minimize wall effects, the rod is "shrink wrapped" in PEEK. The mesopores provide a high surface area of 300 m²/g. The silica monolith matrix has a porosity of 80%, which is higher than the typical 65% porosity of microparticulate-packed columns. This allows monoliths to be operated at high flow rates with modest pressures, with little loss of performance in terms of band broadening. Polymethacrylate and polystyrene-divinylbenzene monoliths are commercially available in both column and disk formats for analytical and preparative chromatography. These materials also contain a bimodal pore structure of macro- and mesopores.

Stationary-Phase Ligand

Silica-based reversed phases are prepared by reacting the silica matrix with a silane reagent carrying the desired hydrophobic ligand. Monomeric phases are synthesized using a monofunctional silane, LR_2SiX, where L is the chromatographic ligand, R is a protective group (e.g., methyl), and X is the leaving group (e.g., chlorine or alkoxy group). The resulting siloxane bond (Si-O-Si) is chemically stable above pH 2. The product of this reaction is a monolayer that exhibits favorable kinetics in the chromatographic binding of polypeptides and so displays high efficiency. An advantage of this attachment chemistry is its convenience and reproducibility. Its limitations are the presence of residual silanols following the bonding process and the susceptibility of the phase to hydrolysis. In the bonding process, the accumulation of hydrophobic ligands blocks access to all of the active silanols by steric hindrance. These residual silanols are, however, accessible to small, polar analytes and are a major cause of tailing, particularly for basic analytes. The level of residual silanols can be reduced by performing a secondary

Table 4.3   Common Reversed-Phase Ligands

Trimethyl
Butyl (C4)
Octyl (C8)
Octadecyl (C12, ODS)
Phenyl
Diphenyl
Cyano

silanization with a silane reagent carrying small alkyl groups (e.g., trimethylchlorosilane) so that the reagent has easier access to the remaining silanol groups. This process is termed *endcapping*, and the use of endcapped columns for polypeptides is recommended to minimize the potential for undesirable interactions of basic amino acid side chains with the silica. Columns prepared using low-activity type B silica and endcapping are termed *base deactivated*, and these are marketed for chromatography of basic drugs. Base deactivated columns are suitable for polypeptide applications if the pore diameters are appropriate.

The stability of monomeric reversed phases can be increased by modifying the structure of the silane to minimize its susceptibility to hydrolytic attack. One approach is to replace the methyl protective groups with more bulky and hydrophobic groups such as isopropyl functions. These serve to restrict access of water to the silica surface, protecting the siloxane bond. Another approach is the use of bifunctional silane reagents that react with the silica to create a two-point attachment of the chromatographic ligand. This "bidentate" attachment of the ligand reduces the likelihood that it will be stripped from the surface by hydrolysis.

Another approach to preparing a stable reversed phase with fewer residual silanols is the use of polyfunctional silanes of the type R_2SiX_2. These react to form a polymeric stationary phase that shields the siloxane bonds and restricts access to residual silanols. Polymer phases have higher carbon loads and are typically more retentive than monomeric phases. However, they are more difficult to synthesize reproducibly and may exhibit batch-to-batch variability in their properties. They also exhibit poorer mass transfer kinetics and so provide poorer efficiency than monomeric phases.

The common reversed-phase ligands are listed in Table 4.3. The most popular are the straight-chain alkyl groups. These exhibit increased retention and stability with increasing chain length. Trimethyl phases are the least hydrophobic and might be considered for chromatography of proteins that would be strongly retained on the longer-chain ligands. However, these phases are easily hydrolyzed under the conditions usually employed for protein separations. Octyl (C_8) and octadecyl (C_{18}) are the most popular phases for chromatography of small molecules because of their retentiveness and stability. They are generally the first choice for proteins and peptides, although for protein applications, octyl phases are often selected in preference to octadecyl because they are less likely to cause denaturation and excessive retention of hydrophobic species. However, chromatography of peptides and proteins is generally carried out using gradient elution with organic modifier concentrations rarely exceeding 60 to 80%. Under these conditions, the alkyl chains probably exist largely in a self-associated state such that the effective hydrophobicity of the different alkyl chains is similar. In fact, the selectivities of butyl, octyl, and octadecyl phases for proteins and peptides are comparable.[13] Phenyl and cyano phases are likely to exhibit different selectivities than the straight-chain alkyl phases and therefore are often selected when C_8 or C_{18} columns fail to provide satisfactory resolution. Cyano is the most polar of the common reversed-phase ligands (CN columns are also used for normal-phase chromatography.), and it is the phase of choice for very hydrophobic species, e.g., membrane proteins and the hydrophobic polypeptides generated by cyanogen bromide cleavage.

Column Dimensions

A key property of column liquid chromatography is its scalability. An analytical method that has been developed on a standard 4.6-mm I.D. column can be scaled up in diameter for preparative isolation or down in diameter for improved sensitivity, analysis of volume-limited samples, or coupling to a mass spectrometer. If the column packing and mobile-phase composition are unchanged, the separation should be the same. However, the flow rate must be scaled in proportion to the difference in column diameter to avoid pressure changes and to achieve the same flow velocity and chromatographic efficiency. Assuming that the sample solvent has approximately the same eluting strength as the mobile phase, the injection volume should also be scaled according to column diameter. If the sample is introduced in a solvent that is weaker than the mobile phase (typically the case because protein and peptide preparations are often in an aqueous buffer or salt solution), large sample injections will be focused on the head of the column. This provides better sample throughput when scaling up to preparative isolation and better sensitivity when scaling down to microbore or capillary columns. Assuming a nominal flow rate of 1 ml/min on a 4.6-mm × 150-mm analytical column, comparable flow rates and sample capacities for larger and smaller columns are listed in Table 4.4.

When scaling a chromatographic separation, some precautions should be noted. When scaling up for preparative work, it is advisable (if possible) to use a column packed with material that has the same chromatographic ligand bonded to the same silica. Changes in the ligand bonding density and the type of silica can affect the chromatography by changing the protein–ligand interactions or secondary retention effects contributed by residual silanols. These changes can be subtle or (in the case of silica variations) dramatic.

Transferring a method from a standard 4.6-mm I.D. column to one of narrower diameter can cause loss of resolution due to extracolumn contributions to peak broadening. As column diameter and flow rate are scaled down, the peak

Table 4.4 Relationship Between Column
Diameter, Flow Rate and Sample Capacity

Column Diameter (mm)	Flow Rate (µl/min)	Sample Capacity (µg)
21.0	21000	1000
8.0	3000	150
4.6	1000	50
3.0	400	20
2.0	200	10
1.0	50	2
0.5	10	0.5
0.1	0.5	0.02

volume will decrease in proportion to the square of the column diameter. For example, scaling to columns of 2, 1, 0.3, and 0.1 mm I.D. will decrease peak volumes by factors of approximately 5, 20, 200, and 2000, respectively. If the HPLC system is not modified to minimize extracolumn volume, these contributions to the total peak broadening can become significant enough to degrade resolution. When scaling to 1- to 3-mm columns, transfer lines should be prepared with small-diameter tubing (e.g., 0.007 or 0.005 in.) and short lengths. Also, compression fittings should be assembled correctly with care not to mismatch components such as ferrules and nuts. Scaling to capillary columns of 0.1 to 0.5 mm I.D. will require additions to the HPLC such as nanoscale injectors, flow splitters, or even a change to a nanoflow solvent delivery system.

The choice of HPLC as a separation technique in biopharmaceutical development may depend upon its scalability, even if its resolving power is less than other techniques. For example, HPLC may be used in preference to gel electrophoresis or capillary electrophoresis, where scaleup is difficult or impossible.

Mobile-Phase Composition

Aqueous Component

In reversed-phase chromatography, the mobile phase consists of an aqueous component (the weak solvent) and an organic modifier (the strong solvent), and analyte retention is regulated by the ratio of the two components. The hydrophobicity of a polypeptide is largely dependent on the ionization state of the amino acid termini and of the ionizable side chains of internal residues. Thus, the pH of the mobile phase can have profound effects on polypeptide retention, and it is necessary to control mobile-phase pH to obtain the desired selectivity and achieve reproducible separations. In practice, separations of proteins and peptides are most often performed under acidic pH conditions employing dilute organic acids as additives. Under these conditions, the ionization of residual silanols on the stationary phase is suppressed, reducing unfavorable interactions with basic amino acid side chains that would cause tailing. Also, the organic acid additive can form ion pairs with the protonated basic side chains. The ion-paired groups will be more hydrophobic than their unpaired counterparts, favoring polypeptide retention. The ionization of terminal and side chain carboxyl groups is suppressed, which also favors retention. The most widely used organic acid in RPLC of peptides and proteins is trifluoroacetic acid (TFA), added to a concentration of 0.05 to 0.1% (approximately 5 to 10 mM). The stronger acidity of TFA enables the beneficial effects of low-pH operation to be achieved at much lower concentrations than required with unfluorinated organic acids. Use of lower additive concentrations reduces mobile-phase contributions to the background absorbance when using UV detection at low wavelengths. An added benefit of TFA is its high volatility, which facilitates solvent removal from collected sample fractions in preparative applications.

The use of TFA as a mobile-phase additive in LC-MS can be problematical when using electrospray ionization. In negative ion detection, the high concentration of TFA anion can suppress analyte ionization. In positive ion detection, TFA forms such strong ion pairs with peptides that ejection of peptide pseudomolecular ions into the gas·phase is suppressed. This problem can be alleviated by postcolumn addition of a weaker, less volatile acid such as propionic acid.[14] This "TFA fix" allows TFA to be used with electrospray sources interfaced with quadrupole MS systems. A more convenient solution to the TFA problem in LC-MS is to simply replace TFA with acetic or formic acid. Several reversed-phase columns are commercially available that have sufficient phase coverage and reduced levels of active silanols such that they provide satisfactory peptide peak shapes using the weaker organic acid additives.[15]

A caution in the use of TFA is the effect of its ion-pairing activity on selectivity. This is clearly illustrated in a study performed by Guo et al.[16] using a series of synthetic peptides containing 0, 1, 2, 4, or 6 basic residues (Figure 4.4). As the TFA concentration was increased, the retention of basic peptides increased and elution order changed, with the more highly charged peptides showing the strongest effect. Analysts should be aware that changes in TFA concentration during routine operation can cause changes in selectivity. This could be a concern when using HPLC systems equipped with on-line degassing units.

When chromatographic resolution of species based on modifications located at the protein surface is desired, it may be advisable to use conditions that favor retention of native conformation.[17] Here, the standard acidic conditions described in the preceding text may be inappropriate, and mobile phases buffered near neutrality may be required. Buffers based on ammonium acetate, ammonium bicarbonate, and triethylammonium phosphate may prove more useful in resolving polypeptide variants with differing posttranslational modifications, amino acid substitutions, or oxidation and deamidation products. The addition of more hydrophobic ion-pairing agents may be needed to obtain polypeptide retention, and a variety of alkyl sulfonates and alkyl amines have been described for specific applications.[17]

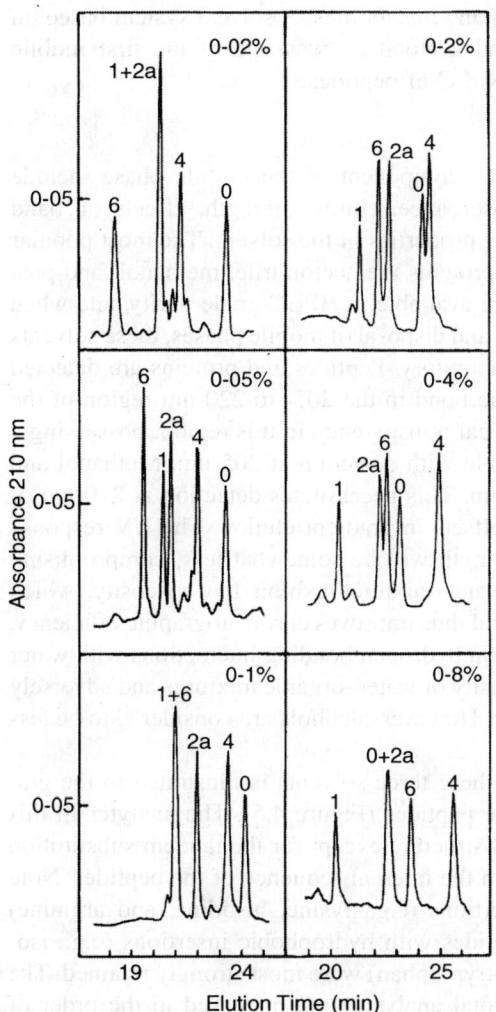

Figure 4.4 Effect of TFA concentration on peptide retention. A series of five synthetic peptides containing 0, 1, 2, 4, or 6 basic residues were separated on an octadecyl reversed-phase column using a 1%/min linear gradient from water to acetonitrile. Both solvents contained TFA at the indicated concentrations. (Reproduced from D. Guo, C.T. Mant, and R.S. Hodges, *J. Chromatogr., 386:* 205 [1987]. With permission from Elsevier Science.)

Triethylamine (TEA) is a common additive used in RPLC, which can have two beneficial effects. It can serve as an ion-pairing agent to promote retention of anionic species, and it can suppress the interaction of basic solutes with residual silanols. In fact, it is most frequently used at concentrations of 5 to 25 m*M* to improve peaks shapes of basic drugs and metabolites. A solvent system based on triethylamine-phosphate (TEAP) and acetonitrile was one of the first mobile phases to be used successfully for RPLC of peptides.[18]

Organic Component

The criteria for selecting the organic component of the mobile phase include solvent purity and toxicity, UV absorbance, eluting strength, effects on band spacing, viscosity, and the denaturing properties of the solvent. The most popular solvents for RPLC of peptides and proteins are acetonitrile, methanol, and propanol or isopropanol. All of these are available in HPLC-grade purity, and when proper care is used in the preparation and disposal of mobile phases, these solvents do not pose severe hazards in the laboratory. Peptides and proteins are detected by the UV absorbance of the peptide bond in the 205- to 220-nm region of the spectrum. Acetonitrile has good optical transparency in this region, possessing a UV cutoff of 190 nm; it is compatible with detection at 205 nm. Methanol and propanol have a UV cutoff at 205 nm. This necessitates detection at 210 nm or above to avoid excessive baseline offsets in gradient elution. The UV response of the analytes at the longer wavelength will be somewhat less, compromising detection sensitivity. Acetonitrile–water mixtures exhibit low viscosity, which reduces resistance to mass transfer and thus improves chromatographic efficiency. The short-chain alcohols participate in hydrogen bonding interactions with water molecules, which increases the viscosity of water–organic mixtures and adversely affects mass transfer and peak shape. However, alcohols are considered to be less denaturing than acetonitrile.

The comparative behavior of these three solvents is illustrated in the gradient elution of a series of synthetic peptides (Figure 4.5). The analytes in this study were octapeptides of identical structure except for the tandem substitution of each of the protein amino acids in the internal sequence of the peptide.[6] Note that peptides with hydrophilic insertions (e.g., lysine, histidine, and arginine) were weakly retained, whereas peptides with hydrophobic insertions (e.g., isoleucine, phenylalanine, leucine, and tryptophan) were most strongly retained. The three chromatograms indicate the total analysis time increased in the order of isopropanol, acetonitrile, and methanol. This demonstrates that methanol is the weakest solvent, isopropanol is the strongest solvent, and acetonitrile is of intermediate eluting power. The best peak shapes and highest efficiencies were observed with acetonitrile, reflecting lower mobile-phase viscosity and better mass-transfer kinetics obtained with this modifer. The variations in peak positions in the three chromatograms demonstrates the different selectivities obtained with the three solvents.

Surfactants

The addition of surfactants to the mobile phase will reduce surface tension and increase eluting strength. For this reason, surfactants might be expected to be useful in the chromatography of polypeptides. This has proven not to be the case

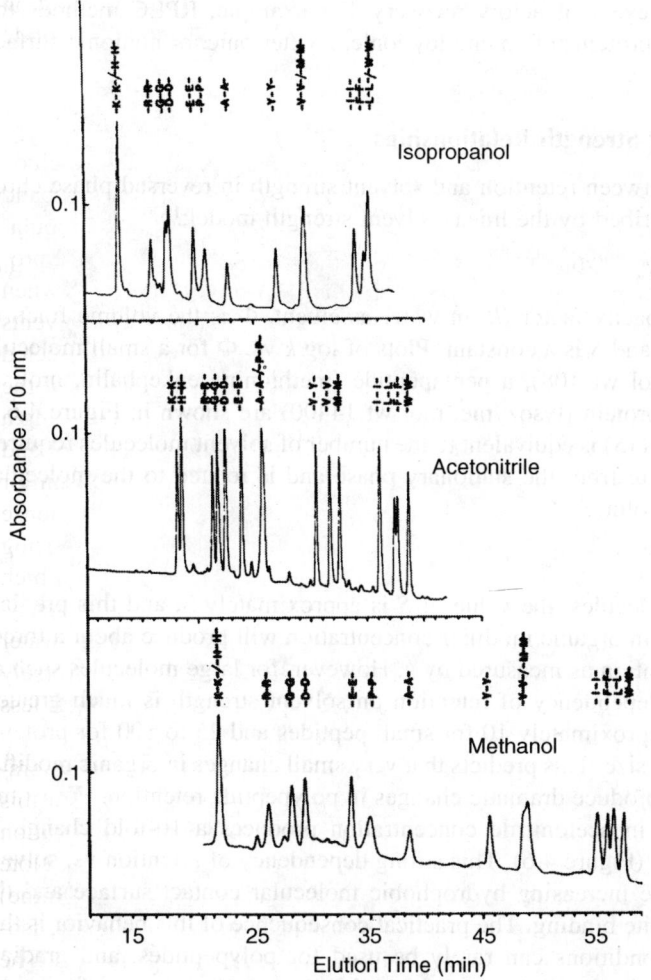

Figure 4.5 Effect of organic modifier on peptide retention. Synthetic octapeptides of the structure described in Figure 4.1 were separated by gradient elution on an octyl column using aqueous–organic systems containing a constant TFA concentration of 0.1%. (Reproduced from D. Guo, C.T. Mant, A.K. Taneja, J.M.R. Parker, and R.S. Hodges, *J. Chromatogr., 359:* 499 [1986]. With permission from Elsevier Science.)

for most separation problems. Surfactants will tend to induce protein unfolding, introducing new hydrophobic contact areas that can complicate the chromatography. Surfactants will also bind to the reversed-phase surface and can permanently change the behavior of the column. Therefore, the use of surfactants is usually reserved for applications where they are required for solubilization of the protein and to achieve satisfactory recovery. For example, RPLC methods for integral membrane proteins often employ ionic, zwitterionic, or nonionic surfactants as additives.

Retention–Solvent Strength Relationships

The relationship between retention and solvent strength in reversed-phase chromatography is described by the linear solvent strength model:[19,20]

$$\text{Log } k = \log k_w - S\Phi \tag{1}$$

where k_w is the capacity factor (k) in water as eluent, Φ is the volume fraction of organic solvent, and S is a constant. Plots of log k vs. Φ for a small molecule (benzyl alcohol, mol wt 108), a pentapeptide (methionine enkephalin, mol wt 573), and a small protein (lysozyme, mol wt 14400) are shown in Figure 4.6.[21] The slope of the plot (S) is equivalent to the number of solvent molecules required to desorb the analyte from the stationary phase and is related to the molecular weight (M) of the solute:[19]

$$S = 0.48M^{0.44} \tag{2}$$

For small molecules, the value of S is approximately 3, and this predicts that a 10% change in organic modifier concentration will produce about a three-fold change in retention as measured by k. However, for large molecules such as polypeptides, the dependency of retention on solvent strength is much greater, with S values of approximately 10 for small peptides and 25 to 100 for proteins of 10 to 100 kDa in size. This predicts that very small changes in organic modifier concentration can produce dramatic changes in polypeptide retention. For example, a 3% change in acetonitrile concentration produces a 10-fold change in lysozyme retention (Figure 4.6). This strong dependency of retention vs. solvent strength reflects the increasing hydrophobic molecular contact surface and the potential for multisite binding. The practical consequence of this behavior is that isocratic elution conditions can rarely be used for polypeptides, and gradient elution is almost universally required for multicomponent mixtures of peptides and proteins. Note that, because S is related to molecular size, the S value for unfolded states will be larger than for native conformations. Also, proteins that occupy hydrophobic environments *in vivo* (e.g., membrane proteins) should be conformationally more stable under reversed-phase conditions and display lower S values. This has been confirmed experimentally.[22]

The plots presented in Figure 4.6 display polypeptide retention data only for organic modifier concentrations of less than 50%. When this type of study is extended to higher organic concentrations, there is a reversal of retention behavior such that polypeptides exhibit enhanced retention (Figure 4.7). This behavior is observed for both peptides and proteins and is interpreted as a transition from hydrophobicity-driven retention at low organic modifier concentration to hydrophilicity-driven retention at high organic concentrations (Figure 4.8). For silica-based

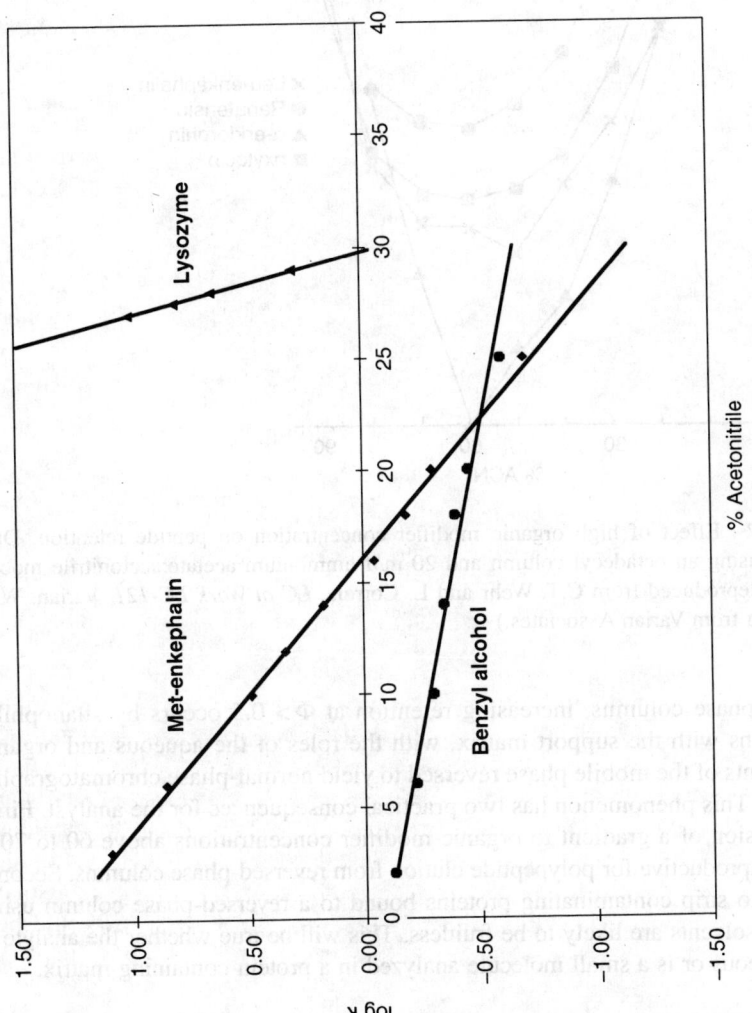

Figure 4.6 Effect of organic concentration on retention. (Reproduced from M.W. Dong, J.R. Gant, and B.R. Larsen, *BioChromatography*, 4: 19 [1989]. With permission from Eaton Publishing.)

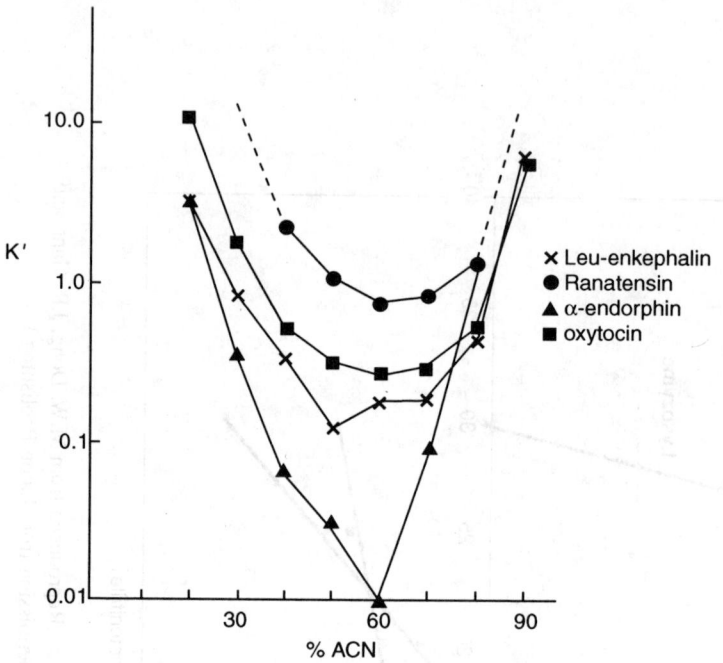

Figure 4.7   Effect of high organic modifier concentration on peptide retention. Data obtained using an octadecyl column and 20 m*M* ammonium acetate:acetonitrile mobile phases. (Reproduced from C.T. Wehr and L. Correia, *LC at Work LC-121,* Varian. With permission from Varian Associates.)

reversed-phase columns, increasing retention at $\Phi > 0.5$ occurs by silanophilic interactions with the support matrix, with the roles of the aqueous and organic components of the mobile phase reversed to yield normal-phase chromatographic behavior. This phenomenon has two practical consequences for the analyst. First, the extension of a gradient to organic modifier concentrations above 60 to 70% will be unproductive for polypeptide elution from reversed-phase columns. Second, attempts to strip contaminating proteins bound to a reversed-phase column using nonpolar solvents are likely to be fruitless. This will be true whether the analyte is proteinaceous or is a small molecule analyzed in a protein-containing matrix.

Gradient Elution

Because of the profound dependency of polypeptide retention on solvent strength, isocratic elution of proteins and peptides is generally impractical. Instead, solvent strength is increased continuously during the analysis, and each analyte elutes from the column at the point when solvent strength minimizes its binding to the stationary phase.

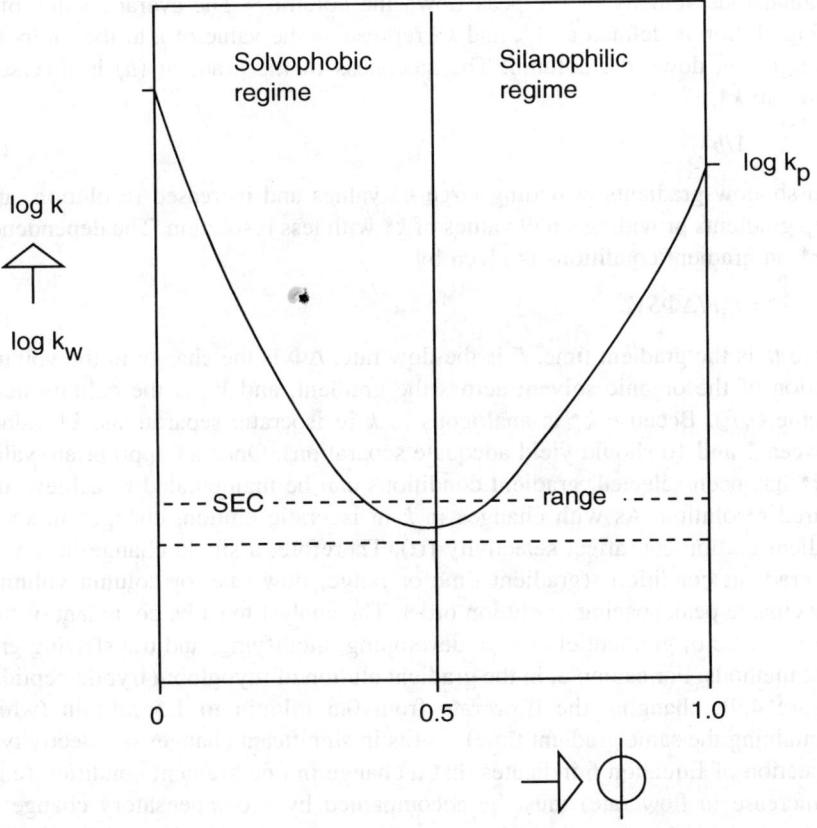

Figure 4.8 Regions of reversed-phase chromatographic mechanisms. (Reproduced from K.K. Unger, R. Janzen, and G. Jilge, *Chromatographia, 24:* 144 [1987]. With permission of Friedr. Vieweg & Sohn.)

In isocratic elution, retention is characterized by the capacity factor, k:

$$k = (t_r - t_0)/t_r \tag{3}$$

where t_r is the retention time of the analyte and t_0 is the elution time of an unretained species. The mobile-phase elution strength is adjusted to provide values of k between 1 and 10. The resolution R of two components is described by:

$$R = \tfrac{1}{4}(N)^{1/2} (\alpha - 1)(k/k + 1) \tag{4}$$

where N is the column efficiency in plates and is the selectivity (calculated as the ratio of the k values of two components). It should be noted that changes in k can sometimes change selectivity.

In gradient elution, the mobile-phase composition in the region of the analyte at any point in time determines an isocratic value of k, which defines the instantaneous velocity of the peak down the column.[20] The average value of k during elution is defined as k^*, and k^* represents the value of k at the midpoint of migration down the column. The steepness of the gradient (b) is inversely related to k^*,

$$k \sim 1/b \tag{5}$$

with shallow gradients providing large k^* values and increased resolution, and steep gradients providing small values of k^* with less resolution. The dependence of k^* on gradient conditions is given by

$$k^* = t_G F/\Delta\Phi S V_m \tag{6}$$

where t_G is the gradient time, F is the flow rate, $\Delta\Phi$ is the change in the volume fraction of the organic solvent across the gradient, and V_m is the column dead volume ($t_0 F$). Because k^* is analogous to k in isocratic separations, k^* values between 1 and 10 should yield adequate separations. Once an appropriate value of k^* has been selected, gradient conditions can be manipulated to achieve the desired resolution. As with changes in k in isocratic elution, changes in k^* in gradient elution can affect selectivity (α). Therefore, a single change in any of the gradient conditions (gradient time or range, flow rate, or column volume) may change peak spacing or elution order. The analyst must be cognizant of this characteristic of gradient elution in developing, modifying, and transferring gradient methods. For example, in the gradient elution of myoglobin tryptic peptides (Figure 4.9), changing the flow rate from 0.5 ml/min to 1.5 ml/min (while maintaining the same gradient time) results in significant changes in selectivity.[25] Inspection of Equation 6 indicates that a change in one gradient condition (e.g., an increase in flow rate) must be accompanied by a compensatory change in another variable (e.g., a collateral decrease in gradient time) to maintain a constant value of k^* and unaltered peak spacing.

Consideration of Equation 6 also suggests strategies for optimizing gradients for proteins and peptides, which have large S values. Reasonable k^* values can be achieved by increasing the gradient volume ($t_G F$), by reducing V_m, or by using a narrower gradient range ($\Delta\Phi$). Reduction in column volume can be accomplished by using a smaller column diameter (d) or a shorter column length (L). With the first approach, it should be noted that the flow rate must be scaled in proportion to the change in d^2 to maintain the same flow velocity to avoid loss in column efficiency. Reducing V_m by decreasing L would be expected to reduce efficiency and thereby compromise resolution. In practice, protein resolution does not seem to be strongly dependent on column length. This is because the negative effect of decreasing N is offset by the positive effect of increasing k^*.[20]

Using the preceding guidelines, a systematic approach to gradient optimization can be followed:[20]

1. Estimate the mean S values for the analytes from Equation 2.
2. Select an appropriate value for k^*, e.g., $k^* = 5$.

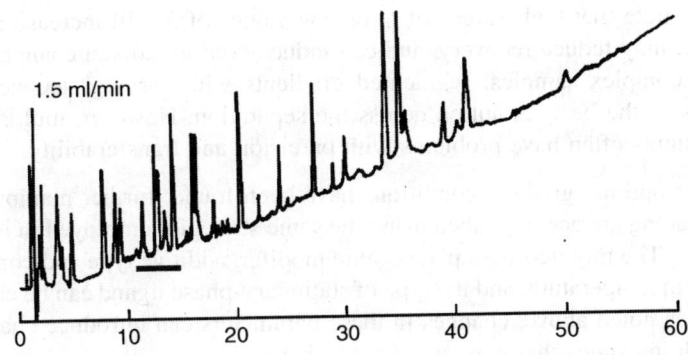

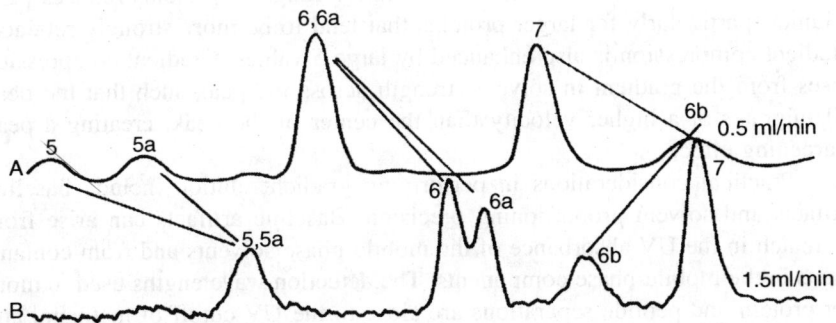

Figure 4.9 Changes in selectivity resulting from a change in flow rate from 0.5 to 1.5 ml/min for the reversed-phase gradient separation of a myoglobin tryptic digest. (A) entire chromatogram obtained at 1.5 ml/min, (B) expanded portion of the peaks eluting in the indicated segment of the chromatogram for both flow rates. (Reproduced from J.L. Glajch, M.A. Quarry, J.F. Vasta, and L.R. Snyder, *Anal. Chem.*, *58:* 280 [1986]. With permission from the American Chemical Society.)

3. Select a solvent range ($\Delta\Phi$) for a scouting gradient.

4. Using the values of S, k^*, and $\Delta\Phi$ of in combination with V_m and F, estimate t_G from Equation 6. For example, for a mean molecular weight of M and a k^* of 5,

$$t_G = 3\Delta\Phi\, V_m\, FM^{0.44}$$

5. Run the scouting gradient, then narrow the gradient range so that the earliest peak elutes after t_0 + dwell time (t_D) and the last peak elutes near the final solvent concentration. The gradient steepness should be maintained in the narrowed gradient.

6. Fine tune t_G, F, and L for best resolution while keeping k^* constant. However, note that high values of t_G or low values of F will increase analysis time, may reduce recovery, and can induce conformational changes.

7. For complex samples, segmented gradients with varying steepness may provide the best resolution across the separation. However, multisegment gradients often have problems with precision and transferability.

Once optimal gradient conditions have been found, further manipulations of peak spacing are accomplished using the same strategies employed in isocratic separations. The mobile-phase pH, organic modifer, additive type and concentration, column temperature, and the type of stationary-phase ligand can be changed. However, as noted above, changes in these parameters can introduce changes in protein folding states that may affect peak shape.

The large S values observed for polypeptides have two beneficial consequences in gradient elution. The narrow elution range of proteins reduces peak volumes, particularly for larger proteins that tend to be more strongly retained. Gradient compression is also enhanced by large S values. Gradient compression arises from the gradient in solvent strength across the peak such that the peak tail migrates at a higher velocity than the center of the peak, creating a peak sharpening effect.

Practical considerations in performing gradient elution include baseline artifacts and solvent proportioning precision. Baseline artifacts can arise from mismatch in the UV absorbance of the mobile-phase solvents and from contaminants in the mobile-phase components. The detection wavelengths used to monitor protein and peptide separations are close to the UV cutoff of methanol and propanol, and these solvents generate continuously rising baselines across the gradient. In addition, the absorbance of TFA increases with increasing organic solvent concentration due to shifts in the absorbance spectrum of the additive in nonpolar environments. To compensate for such solvent mismatches, acetonitrile is often used in preference to alcohols, the TFA concentration in the strong solvent can be reduced, and detection can be performed at 214 to 216 nm instead of 210 nm.

Mobile-phase contaminants often elute as discrete peaks, usually late in the gradient. Contaminants may be present in the buffers and additives or (more often) in the water used to prepare the weak solvent. If these are hydrophobic in nature, they will accumulate on the column in the early part of the gradient and elute as peaks as solvent strength increases. To confirm that artifact peaks arise from mobile-phase contaminants, the weak solvent may be pumped isocratically for a period prior to initiating the gradient. An increase in the height of the artifact peaks is diagnostic of contaminants in the weak solvent. Elimination of artifact peaks may require the use of purer reagents and water, or off-line cleanup of the weak solvent with a reversed-phase stripper column (an expired analytical column is useful for this task).

The proportioning accuracy and precision of the solvent delivery system is more important in the chromatography of proteins and peptides than in the analysis of small molecules. Small deviations in solvent composition can produce significant variances in analyte elution due to the strong dependency of polypeptide retention on solvent strength (e.g., large S values). The use of premixed mobile phases can minimize this problem. For example, a separation requiring a gradient range from 20 to 60% acetonitrile should be performed by proportioning between premixed 80:20 water:acetonitrile (solvent A) and premixed 40:60 water:acetonitrile (solvent B) and executing a gradient from 0 to 100% B.

Temperature

It might be expected that column temperature would affect retention and selectivity in the chromatography of proteins and peptides because changes in temperature can shift the ionization state of ionizable amino acid side chains and can induce conformational changes. In fact, studies with tryptic digests confirm that temperature can be an important variable in controlling peak spacing of peptides in gradient elution.[26] Moreover, selectivity effects due to temperature change are independent of effects due to changes in gradient steepness. This argues that simultaneous variation of temperature and gradient steepness should be useful in optimizing polypeptide separations. This strategy may be more convenient than changing mobile-phase composition or column chemistry because gradient and temperature changes are easily implemented by changing settings in the HPLC system controller. Chromatography simulation software that allows the analyst to find optimal separation conditions for polypeptides with input of a few experimentally established temperature and steepness data is commercially available.[26] A reason for caution in the use of temperature as a chromatographic variable is the instability of many HPLC columns under conditions of elevated temperature and low pH. Fortunately, there are several reversed-phase columns now available with enhanced stability at extremes of pH and temperature.[27,28]

ANOMALOUS BEHAVIOR OF PROTEINS IN RPLC

Many globular proteins in the 10- to 100-kDa range behave as predicted[20] in terms of peak width and shape. However, deviations from expected behavior are often observed with proteins in RPLC, particularly with those larger than 25 kDa. These anomalies include excessive bandwidth, poor recovery, the observation of multiple peaks for a single species, and "ghosting." Ghosting refers to the appearance of protein peaks in a blank gradient following a chromatographic run. Anomalous behavior has been shown to arise from interconversion of the protein between different conformational states.[29,30] If the rate of interconversion during elution is fast, a single narrow band is observed. If the rate of interconversion is slow, the band may be broad and asymmetric. Very slow interconversion rates may generate multiple peaks for the different conformers. Protein unfolding within the pores may lead to entrapment, and the phenomenon of ghosting is thought to represent refolding and subsequent elution of entrapped species. Such anomalous or "nonideal" behavior of proteins in RPLC can be minimized by employing conditions that stabilize a particular conformer, e.g., operation at elevated temperature or extremes of pH. The use of sample pretreatment with denaturing conditions and separation at elevated temperature (60°C) has been used successfully to obtain acceptable peak shapes for "non-well-behaved" proteins.[31,32]

DETECTION TECHNIQUES

Absorbance detection in the low-UV region is the most popular method for polypeptides. Although the peptide bond exhibits strongest absorbance below 200 nm, the use of longer wavelengths (210 to 220) provides an acceptable tradeoff between obtaining satisfactory signal for analytes and minimizing background absorbance from mobile-phase components. Photodiode array (PDA) detectors enable multiwavelength detection (e.g. 215, 260, and 280 nm) and on-line acquisition of absorbance spectra. This information can be useful for confirming peak purity or the presence of aromatic side chains in the peptide primary sequence. For proteins, spectral information can provide clues about the solvent environment of aromatic residues, e.g., whether tryptophan and tyrosine residues are buried in the protein interior or exposed on the protein surface.

Fluorescence is not widely used as a general detection technique for polypeptides because only tyrosine and tryptophan residues possess native fluorescence. However, fluorescence can be used to detect the presence of these residues in peptides and to obtain information on their location in proteins. Fluorescence detectors are occasionally used in combination with postcolumn reaction systems to increase detection sensitivity for polypeptides. Fluorescamine, *o*-phthalaldehyde, and napthalenedialdehyde all react with primary amine groups to produce highly fluorescent derivatives.[33,34] These reagents can be delivered by a secondary HPLC pump and mixed with the column effluent using a low-volume tee. The derivatization reaction is carried out in a packed bed or open-tube reactor.

Multiangle light-scattering detectors are increasingly used to obtain on-line information on protein molecular weight. However, they must be used in combination with refractive index detectors, and so this technique is not compatible with reversed-phase gradient elution.

Mass spectrometry is the fastest-growing detection technique for HPLC in general and for proteins and peptides in particular. Electrospray ionization permits mobile-phase streams to be introduced directly into the MS system at the flow rates used with analytical HPLC columns. Among the chromatographic modes employed for polypeptide separations, reversed phase is ideally suited for ESI-MS detection because the common mobile-phase components are volatile. Virtually every type of mass analyzer (quadrupole, ion trap, time-of-flight, FT-MS, magnetic sector, and hybrid mass spectrometers) have been successfully coupled to RPLC-ESI systems. Electrospray is a soft-ionization technique that provides mostly pseudomolecular ions with little loss to fragmentation. Applications of LC-MS include (1) high-sensitivity detection of particular peptides in complex mixtures using selected ion monitoring; (2) determination of peptide and protein mass using scanning detection; (3) determination of peptide sequence using tandem MS; and (4) investigation of protein conformation and protein–protein interaction.

Although RPLC-MS is a very powerful analytical tool, not all RPLC conditions are compatible with mass spectrometric detection. Nonvolatile buffers, salts, and additives (e.g., phosphates, and nonvolatile ion-pairing agents) are not compatible with electrospray sources. Electrospray systems with orthogonal sprayers can tolerate low concentrations of such mobile-phase components for brief periods, but volatile components are preferred. For low pH operation (pH 2 to 3), phosphate should be replaced with dilute solutions of volatile organic acid (e.g., 0.1 to 1% acetic or formic acid). As discussed in the preceding text, the use of TFA causes ion suppression, and it is usually replaced with acetic or formic acid. For operation at higher pH, volatile ammonia-based buffers can be used: ammonium formate (pH 2.7 to 4.7), ammonium acetate (pH 3.7 to 5.7), or ammonium bicarbonate (pH 5.4 to 7.4). These buffers should be used at concentrations below 50 mM to avoid ion suppression. Volatile ion-pairing agents such as heptafluorobutyric acid (HFBA) can be used for basic analytes. Alkyl amine additives such as TEA, which are often used to suppress tailing of basic analytes in RPLC, will cause strong ion suppression in positive-mode electrospray. The use of surfactants as mobile-phase additives is not recommended as they rapidly contaminate the ionization source.

Although modern electrospray sources can accommodate mobile-phase flow rates of up to 1 ml/min, lower flow rates are preferred for best ionization efficiency. This can be achieved by installing a flow splitter between the column and the ionization source. Because electrospray ionization is concentration sensitive, this has little effect on signal strength. Alternatively, small-diameter columns can be used. Conventional electrospray sources are usually coupled to a 2- to 3-mm I.D. column operated at 200 to 500 μl/min. When limited amounts of sample are available, capillary columns may be preferred to increase sensitivity and minimize on-column loss of sample components. Capillary LC-MS is the preferred technique for the separation of very complex peptide mixtures generated in proteomic studies.[35] These may be proteolytic digests of protein spots obtained from 2-D gels, or digests of proteins extracted directly from cells or tissues. In the latter case, the extraordinary complexity of the samples requires coupling of multiple chromatographic modes. Multidimensional chromatographic techniques generally employ ion exchange as the first dimension, and reversed-phase LC-MS as the second dimension. Multidimensional techniques include off-line first-dimension separations, on-line column switching, or direct coupling of a mixed-bed column to the mass spectrometer.[36] In the last approach,[37] the capillary is packed with a reversed-phase material in the outlet segment and an ion-exchange material at the inlet. Samples are loaded onto the ion-exchange bed and eluted with a series of salt solutions of increasing ionic strength. Each ion-exchange eluate is resolved on the second-dimension bed with a reversed-phase gradient.

APPLICATIONS

The primary analytical applications of RPLC in the development of biopharma-ceuticals are the determination of protein purity and protein identity. Purity is established by analysis of the intact protein, and RPLC is useful in detecting the presence of protein variants, degradation products, and contaminants. Protein identity is most often established by cleavage of the protein with a site-specific protease followed by resolution of the cleavage products by RPLC. This tech-nique, termed peptide mapping, should yield a unique pattern of product peptides for a protein that is homogeneous with respect to primary sequence.

Proteins

Reversed-phase chromatography is a powerful tool for detecting protein structural changes that are introduced during biosynthesis, isolation, or processing. These may include amino acid substitutions, side-chain oxidation and deamidation, and N- or C-terminal cleavage. Protein variants that differ in the number and type of posttranslational modifications such as gycosylation, phosphorylation, acetylation, and sulfaction may also be identified by RPLC. If structural changes are located at the protein surface, the variant will more likely be distinguished using chromato-graphic conditions that favor retention of native conformation.[17,38] The typically harsh RPLC conditions using acid water:acetonitrile:TFA mobile phases and hydrophobic octadecyl or octyl ligands may not be appropriate. Instead, alcohol modifiers (meth-anol, propanol) and neutral pH buffers may be used with short-chain (C4) or polar (cyano) stationary phases. Confidence in the purity of a protein separation is increased by the appearance of a single peak under two different sets of RPLC conditions, i.e., different pH values, temperatures, or stationary phases.[38]

Proper protein folding is a concern in the production of a bioactive protein therapeutic, and it is of interest to verify that the conformation of a recombinant protein is the same as the wild-type molecule. Because retention in RPLC depends on the surface hydrophobic contact area, comparable chromatographic behavior of a recombinant protein with that of the wild-type molecule provides evidence of similar 3-D structure.[17,38]

The literature is rich in publications describing the use of RPLC for protein purification and characterization, and a review is provided in Reference 17. Two examples are given here as illustration.

Interleukin 2 (IL-2) is a hydrophobic immune modulator protein with a mol wt of about 15,500 Da. Recombinant IL-2 expressed in *E. coli* forms insoluble aggregates located in inclusion bodies. These can be easily isolated from host cell extracts but must be solubilized under denaturing conditions and refolded to yield active protein. Reversed-phase chromatography has been used as the prin-ciple purification step for IL-2.[39,40] RPLC has also been used[41] to determine the purity of IL-2 isolated by gel permeation chromatography and to monitor the refolding of denatured IL-2 to the native state under different renaturing condi-tions (Figure 4.10).

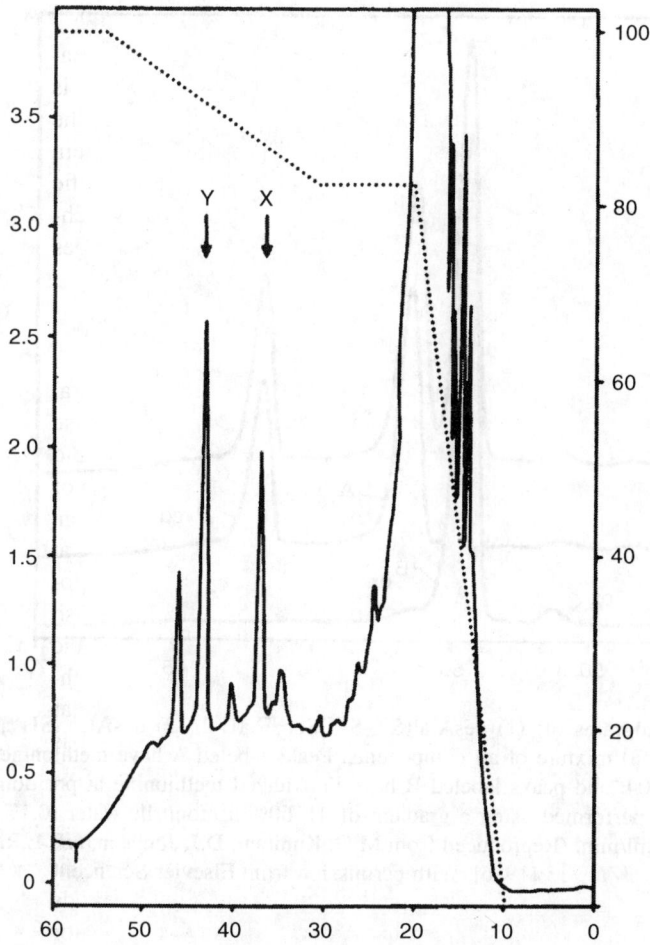

Figure 4.10 Reversed-phase chromatography of refolded IL-2. Peak X is oxidized IL-2 and peak Y is reduced IL-2. (Reproduced from M.P. Weir and J. Sparks, *Biochem. J.,* 245: 85 [1987]. With permission from Portland Press.)

Amino acid variants of IL-2 have been used to investigate the relationship between retention and protein structure in gradient RPLC.[22] The protein contains three cysteine residues in its primary sequence at positions 58, 105, and 125. The two located at positions 58 and 105 are linked in a disulfide bridge in the native molecule. A series of variants in which the three cysteinyl residues were replaced with serines were compared. Substitution with serine at positions 58 or 105 forces the molecule to form an unnatural disulfide between positions 125 and 58 or 105. A methionine residue located at position 104 can also be oxidized to the sulfoxide

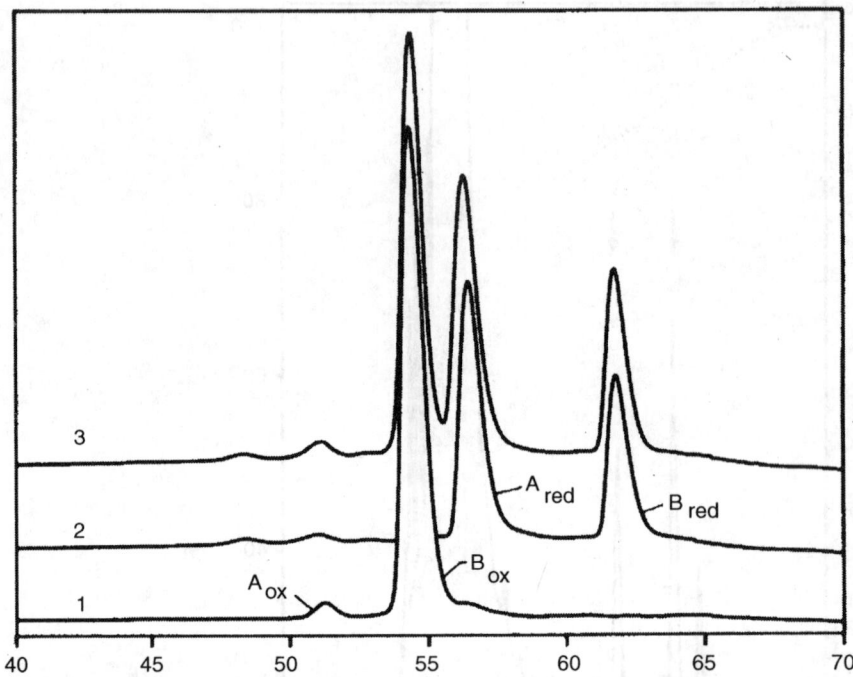

Figure 4.11 Chromatograms of: (1) desAla^1(S^{58}–S^{105})Ser^{1125} IL-2, (2) desAla^1 (SH^{58}, SH^{105})Ser^{125} IL-2, and (3) mixture of all components. Peaks labeled A have methionine sulfoxide at position 104, and peaks labeled B have unoxidized methionine at position 104. Separations were performed using a gradient of 41–60% acetonitrile-water (0.1% TFA) in 60 min at 0.5 ml/min. (Reproduced from M.G. Kunitani, D.J. Johnson, and L.R. Snyder, *J. Chromatogr., 371:* 313 [1986]. With permission from Elsevier Science.)

form. All of these variants have similar molecular weight and hydrophobicity and should be expected to exhibit similar S values in gradient elution. This was not the case; variants that possessed the unnatural disulfide bridge exhibited higher-than-expected S values. This indicated conformational instability in these variants that exposed greater hydrophobic contact areas during gradient elution. Plots of k^* vs. Φ for the oxidized vs. reduced and methionyl vs. methionyl sulfoxide forms obtained by computer modeling studies predicted that these four species could be resolved by manipulation of low rate F and gradient range ($\Delta\Phi$), and this was confirmed experimentally (Figure 4.11).

Although gradient elution is generally required for RPLC separations of proteins, isocratic elution can be successful in some instances. For example, isocratic elution has been used for the determination of purity of production batches of biosynthetic human growth hormone (HGH).[42] The method was used to

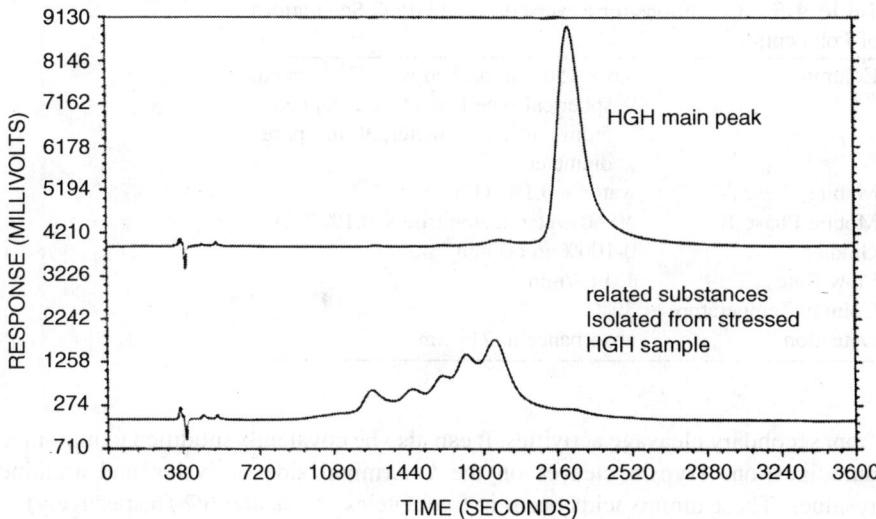

Figure 4.12 Analysis of biosynthetic human growth hormone by isocratic reversed-phase chromatography. To generate all possible degradation products, a production lot of HGH was exposed to 40°C. The profile of the unstressed HGH is shown in the upper trace. (Reproduced from R.M. Riggin, G.K. Dorulla, and D.J. Miner, *Anal. Biochem., 167:* 199 [1987]. With permission from Elsevier Science.)

resolve deamidated species and sulfoxide derivatives from native growth hormone (Figure 4.12), which could not be adequately resolved using reversed-phase gradient elution or anion-exchange chromatography. It is interesting to note that best resolution of degradation products from the parent protein required conditions which promoted the retention of native conformation (C4 stationary phase, 71:29 50 mM Tris buffer, pH 7.5:n-propanol). The use of acidic conditions (sodium phosphate, pH 2) produced poor peak shapes and inadequate resolution.

Peptides

The extraordinary ability of reversed-phase chromatography to resolve peptides with very subtle differences in sequence or structure makes it the method of choice for peptide mapping. It is not uncommon for two peptides differing in a single residue or the presence of a single side-chain modification to be fully resolved. The high efficiency of monomeric, microparticulate reversed-phase columns generates peak capacities in excess of 250, enabling separation of the complex peptide mixtures often encountered in peptide-mapping experiments.

A variety of proteases are used for peptide mapping, but trypsin is the most popular because it is stable and available in high purity with little contamination

Table 4.5 Conditions for Reversed-Phase HPLC Separations of Polypeptides

Column	4.6 × 150 mm packed with C8-bonded spherical type B silica, endcapped, 5 μm particle diameter, 30 nm pore diameter
Mobile Phase A	water + 0.1% TFA
Mobile Phase B	40:60 water:acetonitrile + 0.1% TFA
Gradient	0-100% in 60 min
Flow Rate	1.0 ml/min
Column Temperature	30°C
Detection	absorbance at 210 nm

from secondary cleavage activities. It can also be covalently modified to minimize autodigestion. Trypsin cleaves on the C-terminal side of lysine and arginine residues. These amino acids occur at frequencies of 5% and 6% (respectively) in the proteome, so tryptic peptides are usually in the range of 2 to 20 residues in length. This has several beneficial consequences for tryptic mapping experiments. First, peptides of this size possess little or no secondary structure, so that conformational effects do not cause aberrant peak morphology. Second, molecular diffusion rates are relatively high, so that mass-transfer kinetics are favorable. Third, tryptic peptides are usually within the mass range of quadrupole and ion-trap mass spectrometers. Thus, peptide mass and sequence information is easily obtained with these robust and relatively inexpensive detectors. Fourth, because every tryptic peptide has a C-terminal lysine or arginine in addition to the N-terminal amine group, they produce doubly charged ions in electrospray ionization. This facilitates fragmentation in tandem MS experiments and aids in the interpretation of MS-MS spectra.

The standard RPLC conditions used for polypeptides (Table 4.5) are usually satisfactory for tryptic peptides. The resolution of the 21 tryptic peptides of recombinant HGH[26] was obtained using these mobile-phase and column conditions with elevated temperature (Figure 4.13). Modification of side-chain residues is likely to change the chromatographic properties of a peptide such that the magnitude of that peak is diminished and a new peak appears with the retention time characteristic of the modified peptide. For example, phosphorylation of serine or threonine residues, or oxidation of methionine residues, will shift the retention of a peptide to earlier times. Similarly, deamidation of glutamine or asparagine residues will increase peptide retention.[38] Disulfide bridges can be located by observing peak shifts in digests prepared with and without reducing agents (e.g., dithiothreitol, β-mercaptoethanol). Glycosylated peptides can be recognized by a shift in peptide retention following enzymatic treatment to remove carbohydrate groups, for example, incubation with the enzyme PNGase to remove N-linked carbohydrate moieties.

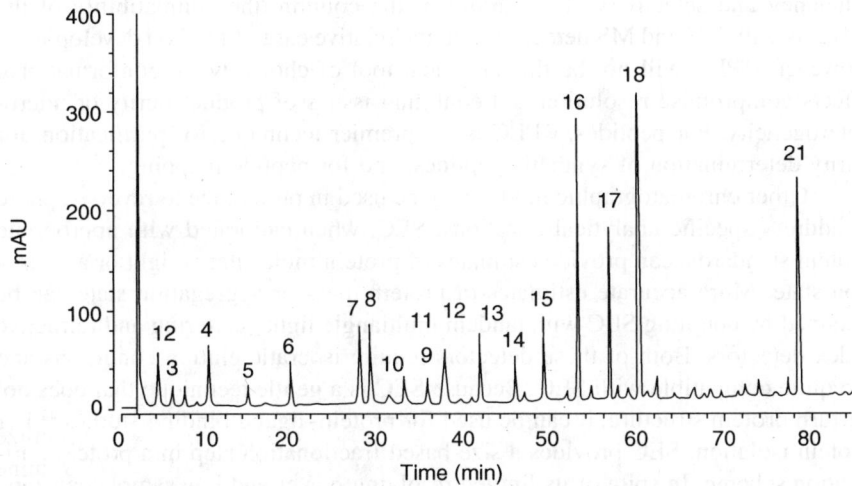

Figure 4.13 Separation of the tryptic peptides from recombinant HGH using a 120-min linear gradient at 60°C from water + 0.1% TFA to 40:60 water:acetonitrile + 0.1% TFA. Column: Zorbax SB-C8, 4.6 × 150 mm, 30-nm pore, 5-μm particle size. (Reprinted from W.S. Hancock, R.C. Chloupek, J.J. Kirkland, and L.R. Snyder, *J. Chromatogr. A, 686:* 31 [1994]. With permission from Elsevier Science.)

COMPARISON WITH ALTERNATIVE TECHNIQUES

Reversed-phase HPLC can be compared in terms of its utility with other modes of chromatography and with other separation techniques such as gel electrophoresis, capillary electrophoresis (CE), and capillary electrochromatography.

Other Chromatographic Modes

Other modes of chromatography that are used for protein and peptide separations include size exclusion chromatography (SEC), ion-exchange chromatography (IEC), HIC, immobilized metal affinity chromatography (IMAC), affinity chromatography, and hydrophilic interaction chromatography (HILIC). All of these tend to be less denaturing than RPLC and will be preferred for preparative isolation of proteins with retention of native conformation and biological activity. Reversed phase can be used as a preparative tool and is most successful for small proteins (<25 kDa), those that retain native conformation under reversed-phase conditions, or proteins that easily refold to the native state following chromatography. Preparative isolation of a protein from a complex sample often requires sequential purification steps with two or more chromatographic modes. Where possible, reversed phase may be included in a purification scheme because it offers a unique chromatographic selectivity. For analytical applications, reversed-phase chromatography is usually the technique of choice because of its high efficiency and selectivity, the stability of the column, the compatibility of the solvents with UV and MS detection, and the relative ease of method development. However, RPLC will not be the analytical tool of choice when conformational effects compromise resolution and confound issues of product purity or micro-heterogeneity. For peptides, RPLC is the premier technique for purification, for purity determination of synthetic peptides, and for peptide mapping.

Other chromatographic modes may be used in preference to reversed phase to address specific analytical questions. SEC, when calibrated with appropriate protein standards, can provide estimates of protein molecular weight or aggregation state. More accurate estimates of protein mass or aggregation state can be obtained by coupling SEC with tandem multiangle light-scattering and refractive index detectors. Both of these detectors require isocratic elution conditions and are quite compatible with SEC. Because SEC is a gentle technique that does not perturb protein structure, it can be used for protein–ligand binding studies.[43] For protein isolation, SEC provides a size-based fractionation step in a protein puri-fication scheme. In spite of its limited resolving power and low sample capacity, the excellent recovery of protein mass and activity from SEC columns favors the inclusion of SEC steps in purification protocols.

IEC is usually carried out under physiological conditions, e.g., at neutral pH with salt solutions. It is therefore widely used for purification of proteins in their native state with high recovery of mass and activity. From an analytical perspective, ion exchange may be used in preference to reversed phase to deter-mine purity or charge heterogeneity in cases where the denaturing conditions of RPLC would yield confusing results. In particular, IEC can provide information about protein variants with differences in surface charge.

HIC, like IEC, is performed under conditions that preserve protein shape and activity. It is used in preparative applications to obtain a selectivity compli-mentary to IEC and akin to RPLC but without the denaturing properties of the latter technique. Although HIC and RPLC share a mechanism based on hydro-phobic partitioning, the actual peak spacing and elution order of the two techniques can be different. This arises from the different hydrophobic contact points presented by the protein under native (HIC) and denaturing (RPLC) conditions. Although not widely used for analytical separations, HIC can be used to answer questions about accessible hydrophobic surface area that cannot be addressed by RPLC.[44]

Affinity chromatography and related techniques (e.g., thiol chromatography and IMAC) are widely used for preparative isolation because they enable a single protein or class of proteins to be selectively purified from very complex mixtures. They may be occasionally used as analytical tools. For example, protein A affinity chromatography has been used for quantitative analysis of immunoglobulins in ascites fluid.[45] Information about surface-accessible histidine and phosphate groups may be obtained using IMAC.

HILIC is a variant of normal-phase chromatography that employs polar stationary phases and RPLC-type mobile phases. Because HILIC separations occur by a normal-phase mechanism, the organic component of the mobile phase (typically acetonitrile) is the weak solvent and the aqueous component (an acidic or neutral pH buffer) is the strong solvent. Elution is generally accomplished with a gradient from low to high percentages of the aqueous component. Polar polymeric stationary phases are used, such as polyhydroxyethyl aspartamide.[46] HILIC offers an alternative to RPLC for very polar species that are poorly retained on reversed-phase columns, such as phosphorylated peptides[47] and hydroxyproline-rich peptides.[48] HILIC elution solvents are also compatible with electrospray–MS. In contrast to RPLC, less-retained analytes from a HILIC column will elute at high organic solvent concentrations that are more easily desolvated in the ionization process.

Gel Electrophoresis

For decades, polyacrylamide gel electrophoresis (PAGE) has been the primary tool of the protein biochemist for characterizing protein mixtures, estimating protein molecular weight and isoelectric point, monitoring protein purification, and determining microheterogeneity. Gel electrophoretic techniques include native PAGE, SDS-PAGE, and isoelectric focusing (IEF). Two-dimensional PAGE couples the latter two techniques in a two-step process. Native and SDS-PAGE offer several advantages compared to column chromatographic techniques. When separations are performed on slab gels, several samples can be run in parallel lanes on the gel, and individual separations can be directly compared with each other and with standards and reference materials run on the same gel. If the PAGE experiment is terminated at the appropriate time, all sample components will be confined between the origin and the tracking dye. This is in contrast to chromatography in which sample components that are strongly retained on the stationary phase will not be detected. The disadvantages of PAGE include the requirement for staining to detect protein bands, the labor involved in gel preparation and processing, and gel-to-gel reproducibility. Stain intensity is either nonlinear with protein mass or linear over a narrow range, and staining intensity varies from protein to protein. Gel electrophoresis is therefore a nonquantitative or, at best, semiquantitative technique. Casting, running, staining, and destaining gels are time consuming, and reproducibility of the results is dependent on the expertise of the analyst. These limitations are minimized by the use of commercial precast gels and robotic systems for processing gels. Also, because a dozen or more lanes can be run on a single gel, the analysis time on a per-sample basis can compare favorably with serial techniques such as chromatography. For all these reasons, PAGE is often the preferred technique for capturing a "snapshot" of the progress of a protein purification process.

For analytical applications, chromatography will be preferred over PAGE where quantitative information is necessary and where automation is desired. As an instrumental technique, all components of a chromatographic analysis including injection, separation, detection, and data analysis are fully automated and under single-point control by the HPLC workstation. The limitation of serial analysis is offset by the ability to analyze a hundred or more samples with no further operator intervention after loading samples in the autosampler. Analytes are detected by their native absorbance with linear dynamic ranges of up to five orders of magnitude, and peak-area precision is typically better than 2% RSD. Reversed-phase separations are highly reproducible, with retention time precision typically 1% RSD or better. When regulatory considerations require precise quantitation of product and impurity levels, chromatography will be the method most likely to return reliable information. Reversed-phase chromatography is almost always the first choice for the analysis of peptides. Peptides do not stain well with the standard PAGE staining techniques, and they can easily be lost by diffusion at the end of the electrophoretic separation.

For accurate determination of protein molecular weight, mass spectrometry and LC-MS have largely displaced SDS-PAGE. However, SDS-PAGE will still be used where estimates of molecular weight suffice or where MS instrument time is limited.

Capillary Electrophoresis and Capillary Electrochromatography

Because it shares the separation chemistries of gel electrophoresis and the instrumental advantages of HPLC, CE provides an alternative when either of these two competing techniques fail to deliver the desired performance.[49] Like gel electrophoresis, CE separations are based on differential migration in an electric field, and CE analogs for each gel electrophoresis mode are available: capillary zone electrophoresis (CZE [native PAGE]), capillary IEF, and SDS-entangled polymer sieving (SDS-PAGE). As an instrumental technique, CE is fully automated and employs on-line optical detection with similar benefits to LC-UV detection. The potential for replacing gel electrophoresis with precise and quantitative methods has spurred the growth of CE in the biotechnology and biopharmaceutical industries.

CE is strictly an analytical technique. The dimensions of the fused silica capillaries limit sample loads to picogram amounts, which precludes its use as a preparative tool. However, this requirement for limited sample amounts is an advantage when using CE as a complement to RPLC. CZE has been remarkably successful for the separation of peptides.[50] Because the separation of peptides in CZE is dependent on differences in mass/charge ratio, the separation selectivity of CZE is distinct from that of RPLC (in which separation is based on differences in hydrophobicity). This orthogonality in separation selectivity predicts that peptides that coelute in RPLC might be resolvable by CZE. This in fact is often the case.[51] Consequently, CZE is used as a companion technique to RPLC to confirm protein identity by peptide mapping. Also, it can be used for rapid assessment of the purity of peaks collected from an RPLC peptide-mapping experiment and to guide the analyst in further characterization of the fraction by mass spectrometric or chemical sequence analysis.

Micellar electrokinetic chromatography (MEKC) and capillary electrokinetic chromatography (CEC) are, as their names imply, chromatographic techniques performed with CE instrumentation. Both are based on hydrophobic partitioning into a nonpolar phase and use electroosmotic flow (EOF) as a means of transporting the mobile phase through the column. In MEKC, the capillary is filled with an aqueous solution of a surfactant, typically SDS. The nonpolar phase is created by the formation of SDS micelles, which are in rapid equilibrium with surfactant monomers. Micelles and monomers move electrophoretically toward the anode, while EOF carries the bulk solution toward the cathode. Separation is accomplished by differential partitioning of analytes into the micellar phase in combination with electrophoretic contributions dictated by the analyte mass and charge state. MEKC has been occasionally used for peptide separations,[52] where it can offer the benefits of nanoscale separations with a selectivity similar to RPLC. MEKC-like conditions have also been used for protein separations,[53] although the mechanism probably includes contributions from micellar partitioning and protein–surfactant complexation.

CEC employs capillaries packed with conventional microparticulate materials. Most CEC separations have been achieved with reversed-phase packings and aqueous–organic mobile phases. EOF provides the driving force for transporting the mobile phase. In contrast to pressure-driven flow, laminar flow band broadening is eliminated and column back pressure is absent. Thus, CEC separations are characterized by high efficiencies compared to pressure-driven HPLC. Peptide separations have been performed by CEC,[54] but the difficulty in performing gradient elution on commercial systems limits the usefulness of this technique in practical applications.

All three of these capillary techniques (CE, MEKC, and CEC) share two limitations that have hindered their adoption as general replacements for conventional RPLC. First, the nanoliter volumes of sample and mobile phase require the use of on-tube detection. The reduction of detector light path from millimeters in HPLC to micrometers in CE compromises detection sensitivity. Second, changes in the state of the wall in CE and MEKC or of the particle surface in CEC will alter the magnitude of EOF, causing poor reproducibility of migration/retention times and peak areas. Sample adsorption to active sites on the wall or packing is the most frequent cause of drifting EOF. Much work has been done to minimize this problem, including the use of coated capillaries, capillary regeneration techniques, and electrolyte additives. To date, CZE separation of peptides has met with the greatest success because the low-pH electrolytes used for peptide separations suppress silanol ionization and peptide adsorption.

References

1. C. Horvath, W. Melander, and I. Molnar, *J. Chromatogr., 125:* 129 (1976).
2. R.F. Rekker, The Hydrophobic Fragmental Constant. Elsevier, Amsterdam, 1977, p. 302.
3. I. Molnar and C. Horvath, *J. Chromatogr., 142:* 623 (1977).
4. J.L. Meek, *Proc. Natl. Acad. Sci. U.S.,* 77: 1632 (1980).

5. J.L. Meek and Z.L. Rossetti, *J. Chromatogr., 211:* 15 (1981).
6. D. Guo, C.T. Mant, A.K. Taneja, J.M.R. Parker, and R.S. Hodges, *J. Chromatogr., 359:* 499 (1986).
7. D. Guo, C.T. Mant, A.K. Taneja, and R.S. Hodges, *J. Chromatogr., 359:* 519 (1986).
8. X.M. Lu, K. Benedek, and B.L. Karger, *J. Chromatogr., 359:* 19 (1986).
9. Lichrospher & Lichroprep Sorbents Tailored for Cost Effective Chromatography, EM Separations, Gibbstown.
10. N.B. Afeyan, N.F. Gordon, I. Mazaroff, L. Varady, S.P. Fulton, Y.B. Yang, and F.E. Regnier, *J. Chromatogr., 519:* 1 (1990).
11. K. Kalghatgi and C. Horvath, Micropellicular sorbents for rapid reversed-phase chromatography of proteins and peptides, in *Analytical Biotechnology, Capillary Electrophoresis, and Chromatography,* C. Horvath and J.G. Nikelly (Eds.), American Chemical Society, Washington, D.C., 1990, p. 162.
12. D. Lubda, K. Cabrera, W. Kraas, C. Schaefer, and D. Cunningham, *LC GC N. Am., 19:* 1178 (2001).
13. M. Kawakatsu, H. Kotaniguchi, H. Freiser, and K.M. Gooding, *J. Liq. Chromatogr., 18:* 633 (1995)
14. A. Apfel, S. Fischer, G. Goldberg, P.C. Goodley, F.E. Kuhlmann, *J. Chromatogr. A, 712:* 177 (1995).
15. T. Wehr, *LC GC N. Am., 18:* 406 (2000).
16. D. Guo, C.T. Mant, and R.S. Hodges, *J. Chromatogr., 386:* 205 (1987).
17. M.T.W. Hearn, Reversed-phase and hydrophobic interaction chromatography of proteins and peptides, in *HPLC of Biological Macromolecules,* 2nd ed., K.M. Gooding and F. Regnier (Eds.), Marcel Dekker, New York, 2002, pp. 172–173.
18. J.E. Rivier, *J. Liq. Chromatogr., 1:* 343 (1978).
19. M.A. Stadalius, H.S. Gold, and L.R. Snyder, *J. Chromatogr., 296:* 31 (1984).
20. L.R. Snyder and M.A. Stadalius, HPLC separations of large molecules: a general model, in *HPLC — Advances and Perspectives,* Vol. 4, C. Horvath (Ed.), Academic Press, New York, 1986, pp. 195–312.
21. M.W. Dong, J.R. Gant, and B.R. Larsen, *BioChromatography, 4:* 19 (1989).
22. M.G. Kunitani, D.J. Johnson, and L.R. Snyder, *J. Chromatogr., 371:* 313 (1986).
23. C.T. Wehr and L. Correia, *LC at Work LC-121,* Varian.
24. K.K. Unger, R. Janzen, and G. Jilge, *Chromatographia, 24:* 144 (1987).
25. J.L. Glajch, M.A. Quarry, J.F. Vasta, and L.R. Snyder, *Anal. Chem., 58:* 280 (1986).
26. W.S. Hancock, R.C. Chloupek, J.J. Kirkland, and L.R. Snyder, *J. Chromatogr. A, 686:* 31 (1994).
27. J.L. Glajch and J.J. Kirkland, *LC GC, 8:* 140 (1990).
28. R.E. Majors, *LC GC, 18:* 262 (2000).
29. S.A. Cohen, K. Benedek, Y. Tapuhi, J.C. Ford, and B.L. Karger, *Anal. Biochem., 144:* 275 (1985).
30. M.T.W. Hearn, A.N. Hodder, and M.-I. Aquilar, *J. Chromatogr., 327:* 47 (1985).
31. W.G. Burton, K.D. Nugent, T.K. Slattery, B.R. Summers, and L.R. Snyder, *J. Chromatogr., 443:* 363 (1988*)*.
32. K.D. Nugent, W.G. Burton, T.K. Slattery, B.F. Johnson, and L.R. Snyder, *J. Chromatogr., 443:* 381 (1988).
33. R. Newcomb, *LC GC, 10:* 34 (1992).
34. S.M. Lunte and O.S. Wong, *LC GC, 7:* 908 (1989).
35. T. Wehr, *LC GC, 19:* 702 (2001).

36. H. Liu, D. Lin, and J.R. Yates, *Biotechniques, 32:* 898 (2002).
37. M.P. Washburn, D. Wolters, and J.R. Yates, *Nature Biotechnol., 19:* 242 (2001).
38. J. Frenz, W.S. Hancock, and W.J. Henzel, Reversed phase chromatography in analytical biotechnology of proteins, in *HPLC of Biological Macromolecules,* K.M. Gooding and F.E. Regnier (Eds.), Marcel Dekker, New York, 1990, pp. 158–168.
39. L.E. Henderson, J.F. Hewetson, R.F. Hopkins, R.G. Sowder, R.H. Neubauer, and H. Rabin, *J. Immunol., 131:* 810 (1983).
40. A.S. Stern, Y.E. Pan, D.L. Urdal, D.Y. Mochizuki, S. DeChiara, R. Blagher, J. Wideman, and S. Gillis, *Proc. Natl. Acad. Sci. U.S.A., 81:* 871 (1984).
41. M.P. Weir and J. Sparks, *Biochem. J., 245:* 85 (1987).
42. R.M. Riggin, G.K. Dorulla, and D.J. Miner, *Anal. Biochem., 167:* 199 (1987).
43. B. Sebille, The measurement of interactions involving proteins by size exclusion chromatography, in *HPLC of Biological Macromolecules,* K.M. Gooding and F.E. Regnier (Eds.), Marcel Dekker, New York, 1990, pp. 585–621.
44. R.E. Shansky, S.-L. Wu, A. Figueroa, and B.L. Karger, Hydrophobic interaction chromatography of biopolymers, in *HPLC of Biological Macromolecules,* K.M. Gooding and F.E. Regnier (Eds.), Marcel Dekker, New York, 1990, p. 138.
45. D. Burke, Affinity chromatography and related techniques, in *Basic HPLC and CE of Biomolecules,* R.L. Cunico, K.M. Gooding, and T. Wehr (Eds.), Bay Bioanalytical Laboratory, Richmond, CA, 1998, pp. 223–242.
46. A.J. Alpert, *J. Chromatogr., 499:* 177 (1990).
47. J.A. Boutin, A.P. Ernould, G. Ferry, A. Genton, and A.J. Alpert, *J. Chromatogr., 583:* 137 (1992).
48. M.J. Kieliszewski, M. O'Neill, J. Leykam, and R. Orlando, *J. Biol. Chem., 270:* 2541 (1995).
49. T. Wehr, R. Rodriguez-Diaz, and M. Zhu, *Capillary Electrophoresis of Proteins,* Marcel Dekker, New York, 1999, pp. 20–21.
50. T. van de Goor, A. Apfel, J. Chakel, and W. Hancock, Capillary electrophoresis of peptides, in *Handbook of Capillary Electrophoresis, 2nd ed.,* J. Landers (Ed.), CRC Press, Boca Raton, 1997, p. 214.
51. R.G. Nielsen, G.S. Sittampalam, and E.C. Rickard, *Anal. Biochem., 177:* 120 (1989).
52. A. Amini, S.J. Dormady, L. Riggs, and F.E. Regnier, *J. Chromatogr. A, 894:* 345 (2000).
53. J.H. Beattie and M.P. Richards, *J. Chromatogr., 664:* 129 (1994).
54. K. Walhagen, K.K. Unger, and M.T. Hearn, *Anal. Chem., 73:* 4924 (2001).

36. H. H. D. Jia and C. R. Yates, *Bioorganic Int.* 73, 498 (2005).
37. V. H. Wysocki, L. Wойерs, and J. C. Yang, *Mass Spectrom.* 35, 1399 (2000).
38. J. Bretz, W. S. Hancock, and V. A. Heller, *Reversed phase chromatography in analytical biochemistry of proteins,* in *J. J. Pingoud, Macromolecules, Laboratory guide*, Weinheim: VCH, New York, Dekker, New York, 1990, pp. 335–345.
39. E. P. Hoefnagel, W. Heyvaert, R. J. Hopkins, R. C. Sundell, R. J. Peterson, and H. Rautela, *J. Anal.* 1213 (1983).
40. A. G. Glover, T. S. Pan, D. Lai, and Y. Morishima, S. Devman, R. Brigitte, J. Morishima, and S. J. Kittle, *Anal. Methods, New York*, 42, 871 (1988).
41. M. P. Wuronov, J. Ipatov, B. Kipper, *Anal.* 1, 213 (1987).
42. H. M. Pugachin, R. Hoshi, and D. G. Miller, *Anal. Biochem.* 71, 199 (1987).
43. P. Sweilo, *The preparation of chimeric protein by shuffling, process by steric solution encompassing*, in *Methods in Enzymology*, New York, 1990.
44. R. E. Smith, S. S., Wu, A. Fortune, and J. Larsson, *Hydrophobic interaction chromatography of biopolymers*, in *Handbook of biopharmaceutical techniques* (S. R., Thomas), L. R. Zaner (eds.), Marcel Dekker, New York, 1990, p. 189.
45. P. turbid Amphi-entrophoresis and related techniques, in *HPLC and*, Ford (ed.), P. L., Cannon, R. M. Morrison, L. V. Wohlgast (eds.), Butterworth, Boston, 1996, pp. 201–230.
46. E. J. Alpert, *J. Chromatogr.* 499, 177 (1990).
47. J. A. Rostichi, J. P. Chang, C. Horv, V. Simon, and M. Schmidt, *Anal. Chem.* 1, 193 (1997).
48. M. T. Stephenson, O. O'Sullivan, Development in Chromatography, Boston, 1990.
49. R. White, R. O. Ashton, and M. Thomas, *Techniques in protein chemistry*, 1995, p. 10.
50. J. V. McCook, A. Ayeh, McCook, and M. Thomas, Carbohydrate determination, in *Handbook of analytical separation*, Boca Raton, CRC Press, Boca Raton, 1990, p. 210.
51. K. H. Hierons, C. J. Strappingham, and F. J. Thomas, *Anal. Biochem.* 177, 420 (1989).
52. A. Amitai, S. J. Brouwer, C. Kipper, and H. E. Reguera, *Anal. Biochem.* 278, 159 (2001).
53. H. Bentz and M. P. Mirande, J. Chromatogr. 1, 204 (1990).
54. R. Stallings, K. K. Thiga, and M. Tallmann, *Anal. Chem.* 29, 1657 (2001).

5

Practical Strategies for Protein Contaminant Detection by High-Performance Ion-Exchange Chromatography

Pete Gagnon

Ion-exchange chromatography (IEC) is a flexible and powerful analytical tool for aiding purification process development, performing in-process monitoring, and documenting final product quality. The primary retention mechanism in IEC is simple: electrostatic interactions. Anion exchangers carry a positive charge. Negatively charged molecules are retained; the more strongly electronegative, the more strongly they are retained. The opposite is true for cation exchangers. In either case, retained molecules can be eluted in order of increasing retention. The most common method of elution is by means of a salt gradient in which an increasing concentration of dissolved ions eventually outcompete the bound analyte for the support. Weakly bound analytes elute first, followed by analytes that are more strongly bound. A less common method of elution is to suspend the attraction between the column and the bound analyte by changing the charge on the analyte. This is accomplished by changing the pH of the mobile phase. In the case of amphoteric molecules such as proteins, for example, reducing the pH reduces negative charge and increases positive charge, thereby weakening the attraction to anion exchangers. The opposite is true for cation exchangers.

IEC presents a number of significant advantages. It is fast, with analysis times generally ranging from 5 to 15 min, and it is readily automatable. These two features endow it with the ability to support high throughput. Although the columns can be expensive, the cost per analysis is low. It does not employ toxic chemicals or require special handling. It is simple and very versatile.

IEC is commonly able to differentiate a protein product from its own variants, such as deamidated forms, terminal and intrachain enzymatic clips, and misfolded variants, as well as from a wide range of nonproduct protein contaminants. Likewise, in the case of protein conjugates, analytical IEC can discriminate and provide relative quantitation of unreacted materials, various conjugate morphs, and other reaction by-products. These extensive capabilities naturally come at a price. That price is paid in two essential installments: minimizing variation in the analytical system and thoroughly evaluating the scope of potential selectivities. This chapter describes practical strategies for addressing both points.

MINIMIZING VARIATION IN THE ANALYTICAL SYSTEM

The primary source of variation in an assay system is usually the sample itself. This is the reason to have an assay in the first place. However, there may be variations in sample composition other than those toward which the assay is directed, and those variations may detract from the performance and reproducibility of the assay. Eliminating, or at least controlling, these influences can significantly improve resolution and sensitivity, as well as reproducibility. An important example is the ability of proteins to form stable complexes with a wide range of other solutes. Such complexes exhibit a range of aberrant retention behaviors that can confound interpretation of analytical results.

Metal complexation — One of the most insidious and widely occurrent sources of analytical variation in IEC is product complexation with metal ions. Most proteins can form complexes with metals, regardless of whether or not they are metalloproteins.[1] Participant metal ions can derive from the cell culture production process, purification process buffers, or even stainless steel chromatography systems. Complexation can alter retention times, create aberrant peaks, and substantially increase peak width. To the extent that metal contamination of your sample is uncontrolled, so too will be the performance of your assay.

A complete systematic description of protein–metal complexation has yet to be presented, but it is apparent that many mechanisms are involved. Some proteins may participate in classical chelation interactions via polycarboxy clusters on their surfaces.[2] Others interact with metals via coordination with polyhistidyl or other aromatic domains.[1,3–5] Still others may interact with metals via sulfhydryl residues.[1,3] The literature on immobilized metal affinity reveals examples of unexplained retention that may involve yet other mechanisms.[1]

Protein polycarboxy sites can interact with a wide range of metals. The interaction partially or wholly neutralizes the charge on involved carboxyl groups, making the protein less electronegative.[6] Calcium and (ferric) iron interact most strongly under mildly acidic conditions. Their effects are most often observed on cation exchangers, where the complexed forms tend to exhibit stronger retention than the native protein. Other metals interact more strongly under neutral and alkaline conditions, in which the reduction of electronegativity weakens retention on anion exchangers.

Among protein aromatic groups, histidyl residues are the most metal reactive, followed by tryptophan, tyrosine, and phenylalanine.[1] Copper is the most reactive metal, followed in order by nickel, cobalt, and zinc. These interactions are typically strongest in the pH range of 7.5 to 8.5, coincident with the titration of histidine. Because histidine is essentially uncharged at alkaline pH, complexation makes affected proteins more electropositive. Because of the alkaline optima for these interactions, their effects are most often observed on anion exchangers, where complexed forms tend to be retained more weakly than native protein. The effect may be substantial or it may be small, but even small differences may erode resolution enough to limit the usefulness of an assay.

Metal decomplexation can usually be accomplished by pretreating the sample with 0.05 M EDTA and 0.05 M imidazole. The sample need not be buffer exchanged prior to analysis so long as the injection volume is kept low. Injections up to 5% of the column volume (CV) may be tolerated by strongly retained proteins. Reduction to 2% CV may be necessary for weakly retained proteins. Inclusion of these dissociating agents in the actual chromatography buffers can be risky because they may prevent binding of weakly charged proteins. This is especially true with EDTA on anion exchangers, where its tetravalency makes it a very strong eluting ion. On cation exchangers at low pH, both agents will interfere with protein binding.

Charge complexation — Simple charge complexes between solutes can also occur. Alkaline proteins, for example, commonly form stable charge complexes with acidic solutes such as DNA, phospholipids, acidic proteins, and even with polyanionic buffer components such as phosphate or citrate.[6] This problem tends to be most severe at lower pH values where proteins carry the strongest positive charge and is consequently observed most often on cation exchangers. Strong charge complexes may survive the ion-exchange process and create a variety of retention of artifacts. Weak charge complexes may dissociate spontaneously in the presence of strong charge competition by the ion-exchange surface.

The most problematic part of charge complexation in IEC assays is sample loading. This is because sample loading in IEC is typically done at ionic strengths that may be too low to fully dissociate charge complexes. Depending on the charge properties of the cocomplexant, a proportion of the product may be neutralized and fail to bind. If cocomplexant content is uncontrolled from sample to sample, then assay reproducibility will be sacrificed as well.

Even strong charge complexes can be dissociated with 1 M NaCl, and most can be dissociated with half that or less. Ion exchangers will generally tolerate salt levels of this magnitude in a sample so long as the injection volume is kept at 2% or less of the CV. Be aware that complexes formed initially by simple charge attraction may be stabilized secondarily by hydrophobic interactions and hydrogen bonding. The inclusion of 2 M urea will help to dissociate both these mechanisms. Experimentation may allow you to reduce the salt concentration, dispense with the urea, and increase injection volume.

In general, charge cocomplexants exhibit dramatically different retention characteristics than the product of interest and will not reassociate spontaneously after a sample has been injected. The blind spot in this generalization is the potential for postelution reformation of complexes between the product and buffer components, chiefly phosphates and citrates with strongly alkaline proteins. If you are going to use such buffers, at least run a parallel control experiment with a monovalent buffer.

Affinity complexation — Product complexation with leached affinity ligands from a previous purification step is one of the most difficult complexation problems, mainly because of the strong association between the complexants. The charge properties of the leached ligand can influence product retention either by direct charge alteration, or — if the ligand is multivalent — by creating product aggregates ranging from dimers to much larger assemblies.

IgG complexation with leached protein A is a good example. Native protein A has five IgG-binding domains, each with one Fc-binding site and one FAb-binding site.[7] This gives each domain the ability to bind two IgG molecules. The domains are linked in a linear series by sections of random coil. Leaching occurs mainly due to proteolysis of the random coil sequences. Thus the leachate contains a cocktail of degraded forms ranging from one to five binding domains, usually dominated by smaller fragments. The pIs of these domains are strongly acidic (~5.1), making complexes with IgG more electronegative than free IgG.[7] The degree of aberration is proportional to the number of binding domains to which an individual IgG molecule is complexed. This causes broadening to the tailing side of the IgG peak on anion exchangers and toward the leading side on cation exchangers. Some of the IgG is also converted to dimers, soluble aggregates, and precipitates that can further complicate the elution profile.

On cation exchangers operated at pH levels low enough to dissociate the complexes, protein A is displaced and elutes substantially in advance of IgG.[7] The entire run need not be conducted at a dissociative pH. Column pH can be reduced with a low pH wash following sample injection, then restored prior to elution at the pH that supports the best analytical performance. This strategy protects the product from excessive exposure to the potentially denaturing wash conditions in two different ways. First, it minimizes exposure time to the harsh pH environment. Second, experimental evidence suggests that a protein immobilized on a cation exchanger is physically constrained from major conformational rearrangements, such as those that typically accompany denaturation.[8] Additional protection can be obtained by the inclusion of 1 *M* urea and 50% ethylene glycol in the dissociative wash. These additives will suspend hydrogen bonds and hydrophobic interactions that stabilize the complex, and may allow you to moderate the pH of the dissociating wash by 1 or 2 units.[7] The same conceptual approach can be applied for dissociation of many leached affinity ligands.

Compound complexation — One of the most challenging aspects of dealing with complexation is that several different complexation mechanisms may be operating simultaneously. Affinity complexation is one example, as discussed in the

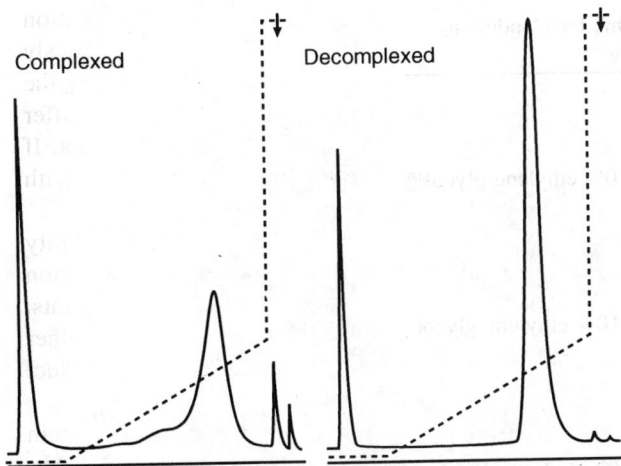

Figure 5.1 Comparison of IgM complexed with DNA versus IgM decomplexed by pretreatment with 1.0 M NaCl and 0.05 M EDTA. Sample: purified IgM plus partially degraded DNA extracted from the host cell line. Dotted trace marks the salt gradient. Arrow marks introduction of 5 mM EDTA to the salt strip. Mono-S HR5/5, pH 5.5. See text for explanation. (Data from P. Gagnon, 1996, Special weapons and tactics for removal of product-bound DNA, oral presentation, BioEast '96, Washington, D.C.)

preceding text, but combinations of nonspecific mechanisms can also occur. Addressing them individually may not be sufficient to achieve quantitative dissociation.

Product complexation of DNA with an IgM provides an example (Figure 5.1). When fragments of DNA bind to positive charge domains on proteins, they can neutralize those domains and make them unavailable for binding to a cation exchanger.[6,9] Such complexes flow through the column immediately after sample application. Other complexes may exist in which the positive charge domains on the protein are only partially occluded. This causes charge heterogeneity of the product, which is usually observed as peak broadening. Very large heteroaggregates may fail to enter the column and be eluted only when the column is stripped under extreme conditions. Complexes with similar behavior may occur when a calcium or ferric ion bound to a polycarboxy site on a protein binds with a phosphoryl moiety on a DNA or phospholipid molecule. It is necessary to address both mechanisms at once to achieve full dissociation.

EVALUATING SELECTIVITY

Part of the reason that ion exchange is so versatile is that proteins are amphoteric; their charge characteristics vary according to the pH environment. In addition, each species of protein responds differently, so that a group of proteins eluting in the order A,B,C at one pH may elute in a different order or with different

Table 5.1 Buffer Systems for Conducting
Initial Selectivity Screening

pH 5.5 binding buffer
0.05 *M* MES
pH 5.5 eluting buffer
0.05 *M* MES, 1 *M* NaCl, 10% ethylene glycol[a,b]

pH 7.0 binding buffer
0.05 *M* Hepes
pH 7.0 eluting buffer
0.05 *M* Hepes, 1 *M* NaCl, 10% ethylene glycol

pH 8.5 binding buffer
0.05 *M* Tris
pH 8.5 eluting buffer
0.05 *M* Tris, 1 *M* NaCl, 10% ethylene glycol

[a] The ethylene glycol is optional. It improves peak
sharpness. See text for discussion.
[b] 1 *M* urea may be substituted for ethylene glycol.

resolution at another pH. The elution order also differs frequently on anion vs.
cation exchangers. Accordingly, thorough screening for assay conditions must
include both anion and cation exchangers over a range of pH values.

The first phase of screening is to conduct salt gradients at a range of fixed
pH values. Table 5.1 describes a series of three buffer systems that can be used
on both anion and cation exchangers. Your screening efforts may benefit by
including a neutral chaotrope such as 1 *M* urea or neutral organic solvent such
as 10% ethylene glycol in your buffers to reduce nonspecific hydrophobic interactions between proteins and the exchanger surface. You may consider nonionic
or zwitterionic detergents for the same purpose, but they require development of
rigorous column-cleaning procedures not necessary with either urea or ethylene
glycol. Whatever additive you choose, you should expect to obtain sharper peaks.
If any of your sample components are extremely hydrophobic, you may also
observe significant reductions in their retention time.

Table 5.2 describes a linear gradient configuration that can be used for
initial screening. Although the gradient segment is conspicuously long, its average
slope of 0.01 *M* NaCl per CV has proven to provide a good initial balance between
separation potential and dilution. If promising results are encountered, the gradient interval can be optimized. Sample volumes up to about 5% of the CV can
be injected without preequilibration when the sample is at more or less physiological conditions. If a sample contains a high concentration of salt (0.5 to 1.0
M) it may still be injected without preequilibration, although at a lower proportional volume, for example about 2% CV.

Table 5.2 Initial Screening Gradient Configuration

Equilibrate column: 10 CV binding buffer
Inject sample: 2–5% unequilibrated sample[a]
Wash: 5 CV binding buffer
Elute: 40 CV linear gradient to 40% eluting buffer
Strip: 5 CV eluting buffer

[a] 5% CV if sample is at roughly physiological ionic strength,
2% if ionic strength is above 0.5 M.

Fractions of 1 CV or less may be collected across the gradient. Fractions from the main product peak and from each side can be analyzed by SDS-PAGE with silver stain. Contaminants can then be identified by their molecular weights. Extrapolating the eluting salt concentration at peak center for each component of interest will allow you to map their relative retention characteristics as a function of both pH and salt concentration.

Initial screening results may suggest that an intermediate or more extreme pH could provide better results, but before you invest substantial resources in optimization, there is another dimension of pH screening that frequently produces selectivities that are difficult or impossible to recreate with salt gradients at fixed pH. This second phase involves pH gradients at fixed salt concentrations (descending pH gradients on anion exchangers, ascending pH gradients on cation exchangers). Whereas salt gradients at fixed pH separate proteins based on quantity of charge regardless of their isoelectric points, pH gradients at fixed ionic strength separate proteins based on their pH titration characteristics.

Results from the first phase of screening will have substantially narrowed the range of conditions you need to evaluate. They would have identified the pH values required for your protein to bind to the respective ion exchangers, and they will have limited the pH interval of the gradient to about 1.5 pH units. For example, if your product binds to a cation exchanger at pH 5.5, but not at pH 7, a 1.5-pH unit gradient from 5.5 to 7 will be sufficient. If your product binds to an anion exchanger at pH 8.5 but not pH 7, then begin with a gradient from pH 8.5 to 7. If it binds at pH 7 but not 5.5, then drop the interval accordingly.

There are two ways to generate pH gradients. One is to use a mechanical gradient mixer. For the pH interval of pH 5.5 to 7, make a buffer containing 0.05 M (each) of the so-called biological buffers MES, ADA, and BES. Titrate half of the buffer to pH 5.5 and the other half to pH 7. For cation exchangers, run a 30-CV gradient from 5.5 to 7. For anion exchange, reverse the order. For the pH interval from 7 to 8.5, use a corresponding buffer system containing 0.05 M BES, Hepes, and Tris.

The other way to create a pH gradient is to employ commercial chromatofocusing buffers (polybuffers, Amerham Biosciences [Piscataway, NJ]). For anion exchange, you will preequilibrate the column to its upper pH limit, then apply a dilution of Polybuffer 94 that has been pretitrated to the lower pH limit. Product specifications accompanying the buffer suggest dilution ranges to achieve particular gradient slopes. Select a dilution that will create a gradient length of about 30 CV. For cation exchange, you will titrate the column to its low pH starting point and apply polybuffer that has been equilibrated to the high pH end point of the gradient. Use the same dilution suggested for anion exchange.

It is unlikely that either of these strategies will provide you with a perfectly linear pH gradient, and in practice you will find that the same buffer systems will give a different gradient shape on every column it is applied to. This is true not only for cation exchangers compared with anion exchangers, but also for different columns within either class. If the deviation from gradient linearity is substantial, it will be necessary to make compensatory adjustments. The former method of gradient creation offers more opportunities in this regard and is also more· economical. If your gradient dips in the midrange, increase the relative concentration of the midrange buffer or decrease the concentration of one or both of the end point buffers. If the gradient rises out of linearity in the midrange, take the opposite approach. Note that these experiments can be performed in the absence of sample.

Once you have established roughly linear pH gradient conditions for your cation and anion exchangers, conduct an experiment with sample. Follow this with a parallel experiment in which 0.05 M NaCl has been added to the buffers. If the proteins of interest are still retained in the 0.05 M NaCl experiment, conduct another pH gradient in 0.1 M NaCl. Collect and analyze fractions as in the previous experiments.

SELECTIVITY WILD CARDS

It often happens that the best separation you can achieve with conventional approaches fails to completely satisfy your goals or possibly fails altogether. In such cases there are many methods by which protein interactions with an IEC column can be systematically modified to alter the separation. It would be too generous to say that these methods offer the ability to fine-tune a given separation. They can just as easily fail to produce any useful change or even degrade the separation. However, it would be fair to say that these methods offer a rich opportunity for encountering lucky hits. All of the options described in the following text assume that you use your best protocol from initial screening as a baseline.

Differential Charge Modification

IEC is very sensitive to additives that directly alter the charge of sample components. Several classes of additives provide a range of opportunities to exploit this approach. All involve complexants that bind to the proteins in a sample.

cis-**Boronate derivatives** — These have the ability to form covalent bonds with *cis*-diol sugars at alkaline pH. Unlike lectins, which typically bind only the terminal sugar in a complex carbohydrate, boronates can bind any *cis*-diol that is physically accessible.[10-12] When this occurs, the charge characteristics of the particular boronate derivative are conferred on the complex. To the extent that the protein components of a sample are differentially glycosylated, boronate coupling can provide a unique tool for exploiting those differences.

Because of the requirement for alkaline conditions, this technique is practically limited to anion exchangers. Glycoproteins may be either positively or negatively charge enhanced. To increase electronegativity, equilibrate the sample to at least 1 mM boric acid at a pH of 8 to 8.5. To increase electropositivity, substitute *m*-aminophenyl boronic acid. It is not necessary to buffer exchange the sample prior to IEC. If you observe a significant effect, you can optimize the gradient in the region of interest.

Controlled metal complexation — While uncontrolled metal complexation can be a major source of problems, controlled complexation is a potentially useful selectivity modifier. The key is to evaluate the effects of only one metal at a time, which in turn requires that your sample be stripped of metals in preparation for your experimental treatment. The EDTA–imidazole treatment described above can be used for this purpose. For evaluating the effects of ferric iron or calcium, buffer exchange the treated sample into 0.05 M MES, pH 6, then add the metal salt of choice to a concentration of 5 mM. For other metals, buffer exchange the treated sample into 0.05 M Tris, pH 8, then add the metal salt of choice to 5 mM.

Note that it is important to avoid carboxy cation exchangers. A proportion of the charged groups on a carboxy cation exchanger exist in a relative proximity that endows them with the ability to chelate metal ions. These chelating sites may competitively displace metals from proteins in the sample, creating a source of assay variability. Use a noncarboxy cation exchanger such as sulfomethyl, ethyl, or propyl.

Ionic detergents — These are not routinely explored as selectivity modifiers in IEC, but present some interesting possibilities. Use a detergent with the same charge as the ion exchanger; otherwise, the detergent will simply bind to the column. To the extent that the detergent binds to hydrophobic patches on a protein, it will reduce retention both by reducing hydrophobic interactions and by adding a column-repellant charge to the protein. Start with a detergent concentration of about 10% of the critical micelle concentration, added to both the sample and the buffers. If you observe a useful result, check to see if it can be achieved by adding detergent to the sample only.

Affinity complexation — Many proteins have affinities for other molecules that can be exploited to alter their retention characteristics in IEC. For example, some enzymes may be combined with synthetic substrates, cofactors, or products.[13-15] The same principle can be applied to other protein/receptor systems. One well-characterized example is the change in chromatographic behavior of fructose 1,6-diphosphatase in the presence of its negatively charged substrate fructose 1,6-diphosphate.[13] The enhancement of electronegativity prevents the complex from binding to cation exchangers. To the extent that an assay is seeking to discriminate a product from nonproduct-derived contaminants, this approach may be helpful. It can also be useful in discriminating active from inactive product.

Differential Hydration

Differential hydration of proteins has been little exploited as a selectivity factor in ion exchange, but it is simple to evaluate and can produce useful results. This technique relies on the preferential exclusion of certain solutes from protein surfaces to produce an exclusionary effect and favor their interaction with the column. Protein hydration is generally proportional to protein size and solubility. Among proteins of similar size, this predicts that retention will increase with protein solubility. Among proteins of similar solubility, retention increases with protein size.[16]

Polyethylene glycol (PEG) is an effective additive for evaluating this technique. PEG is nonionic and thereby neutral to electrostatic interactions. Start with 5% PEG-6000 in your mobile-phase buffers. If the results are promising but fall short of your needs, try 10%. Technically speaking, you can go as high as 20%, but in general you will benefit from using the lowest PEG concentration possible. The viscosity of PEG solutions reduces protein diffusivity, which translates into broader peaks. High viscosity also elevates column backpressure, and flow rates will probably have to be reduced. Even perfusive chromatography media may be able to support flow rates no greater than 300 to 600 cm/h. Viscosity also affects buffer preparation. Concentrated PEG solutions dissolve and filter very slowly.

Differential Denaturation

Gross conformational changes in proteins may expose additional charge sites that alter the retention of one or more sample components. Add urea to final concentration of 8 M in your sample and allow it to incubate for at least an hour. Add the same concentration of urea to your running buffers and run the same gradient configuration as your nonurea control. You can increase the urea concentration to as high as 10 M, but reducing it can be risky. If protein conformation is in a state of flux during the separation, it may manifest as assay variability.

Different Elution Salts

A number of studies have reported the abilities of different eluting salts to produce both different resolution and different selectivities.[17–23] As expected, polyvalent ions are stronger eluters than monovalent ions. Some studies have suggested a correlation between eluting ability and ranking in the Hofmeister series. Others fail to observe this correlation, but nevertheless note differences in elution behavior. Overall, it is impossible to predict which eluting ions, if any, may produce a useful change in selectivity for a particular application. If you wish to explore the possibilities, start by examining the differences between iodide, bromide, chloride, acetate, and phosphate ions on anion exchangers. Investigate the effects of magnesium, lithium, sodium, potassium, and ammonium ions on cation exchangers.

Column Differences

Different ion-exchange columns are well known to support different selectivities, even when they use the same ion-exchange group.[6,20] This is believed to result from a range of factors including ligand density, ligand accessibility, and hydrophobicity of the support. Unfortunately, ion exchangers are not indexed in any way that permits users to evaluate these variables systematically. .

In general, the most fruitful column variants to explore are those that represent different steric presentation of the ligand. Because charges are distributed unevenly on protein surfaces, any given protein may be bound by a different subset of charged residues depending on ligand accessibility. For practical purposes, there are three basic ligand configurations. The first and most common is represented by chromatography media on which the ion-exchange ligand is bonded to the surface by a short spacer. The second configuration, sometimes referred to as a deep-layer exchanger, is exemplified by so-called tentacle-type exchangers, in which branched clusters of ligand extend out into the mobile-phase space. This potentially allows ligand interactions with a larger subset of a given protein's charges.

The third configuration is represented by so-called gel-in-a-shell exchangers, where the ligand is bound at very high density to a permeable gel that surrounds the protein on all sides. This potentially allows all of a given protein's charged groups to interact with the exchanger. Intuitively, you might expect gels that support greater protein-surface interaction to exhibit proportionally greater retention, but they also interact more with potentially repellant charges that can exert a contrary influence. As a result, and as with other process development options, the results of experimenting with different ligand configurations are unpredictable.

CONCLUSIONS

It is impossible to recommend how far you should pursue the possibilities discussed in this chapter. Essentially, all analytical process developers routinely screen both anion and cation exchangers with salt gradients over a range of fixed pH values. A smaller subset routinely evaluates pH gradients as well. Exploration into more exotic territory is usually undertaken only when conventional approaches fail. This is probably as it should be. As powerful and versatile as IEC is, it is only one of a suite of proven chromatography techniques for protein analysis. If investigating the basics in IEC does not yield the results you seek, it makes sense to investigate the basics with SEC, HIC, and RPC before investing significant resources in a blind search of unproven variables.

Despite the obvious merit of this conventional wisdom, selectivity wild cards often prove essential, and there is real value in developing a repertoire of special weapons and tactics. Even though there is no way to predict which, if any, will produce the effect you seek, the options are limited, well defined, and the investment of resources reasonably modest. By the time you get to the point of evaluating wild cards, you will probably be sufficiently familiar with your analytical system to discern a useful result from a chromatogram without extensive secondary testing. The principal investment will be buffer preparation and the time to run the analyses.

Whatever the scope of your investigations into selectivity, it is critical that you address the issue of product complexation in the sample, particularly complexation with metals. Complexation problems are by no means universal, but they are more common than generally realized. If you confirm the occurrence of product complexation, it is important to discover and characterize its source. Even though this may not fall strictly within the usual bounds of assay development responsibilities, the potential for the problem to affect purification, formulation, and even pharmacokinetics demands that this potential source of variation be addressed. Seek to eliminate it. If you cannot eliminate it, then reduce it. If you cannot reduce it, then at least try to maintain it within defined limits.

REFERENCES

1. L. Kagedal, 1989, in *Protein Purification: Principles, High Resolution Methods, and Applications,* J.-C. Janson and L. Ryden, Eds., p. 227, VCH, New York.
2. M. Gorbunoff, 1984, *Anal. Biochem.,* **136**, 433.
3. J. Porath et al., 1975, *Nature,* **258**, 598.
4. J. Porath and B. Olin, 1983, *Biochemistry,* **22**, 162.
5. J. Hale and D. Biedler, 1994, *Anal. Biochem.,* **222**, 29.
6. P. Gagnon, 1996, Purification Tools for Monoclonal Antibodies, *Validated Biosystems,* p. 57, Tucson, AZ.
7. P. Gagnon, 1996, Purification Tools for Monoclonal Antibodies, *Validated Biosystems,* p. 155, Tucson, AZ.
8. M. Carlsson et al., 1985, *J. Immunol. Metab.,* **79**, 89.
9. P. Gagnon, 1996, Special weapons and tactics for removal of product-bound DNA, oral presentation, BioEast '96, Washington, D.C.
10. A. Foster, 1957, *Adv. Carbohydr. Chem.,* **12**, 81.
11. S. Narasimhan et al., 1980, *J. Biol. Chem.,* **255**, 4876.
12. R. Rothman and L. Warren, 1988, *Biochim. Biophys. Acta.,* **955**, 243.
13. R. Scopes and E. Algar, 1979, *FEBS Lett.,* **106**, 239.
14. R. Scopes, 1981, *Anal. Biochem.,* **114**, 8.
15. R. Davies and R. Scopes, 1981, *Anal. Biochem.,* **114**, 19.
16. P. Gagnon et al., 1996, *J. Chromatogr.,* **743**, 51.
17. W. Kopaciewiecz et al., 1983, *J. Chromatogr.,* **266**, 3.
18. K. Gooding and M. Schmuck, 1984, *J. Chromatogr.,* **296**, 321.

19. K. Gooding and M. Schmuck, 1984, *J. Chromatogr.,* **283**, 37.
20. L. Soderberg et al., 1982, in *Protides of the Biological Fluids,* H. Peeters, Ed., Vol. 30, p. 629, Pergamon Press, Oxford.
21. E. Karlsson and L. Ryden, 1989, in *Protein Purification: Principles, High Resolution Methods, and Applications,* J.-C. Janson and L. Ryden, Eds., p. 107, VCH, New York.
22. M.A. Rounds and F.E. Regnier, 1984, *J. Chromatogr.,* **283**, 37.
23. G. Malmquist and N. Lundell, 1992, *J. Chromatogr.,* **627**, 107.

6

Practical Strategies for Protein Contaminant Detection by High-Performance Hydrophobic Interaction Chromatography

Pete Gagnon

Hydrophobic interaction chromatography (HIC) occupies a unique niche in the field of analytical chromatography. A particular advantage of HIC is its unique selectivity. Whereas ion-exchange chromatography (IEC) principally reveals differences based on the surface charge of native proteins, HIC reveals differences based principally on their surface hydrophobicity. HIC is complementary to reversed-phase chromatography (RPC) in a different sense. Whereas HIC discriminates primarily on the basis of surface hydrophobicity, RPC principally reveals differences based on total hydrophobicity of all the hydrophobic residues of denatured proteins.

The retention mechanism of HIC is complex, involving hydrophobic interactions, which are in themselves strongly dependent on the character of the ligand, and the contribution of solvent–protein interactions. In spite of its mechanistic complexity the technique is simple to conduct. The column is generally loaded in a solution of high salt concentration, and elution is accomplished by reducing the salt concentration. Like both IEC and RPC, HIC is a high-resolution method, able under some circumstances to discriminate between proteins that differ by only a single hydrophobic residue.[1,2] These features make HIC an essential member in this triumvirate of powerful analytical technologies.

On a practical basis, analytical HIC is useful for aiding purification-process development, performing in-process monitoring, and assessing final product quality. It is commonly able to differentiate a protein product from its own variants, such as deamidated forms, terminal and intrachain enzymatic clips, and especially misfolded variants, as well as from a wide range of host cell protein contaminants. Likewise, in the case of protein conjugates, analytical HIC can discriminate and provide relative quantitation of unreacted materials, various conjugate morphs, and other reaction by-products. Obtaining the best performance from analytical HIC can be approached in two steps: minimizing variation in the analytical system and thoroughly evaluating the scope of potential selectivities. This chapter offers practical strategies addressing both sets of issues.

MINIMIZING VARIATION IN THE ANALYTICAL SYSTEM

Column Reproducibility

Ligand density is a primary determinant of selectivity in HIC.[3–5] This has two important ramifications: (1) minor variations in ligand density can significantly alter selectivity and (2) reproducibility of column selectivity is more difficult for manufacturers to achieve than it is for nonhydrophobic retention mechanisms. This makes it necessary for assay developers to document adequate lot-to-lot reproducibility of a given column medium before investing major resources in assay development.

For any column you are considering, obtain samples from at least three different production lots. Some column manufacturers set aside samples for this specific purpose, but you may obtain a better estimate of variability by simply purchasing columns over a period of time. An effective way to evaluate reproducibility among the columns is to run a standardized separation of a multicomponent sample cocktail. The cocktail should contain at least two different components that elute completely resolved from one another through the course of a shallow linear gradient. Column documentation from the manufacturer usually identifies an analytical cocktail with relative concentrations for each component.

Begin by specifying one of the column lots as the reference. Prepare enough buffer so that all runs can be conducted with the same buffer lot. Make sure that the chromatograph has been on long enough so that its internal temperature has stabilized. Run the reference and test columns in the shortest reasonable period of time. If your laboratory has tight temperature controls, you have more latitude with the length of time over which you conduct the runs. Otherwise, most laboratories warm up during the course of the day and at different rates and ranges according to the season. Also be aware that buffers will not equilibrate to temperature as rapidly as their surroundings. Selectivity in HIC is very sensitive to temperature, and uncontrolled variations may add a stratum of variation that can be misassigned to column variability.[5,6]

You can obtain an initial estimate of media reproducibility simply by overlaying the chromatograms. Align them first along the gradient trace. On columns with excellent reproducibility, the eluting peaks will overlap very closely. Vertical lines drawn through the respective peak centers will intersect the gradient trace at the same points. On columns with lesser reproducibility, the profiles may be shifted to the left or right. In this case, realign the chromatograms on the first eluting peak. There is no cause for alarm so long as the gradient offset is modest and the relative separation is maintained. If the offset between the gradient-aligned profiles is substantial and the quality of separation is significantly affected, then that particular column may not be qualified for analytical use.

For validation of columns that will be used for an official assay and to provide an unambiguous standard for qualifying future media lots, it is useful to employ more measurable comparative criteria than a simple overlay. Resolution, plate heights, and peak symmetry, as calculated by the classical formulae, should match very closely among test and reference columns (Figure 6.1, Figure 6.2).

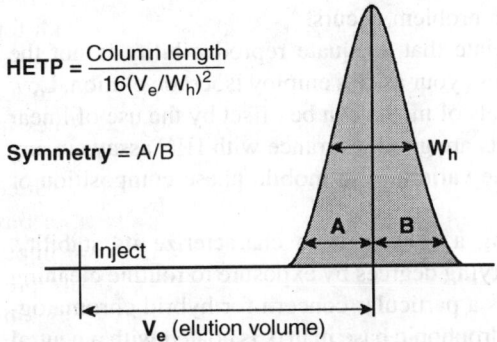

$$HETP = \frac{Column\ length}{16(V_e/W_h)^2}$$

$$Symmetry = A/B$$

Figure 6.1 Calculation of HETP and symmetry. W_h equals width at half of the peak height. A and B are measured at 10% of peak height.

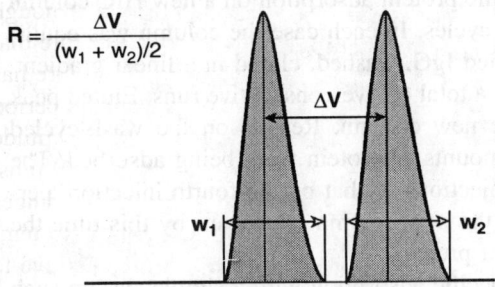

$$R = \frac{\Delta V}{(w_1 + w_2)/2}$$

Figure 6.2 Calculation of resolution.

Plate height and symmetry calculations will provide an index of consistency in particle size, porosity, and packing quality. Resolution is affected by these variables, but will also provide an estimate of reproducibility for ligand density. Plate height and peak symmetry values calculated from gradient-eluted protein peaks will not be directly comparable with values calculated from the passage of nonadsorbed low-molecular weight solutes and should not be interpreted in the same way. They are nevertheless influenced by the same physical variables and will provide you with an unambiguous means of comparing the performance of different lots of chromatography media.

The degree of lot-to-lot reproducibility you require from a column is ultimately a function of the needs of a particular assay, which makes it impossible to state definite limits that will be appropriate in every case. Whatever the level of variation, it is important that it be documented. As new lots of media are brought into use over the course of years, their performance vs. the established reference should be included in a master database begun with the original qualification testing. Among other factors, this will allow you to track the column manufacturer's performance over time and possibly detect trends that could affect your assay performance — before a problem occurs.

It is also important to appreciate that adequate reproducibility is not the same as absolute reproducibility unless your assays employ isocratic elution. Low levels of variation among different lots of media can be offset by the use of linear gradient separations. Linear gradients are good insurance with HIC assays in any case because they also buffer routine variations in mobile-phase composition or ambient temperature.

The second step of qualifying a column is to characterize its stability. Different columns are affected to varying degrees by exposure to routine cleaning and maintenance procedures. This is a particular concern for hybrid chromatography media where an extremely hydrophobic base matrix is coated with a neutral material and then derivatized to incorporate hydrophobic groups.[7] Such compound construction creates more complex degradation pathways than encountered with homogeneous media, the most troublesome of which is loss of coating and subsequent exposure of the base matrix.

Figure 6.3 compares nonspecific protein adsorption on a new HIC column vs. the same column after 50 wash cycles. In each case the column was equilibrated, injected with 10 μg of purified IgG, washed, eluted in a linear gradient, and then cycled four more times for a total of five consecutive runs. Eluted peak height was fairly consistent on the new column. Results on the wash-cycled column indicated that significant amounts of protein were being adsorbed. The effect diminished with successive injections so that by the fourth injection, performance was roughly on par with the new column. However, by this time the column had adsorbed about 15 μg of protein.

In addition to assessing nonspecific adsorption with a simple system such as just described, it is recommended that postwash-cycle testing be performed with the same qualification methods and criteria used to evaluate new column

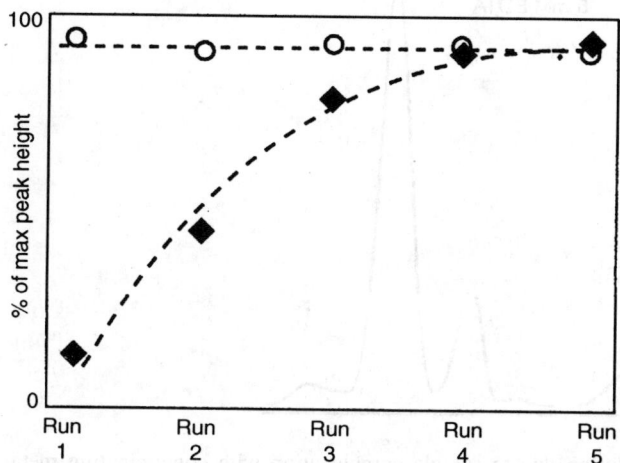

Figure 6.3 Nonspecific protein adsorption as a result of column degradation. White circles indicate eluted column peak heights of a new column. Black diamonds indicate eluted peak heights after the column was treated with 50 wash cycles. See text for discussion. (Data from P. Gagnon, 1997, *Validated Biosystems Quarterly Resource Guide for Downstream Processing*, 2(1), 1, http://www.validated.com/revalbio/library.html.)

reproducibility: resolution, plate height, and peak symmetry. Consult manufacturers for limitations on column chemical exposure, then apply the most severe conditions the columns are certified to withstand. The sample need not be applied during the wash cycling; only the cleaning agents need be added. Automated chromatographs allow this to be performed on an overnight basis. If possible, monitor system pressure throughout the treatment, as increased pressure may also be a sign of media degradation.

A short stability curve is not necessarily a reason to reject a particular column. If the unique selectivity of a given column is essential for achieving a particular analytical separation, it can be used so long as its performance is validated to persist for a specified number of runs and a column log is maintained to document that its usage is limited accordingly. Periodic analyses to document that it is still within functional specification may also be prudent. The point is to ensure that assay performance does not fall victim to an undetected source of progressive variation.

Controlling Sample Variation

Variations in sample composition other than those toward which an assay is directed may detract from the performance and reproducibility of HIC assays. Controlling these influences can significantly improve resolution and sensitivity as well as reproducibility. As with IEC, product complexation with other solutes can be an important source of aberrant retention behavior.

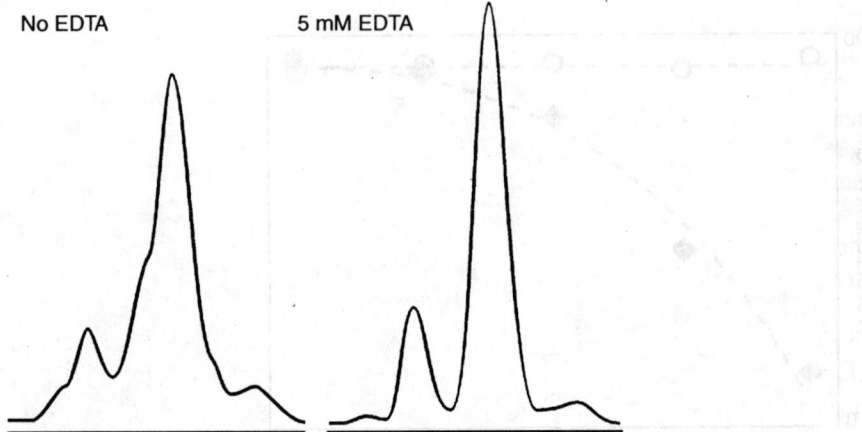

No EDTA   5 mM EDTA

Figure 6.4 Loss of resolution due to sample complexation with contaminating metal ions. Profiles of a 30 KD protein on Tosoh Phenyl 5pw. Elution in ammonium sulfate. EDTA was added to the sample and both buffers.

HIC has a major advantage over IEC in that product binding tolerates high salt concentrations. Charge complexes are dissociated as a by-product of loading conditions. Protein–metal complexes, however, easily survive high salt concentrations and if not addressed can substantially degrade analytical performance (Figure 6.4). Fortunately, HIC's tolerance of high salt provides a way to eliminate most metal complexation problems as well. Metal-scavenging agents such as EDTA and imidazole can be added to the sample prior to loading.

The inclusion of these agents in buffers is also strongly recommended. Many of the binding salts commonly used in HIC are highly contaminated with metals. This creates a risk of protein–metal complex reformation after elution. Inclusion of these agents in buffers is also valuable for scavenging metals that may leach from stainless steel components of the chromatography system. The high salt concentrations used in HIC are notorious for their corrosive effects. Whether or not metal-scavenging agents are included in buffers, stainless steel chromatography systems used for critical analytical procedures should be passivated periodically as a part of their routine maintenance. Passivation is believed to work by creating a thin corrosion-resistant oxide film on exposed stainless steel surfaces. Run a solution of 0.2 to 0.5 *M* citric acid through the system at a low flow rate for 10 to 20 min. Rinse with water, then 0.01 *M* NaOH, and then the storage solution normally used. The obvious alternative is to avoid stainless steel chromatography systems for HIC.

EVALUATING SELECTIVITY

Overall selectivity in HIC is a composite effect created by several different mechanisms, each of which affect protein retention in a different way. Different

HIC supports manifest different balances among these mechanisms. This makes column selection a fundamental aspect of your assay development strategy.

One of the component mechanisms is, of course, hydrophobic interactions. Retention is proportional to column hydrophobicity, and elution order is expected to generally follow solute hydrophobicity. However, it is important to keep in mind that proteins bind preferentially to columns by their dominantly hydrophobic surface. Two proteins with very similar average surface hydrophobicity may exhibit very different retention characteristics due to differences in their respective distribution of hydrophobic residues.[1,2]

Interactions between proteins and salts in the binding buffer are also a major determinant of selectivity. Salts that are strong retention promoters in HIC are excluded from protein surfaces by repulsion from their hydrophobic amide backbones and hydrophobic amino acid residues.[8,9] This causes the mobile phase to exert an exclusionary pressure that favors the association of proteins with the column, regardless of stationary-phase hydrophobicity.[10–12] Because this mechanism involves the entire protein surface, the degree of exclusion is proportional to average protein hydrophobicity, regardless of the distribution of hydrophobic sites.

The balance between stationary-phase hydrophobicity and solvent exclusionary forces is the major determinant of selectivity on nonphenyl columns. On strongly hydrophobic columns, hydrophobic interactions between the proteins and the column are most likely to be the dominant determinant of selectivity. The low levels of binding salts required to achieve retention exert relatively less influence. On weakly hydrophobic columns, the high levels of binding salts cause solvent exclusionary effects to exert a greater influence.

Phenyl columns represent a third mode of selectivity. Pi–Pi bonding, properly speaking, is independent from hydrophobic interactions, but it has been shown to exert substantial influence on retention. Solutes with accessible ring structures are retained much more strongly than they are on similarly hydrophobic nonphenyl columns.[5,13]

These selectivity differences provide a rational basis for a three-column initial screening strategy: a strongly hydrophobic column, a weakly hydrophobic column, and a phenyl column. Commercially available columns are not indexed in a manner that easily allows column selection based on hydrophobicity; however, butyl columns are good candidates for strongly hydrophobic columns. Columns with names like ether-, iso-, and isopropyl are good candidates for weakly hydrophobic columns.

Initial screening conditions are suggested in Table 6.1. Multiple pH values are included because mobile-phase pH can significantly affect retention. Major selectivity shifts such as transpositions in elution order are fairly common; changes in resolution are much more so.[2,14–16] Changes in retention due to pH variation relate to protein hydration. Proteins are minimally charged at their isoelectric points (pIs). This means that they carry the minimum of electrostricted hydration water. Both protein surface hydrophobicity and HIC retention should therefore reach their maximum at a protein's pI.[6] As pH is either increased or

Table 6.1 Buffers for Initial Selectivity Screening

pH 8.5, binding buffer
0.05 M Tris, 2 M potassium phosphate, 5 mM EDTA, 5 mM imidazole
pH 8.5, eluting buffer
0.05 M Tris, 10% ethylene glycol[a,b]

pH 7.0, binding buffer
2 M potassium phosphate, 5 mM EDTA, 5 mM imidazole
pH 7.0, eluting buffer
0.05 M potassium phosphate, 10% ethylene glycol

pH 5.5, binding buffer
0.05 M MES, 2 M potassium phosphate, 5 mM EDTA, 5 mM imidazole
pH 5.5, eluting buffer
0.05 M MES, 10% ethylene glycol

[a] The level of ethylene glycol may be increased if necessary on strongly hydrophobic columns.
[b] 1–2 M urea may be substituted for ethylene glycol.

decreased, protein charge increases and electrostrictive hydration with it. Retention of that protein should decrease regardless of the direction of pH change.

Note that all of the eluting buffers in Table 6.1 contain 10% ethylene glycol. This has little effect on protein elution from weakly hydrophobic columns but can substantially improve peak sharpness on stronger HIC columns. This is especially true of later eluting peaks. You may also find that strongly retained proteins do not elute quantitatively from strongly hydrophobic media in the absence of competing additives. Thus the inclusion of ethylene glycol may improve not only resolution, but also sensitivity and accuracy.

Note also that ammonium sulfate is absent from the recommended binding buffer formulations despite its general popularity in the field. Ammonium ions become fully titrated at alkaline pH and convert to ammonia gas. Buffer pH may become unstable as a result and causticity of the free ammonia may partially hydrolyze the proteins in a sample, creating a source of assay variability.[6,17,18] At small buffer volumes used for analytical applications, liberated ammonia gas may not be a significant health hazard, but precautions may still be necessary to meet regulations. For all these reasons, ammonium salts are best avoided at alkaline pH.

Table 6.2 describes a linear gradient configuration that can be used for initial screening. Although the gradient segment is conspicuously long, this configuration has proven to provide a good initial balance between separation potential and dilution. If promising results are encountered, gradient slope and interval can be optimized to support the best results. Note that sample volumes up to about 2% of the column volume can be injected without preequilibration when the sample is at roughly physiological conditions. You can often increase the relative sample volume by adding dry NaCl directly to the sample. Most proteins

Table 6.2 Separation Conditions for Initial Screening

Equilibrate column: 5 column volumes (CV) 100% binding buffer[a]
Inject sample: 2% CV unequilibrated sample[b]
Wash: 2 CV binding buffer
Elute: 20 CV linear gradient to 100% elution buffer
Strip: 5 CV elution buffer[c]

[a] Starting salt concentration can be reduced to 1 M for strongly hydrophobic columns.
[b] Larger volumes can be applied in some circumstances. See text.
[c] If elution is incomplete, increase concentration of ethylene glycol.

will remain soluble up to at least 4 M NaCl. For strongly hydrophobic columns, 1 M NaCl added to the sample may allow injection volumes of 5% of the column, 2 M twice that. For weakly hydrophobic columns more salt will be required to achieve the same effects. Salts that are stronger retention promoters can achieve the same effects at lower concentrations, but involve a much higher risk that the protein in the sample will precipitate.

SELECTIVITY WILD CARDS

There are a number of methods by which you can modify the surface hydrophobic characteristics of proteins or otherwise modulate hydrophobic interactions. How they will affect a given separation is unpredictable, but they may be useful.

BINDING SALTS

The ability of various salts to mediate HIC retention for any given protein is well known to correlate strongly with the additive ranking of a salt's component ions in the Hofmeister series.[19–22] However, this does not mean that all proteins exhibit the same magnitude of response to different binding salts. Out of a given group of proteins, some may be more or less responsive to a particular salt. As a result, it frequently occurs that different binding salts produce significant changes in resolution, and sometimes produce transpositions in elution order.[6] Because any differential effects are dependent on the individual proteins, there is no way to predict which salts, if any, may produce useful results in a given situation.

Surveying different salts is simple but can be time consuming. For phenyl and butyl columns, because of the modest salt concentrations required to achieve retention for most proteins, the potential range of candidates includes dozens of options. The range is somewhat more restricted for weakly hydrophobic columns because fewer salts will be able to achieve good retention; however, the range is still substantial. Even NaCl may serve at concentrations of 4 to 5 M.

As a baseline, choose the column and operating pH that have so far provided the most promising selectivity. Prepare a nearly saturated solution of the salt you wish to evaluate and run a 20-CV linear gradient, with your sample, from 100 to 0%. Determine the minimum concentration of salt necessary to ensure retention on that particular column. To determine if the particular salt will give you the selectivity you seek, run another gradient from that concentration to 0%. Use the same gradient length as your baseline run.

Boronate Complexation

Cis-boronate derivatives have the ability to form covalent bonds with *cis*-diol sugars at alkaline pH. Unlike lectins, which typically bind only the terminal sugar in a complex carbohydrate, boronates can bind any physically accessible *cis*-diol on a glycoprotein.[23–25] When hydrophobic boronate derivatives are used, the hydrophobicity of each complexation site is enhanced. To the extent that the protein components of a sample are differentially glycosylated, this can provide a tool for exploiting those differences.

A wide variety of substituted boronates are available commercially. For phenyl supports, use a phenyl or biphenyl (napthalene) derivative. For nonphenyl supports, evaluate both weakly and strongly hydrophobic alkyl derivatives (straight chain or branched). Raise sample pH to at least 8 and add the boronate derivative of choice to a concentration of at least 1 mM. The sample need not be buffer-exchanged prior to injection. Column buffers should be at least pH 8.

Organic and Inorganic Modifiers

As explained above, ethylene glycol was added to all of the initial screening buffers to improve peak sharpness. However, it also exerts an effect on selectivity. Other additives can achieve the peak-sharpening effect, but alter selectivity in different ways. Urea (1 to 2 M) in the eluting buffer occasionally gives selectivities different from ethylene glycol.[6] This may relate to the fact that urea is a chaotrope that binds to proteins and reduces their surface hydrophobicity, whereas ethylene glycol is excluded from protein surfaces but reduces solvent polarity and thereby weakens hydrophobic interactions in the system as a whole.[6,26–28]

Ethylene glycol is unusual among organic solvents because it is protein-stabilizing up to concentrations of about 50%.[26,29] Other organic solvents such as alcohols may alter selectivity in different ways, but with a coincident risk of protein precipitation. The addition of chaotropic salts to the elution buffer may alter selectivity in yet different ways, but with a different risk. Protein precipitation is unlikely to be an issue, but salt precipitation may occur. Mix aliquots of your binding and elution buffers at a 1:1 ratio in a test tube before you commit them to the chromatograph. If salt crystals precipitate due to a low solubility product constant for some combination of the ions in your buffers, then look for a different chaotrope. If there is some compelling reason why you still want to evaluate that particular salt, perform a more complex set of mixing experiments in which you reproduce a series of points in the range of mixing proportions that the buffers will experience through the course of the intended gradient. If salts still precipitate, you can also try reducing the concentration of the chaotrope.

Detergents are a unique class of organic modifiers. They have occasionally been used with HIC for special applications such as purification of membrane proteins, but they have not been evaluated systematically as selectivity modifiers.[30] Nevertheless their ability to alter protein surface hydrophobicity makes them potential candidates for this application. It is likely that different classes of detergents will have different effects on selectivity. For example, anionic detergents should have a higher affinity for hydrophobic sites that include or are adjacent to positively charged residues. Cationic detergents should prefer hydrophobic sites of electronegative character. Detergents with ring structures may show a preference for aromatic residues, whereas alkyl detergents may show a preference for alkyl protein residues.

If you decide to pursue these possibilities there are several cautionary points you should take into consideration. First, detergents also bind to HIC supports, especially strong HIC supports. Unless the interaction between your proteins and detergent is extremely strong, you will need to include detergent in your binding buffer. Otherwise, the column may competitively displace it from your proteins, creating a source of uncontrolled variation in your assay. It will also be necessary to extend column equilibration until the concentration of detergent exiting the column equals the concentration entering the column. Second, high salt concentrations severely depress critical micelle concentrations (CMC).[31] In order to prevent micelle formation you will need to keep detergent concentrations very low. How low will depend on the individual detergent and the high salt buffer composition. Consult with detergent manufacturers about the micellar properties of their products. Third, detergents may be difficult to clean from your column. An organic solvent wash will certainly be necessary. If you are evaluating several different detergents you will need to clean the column thoroughly between each treatment because each may confer a different selectivity. Fourth, some nonionic detergents such as Tweens and Tritons are heterogenous in composition and are often contaminated with peroxides that can damage proteins and affect assay results. If the detergent solution is anything less than water-clear, do not use it. Ultrapure detergents are available from a number of suppliers.

There are several formats in which you can conduct your experiments, each of which may produce a different selectivity. The simplest is to add detergent to the eluting buffer, as with other agents as discussed previously. Another is to conduct the entire run at a fixed low concentration of detergent. Contrary to expectations, some proteins actually bind more strongly under these conditions than they do in the absence of detergent.[32,33] A more complex approach is to elute with an increasing detergent gradient while the salt concentration is held constant. In order to minimize the risk of micelle formation during elution, you should first reduce the salt concentration to a level just sufficient to retain the proteins of interest in your assay.

TEMPERATURE

Hydrophobic interactions are very sensitive to temperature. Retention of most proteins increases with temperature, but for some the opposite is true, and the magnitude of response is highly individual in any case.[34-36] If you elevate temperature sufficiently (56°C and above) you may begin to denature proteins in the sample. This may expose more hydrophobic sites and alter selectivity to a greater degree. Whether or not you exploit temperature as a selectivity factor, good temperature control is essential for assay reproducibility.

COLUMN DIFFERENCES

Different HIC columns are well known to support different selectivities, even when they use the same ligand. Among same-ligand supports (for example, among phenyl columns), selectivity differences are chiefly attributed to differences in ligand density. Differences in the surface hydrophobicity of the support matrix and spacer also contribute to selectivity. The differences you observe among same-ligand supports can be characterized generally as frame shifts. Elution order typically remains fairly consistent. The primary exception to this pattern lies with supports that present the ligand in a different steric configuration. Tentacle-type supports, for example, may allow a given protein to bind by a larger subset of its hydrophobic residues. This may alter resolution and elution order among a series of sample components.

CONCLUSIONS

HIC is more challenging than some analytical chemistries. It requires more attention to environmental controls, to equipment maintenance, and to column qualification. The retention mechanism is both more complicated and less intuitive than other methods. And, there are fewer practical guidelines concerning how to exploit the technique to its full potential. Despite these limitations, the bottom line is that HIC provides selectivities that are not obtainable with any other analytical method.

Begin by constructing a foundation based on evaluation of the three-column/multiple-pH strategy outlined above. Optimize the gradient interval and slope of the most promising candidates. Before proceeding with more exotic variables, compare your initial results with those obtained from alternative analytical technologies, such as IEC and RPC. If a different method offers more promising opportunities, start there. If you do come to the point of exploring selectivity wild cards with HIC, there is no way to predict which, if any, will produce the selectivity that you want. Most will probably fail to do so. On the other hand, it is fairly certain that you will encounter substantial variations in selectivity and that you will learn a great deal. Whether or not the results serve your immediate needs, the insight you gain from the exercise is likely to prove valuable in future assay development projects.

REFERENCES

1. J. Fausnaugh and F. Regnier, 1986, *J. Chromatogr.*, 359, 131.
2. R. Chicz and F. Regnier, 1990, *J. Chromatogr.*, 500, 503.
3. J. Rosengren et al., 1975, *Biochim. Biophys. Acta*, 412, 51.
4. T. Miller et al., 1984, *J. Chromatogr.*, 316, 519.
5. K-O. Eriksson, 1989, in *Protein Purification, Principles, High Resolution Methods, and Applications*, J-C. Janson and L. Ryden, Eds., p. 207, VCH, New York.
6. P. Gagnon and E. Grund, 1996, *BioPharm*, 9(5), 55.
7. P. Gagnon, 1997, *Validated Biosystems Quarterly Resource Guide for Downstream Processing*, 2(1), 1, http://www.validated.com/revalbio/library.html.
8. T. Arakawa and S. Timasheff, 1982, *Biochemistry*, 21, 6545.
9. T. Arakawa and S. Timasheff, 1984, *Biochemistry*, 23, 5912.
10. B. Roetger et al., 1989, *Biotechnol. Prog.*, 5, 79.
11. T. Arakawa and S. Timasheff, 1984, *J. Biol. Chem.*, 259, 4979.
12. L. Narhi et al., 1989, *Anal. Biochem.*, 182, 266.
13. S. Cramer, 1999, oral presentation, First International Symposium on HIC and RPC, Phoenix.
14. S. Hjerten et al., 1986, *J. Chromatogr.*, 359, 99.
15. M. Schmuck et al., 1984, *J. Liq. Chromatogr.*, 7, 2863.
16. T. Miller and B. Karger, 1985, *J. Chromatogr.*, 326, 45.
17. Y. Kato et al., 1984, *J. Chromatogr.*, 298, 407.
18. H. Schultze and J. Heremanns, 1966, *The Molecular Biology of Human Proteins*, Vol. 1, Elsevier, New York.
19. F. Hofmeister, 1888, *Arch. Exp. Pathol. Pharmakol.*, 24, 247.
20. D. Gooding et al., 1984, *J. Chromatogr.*, 296, 107.
21. W. Melander and C. Horvath, 1977, *Arch. Biochem. Biophys.*, 183, 200.
22. S. Pahlman et al., 1977, *J. Chromatogr.*, 131, 99.
23. A. Foster, 1957, *Adv. Carbohydr. Chem.*, 12, 81.
24. S. Narasimhan et al., 1980, *J. Biol. Chem.*, 255, 4876.
25. R. Rothman and L. Warren, 1988, *Biochim. Biophys. Acta*, 955, 243.
26. S. Timasheff and H. Inoue, 1978, *Biochemistry*, 17, 2501.
27. J. Lee and S. Timasheff, 1974, *Biochemistry*, 13, 257.
28. H. Inoue and S. Timasheff, 1968, *J. Am. Chem. Soc.*, 90, 1890.
29. C. Tanford, 1968, *Adv. Prot. Chem.*, 23, 121.
30. D. Hereld et al., 1986, *J. Biol. Chem.*, 261, 13813.
31. G. Kreshek, 1975, in *Water, a Comprehensive Treatise*, F. Fairbanks, Ed., p. 95, Plenum Press, New York.
32. D. Wetlaufer, and M. Koenigbauer, 1986, *J. Chromatogr.*, 359, 55.
33. A. Berggrund et al., 1994, *Process Biochem.*, 29, 455.
34. P. Strop et al., 1978, *J. Chromatogr.*, 156, 234.
35. S. Goheen, and S. Englehorn, 1984, *J. Chromatogr.*, 317, 55.
36. R. Ingraham et al., 1985, *J. Chromatogr.*, 327, 77.

7

Use of Size Exclusion Chromatography in Biopharmaceutical Development

Tim Wehr and Roberto Rodriguez-Diaz

INTRODUCTION

Size exclusion chromatography (SEC, earlier referred to as gel filtration chromatography) has several uses in the development of biopharmaceuticals. It is used as a preparative tool to isolate biologically active species, often in concert with other chromatographic techniques in a multistage purification process. It is also used as an analytical tool to obtain information about analyte molecular sizes or shapes and aggregation states, or to determine the extent or kinetics of ligand–biopolymer binding. Preparative SEC often employs soft gels such as dextrans, agarose, or polyacrylamide.[1-3] These are compressible and only compatible with elution by the use of gravity or low-pressure pumps. They may be stabilized by cross-linking, in which case they can be eluted at higher flow rates using modest pressure. Analytical SEC is most often performed using rigid supports such as inorganic silica or cross-linked organic polymers. These materials are mechanically stable at high flow rates and pressures, and are used with HPLC systems.[4] This chapter will focus on analytical applications of SEC.

RETENTION MECHANISMS IN SEC

Size exclusion is the simplest form of chromatography, in which retention depends only on the permeation of analyte into and out of the pore system of the stationary phase. In contrast to other modes of chromatography, such as reversed phase or ion exchange, in which analytes are retained by chemical interactions with the stationary phase, SEC (under ideal conditions) operates only by a molecular sieving mechanism. Molecules that are too large to enter the pores all elute in a volume of mobile phase equal to the interstitial volume between the stationary-phase particles (V_0). Molecules that are small enough to freely enter the pores all elute in a volume equal to the interstitial volume plus the volume of the pore system (V_i). Molecules of intermediate sizes sample different amounts of the pore system, depending on their size or shape, and elute between V_0 and $V_0 + V_i$. The total mobile-phase volume V_M can be expressed as the sum of the interstitial volume and the pore volume:

$$V_M = V_0 + V_i \tag{1}$$

The extent to which an analyte can penetrate the pore system is governed by its distribution coefficient K_D, which is related to its elution volume V_R by:

$$K_D = V_R - V_0/V_M - V_o \tag{2}$$

These equations can be combined:

$$V_R = V_0 + K_D V_i \tag{3}$$

From this it is readily apparent that molecules too large to enter the pores will all have K_D values of zero and will all coelute at V_0. Similarly, all molecules small enough to freely penetrate the pore system will have K_D values of one and coelute at V_M. Molecules of intermediate size will have K_D values between zero and one and will be separated according to size, with larger molecules eluting before smaller molecules.

The relationship between molecular size and elution behavior can be used to estimate the molecular weight of an analyte. A calibration plot of log molecular weight (MW) vs. retention volume (or K_D) exhibits a linear segment between V_0 and V_i (Figure 7.1). If the plot is constructed with standard proteins with shapes similar to that of an analyte protein, the molecular weight of the analyte can be estimated by interpolation of its retention volume on the plot. The relationship between log MW and K_D is linear for K_D values between about 0.2 to 0.8.

Although SEC is often used to estimate protein molecular weight, it should be understood that retention is governed by the hydrodynamic volume of the solute, which may not be closely related to molecular weight. Hydrodynamic volume is affected by the degree of solute hydration and molecular shape. The effective molecular size of a protein in solution depends on its radius of gyration or Stokes radius. Two proteins with similar molecular weights but different shapes (e.g., spherical vs. oblate vs. rod-like) will "carve out" different hydrodynamic volumes and could display significantly different retention volumes (Figure 7.2). To obtain accurate molecular weight estimates with SEC, it is necessary that the proteins used to construct the calibration plot and the analytes all have similar shapes. An alternative approach is to perform calibration and analysis under

Calibration Curve

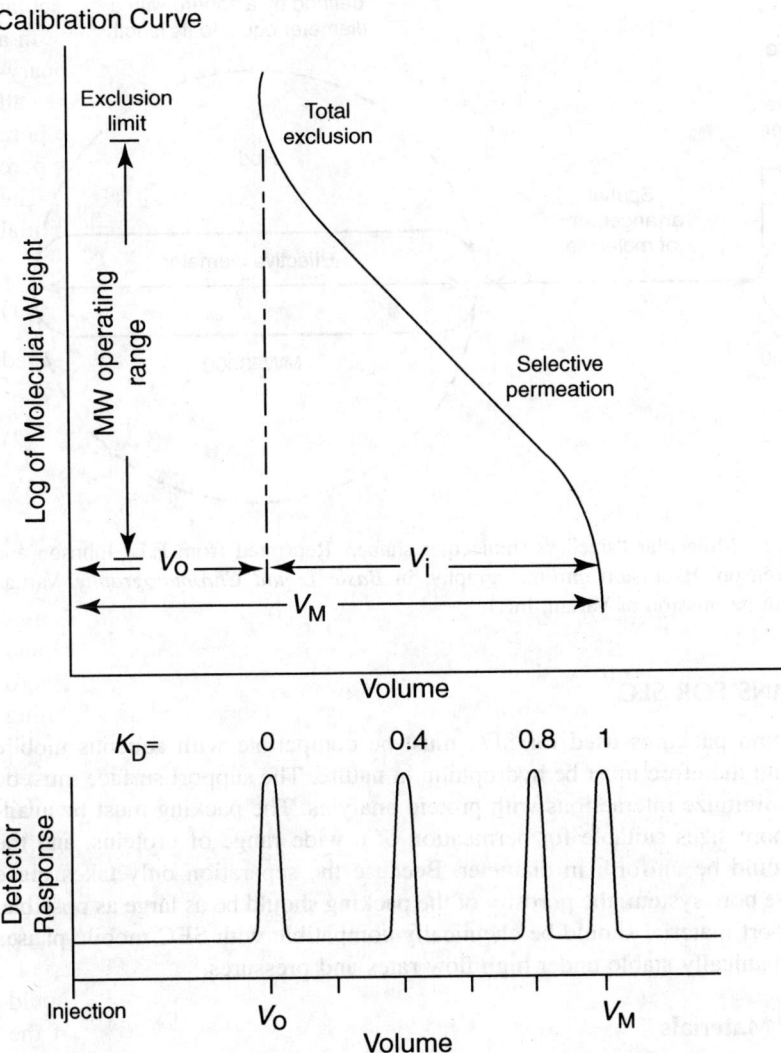

Figure 7.1 SEC chromatogram and calibration curve. (Adapted from E.L. Johnson and R.L. Stevenson, Exclusion chromatography, in *Basic Liquid Chromatography*, Varian, 1978. With permission of Varian, Inc.)

denaturing conditions so that both calibrant and analyte proteins are converted to linear random-coil conformations such that elution behavior is correlated to molecular weight.

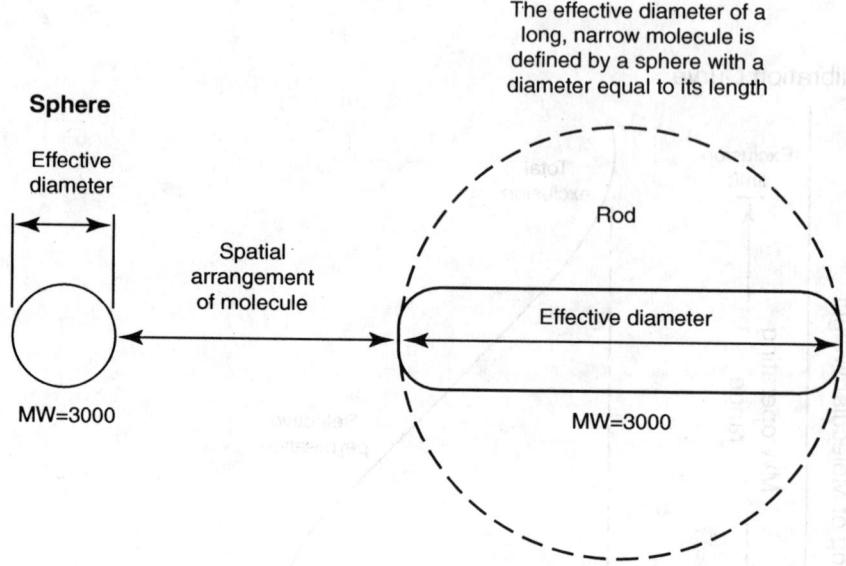

Figure 7.2   Molecular "size" vs. molecular shape. (Reprinted from E.L. Johnson and R.L. Stevenson, Exclusion chromatography, in *Basic Liquid Chromatography*, Varian, 1978. With permission of Varian, Inc.)

COLUMNS FOR SEC

The column packings used for SEC must be compatible with aqueous mobile phases and therefore must be hydrophilic in nature. The support surface must be inert to minimize interactions with protein analytes. The packing must be available in pore sizes suitable for permeation of a wide range of proteins, and the pores should be uniform in diameter. Because the separation only takes place within the pore system, the porosity of the packing should be as large as possible. The support material should be chemically compatible with SEC mobile phases and mechanically stable under high flow rates and pressures.

Support Materials

Two types of materials are used for SEC columns: bonded silicas and hydrophilic organic polymers. Silica is the most widely used material for HPLC packings in general, due to its good mechanical stability, high porosity, and availability in a range of pore sizes with closely controlled pore diameters. However, silica surfaces are highly interactive with proteins and must be derivatized to eliminate interactions. The most common approach is to react surface silanols with an organosilane reagent to introduce a diol-type or carbohydrate-like coating covalently attached to the silica. A limitation of silica-based SEC packings is their instability under alkaline conditions. Silica dissolves at pH values above 8, leading to reduced column lifetime. One product (the Zorbax GF SEC columns from Agilent Technologies) uses a zirconyl cladding to stabilize the silica support.

Because of the limitations of silica, several manufacturers offer SEC columns based on hydrophilic organic polymer. These include polymethacrylate supports, proprietary hydrophilic polymers, and semirigid cross-linked agaroses and dextrans. These materials are more stable under high-pH operation.

Pore Size and Porosity

High-performance SEC packings are available in pore sizes ranging from 10 to 400 nm. A column should be selected with a pore size so that the analytes elute within the linear portion of the calibration plot, e.g., with K_D values between 0.2 and 0.8. Manufacturers provide calibration plots in their product literature for this purpose (Figure 7.3). However, the calibration plots used for column selection should be obtained using calibrants and elution conditions appropriate for the analysis. For example, calibration plots constructed with native proteins should be used for SEC of analyte proteins under physiological conditions. Similarly, column selection for chromatography of proteins under denaturing conditions should be done using calibration plots of denatured proteins or linear hydrophilic polymers (e.g., polyethylene glycols or sulfonated polystyrenes).

Columns with a narrow pore-size distribution will be characterized by high resolution over a narrow fractionation range, i.e., they exhibit a calibration plot with a shallow slope. Columns with a wide pore-size distribution will be characterized by a wider fractionation range, but poor resolution across that range (i.e., a steep calibration plot).

The porosity of an SEC column can be characterized by its phase ratio (V_i/V_0). Soft-gel SEC packings have high porosities with phase ratios of 1.5 to 2.4.[5] High-performance SEC packings have more modest phase ratios of 0.5 to 1.5.[6] This limitation is offset by the high efficiencies and rapid analysis times of high-performance SEC. It should be evident that as support pore diameter and pore volume increase, the amount of solid material in the particle will be reduced, compromising the mechanical strength of the support matrix.

Particle Diameter

As in interactive modes of chromatography, reduction in particle diameter reduces mass transfer effects and improves column efficiency in SEC. Column packings with particle diameters of 10 to 12 μm are available for less demanding applications, whereas SEC packings with particle diameters of 4 to 5 μm can be used for applications requiring higher resolution.

MOBILE PHASES FOR SEC

In contrast to interactive modes of chromatography where the mobile phase is an active participant in the separation process, the mobile phase in SEC is simply a

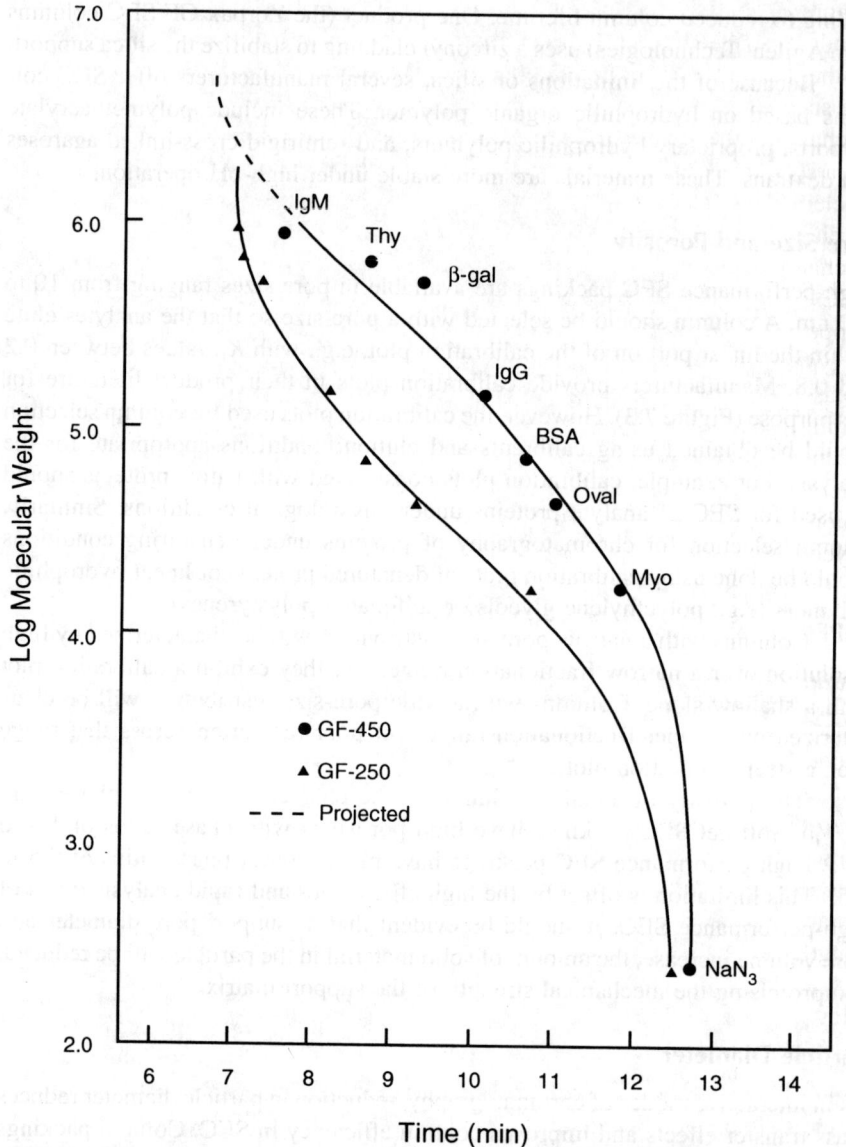

Figure 7.3 Calibration curves of proteins using a 0.2 *M* sodium phosphate (pH 7.5) mobile phase. (Printed with permission of Agilent Technologies, Inc.)

carrier to transport analytes through the column. In principle, the mobile phase chemistry is designed to keep the analyte in solution and in the appropriate conformation and to maximize column lifetime. In practice, it may contain additives to suppress "nonideal" interactions of the analyte with the support matrix or the bonded stationary phase. These interactions may be electrostatic in nature and, in the case of silica-based columns, are most often due to residual silanols on the support. For cationic analytes such as basic proteins, this results in a cation-exchange contribution to retention that causes analytes to elute later than predicted by a solely sieving mechanism. For anionic analytes such as acidic proteins and nucleic acids, this results in an ion-exclusion phenomenon that causes analytes to elute earlier than predicted. In severe cases, analytes may elute after V_M (ion exchange) or before V_0 (ion exclusion). A second type of nonideal behavior in SEC is hydrophobic interaction. This may be due to hydrophobic sites on the support (polymer-based columns) or on the bonded phase (silica-based columns). A typical mobile phase for SEC is 100 mM potassium phosphate + 100 mM potassium chloride (pH 6.8). If nonideal behavior is observed, it can often be minimized by adjusting the salt concentration: increasing ionic strength reduces electrostatic interactions, and decreasing ionic strength reduces hydrophobic interactions. Hydrophobic interactions can also be reduced by adding a small amount (e.g., 5 to 10%) of an organic modifier (e.g., methanol, ethanol, and glycerol).

OPERATIONAL CONSIDERATIONS

Sample Capacity

The loading capacity of SEC columns is quite modest compared to interactive modes of chromatography. A rule of thumb dictates that the sample volume capacity is about 2% of the column volume. A typical analytical SEC column with dimensions of 8 × 300 mm has a V_M of 10 to 11 ml, providing a sample volume limit of about 200 µl. The mass loading limit for such a column is about 1 to 2 mg. Above these volume and mass limits, resolution will be compromised. Sample capacity will scale in proportion to column volumes for different column lengths and diameters.

Flow Rate

The low molecular diffusion coefficients of proteins and other biopolymers reduces the efficiency of mass transfer and compromises efficiency as flow rate is increased. Therefore, high-performance SEC columns are usually operated at modest flow rates, e.g., 1 ml/min or less. However, operation at very low flow rates is undesirable due to excessive analysis times, loss of efficiency due to axial analyte diffusion, and the risk of poor recovery due to analyte adsorption.

Use of Denaturants

As discussed above, accurate estimation of molecular weights may not be achieved under native conditions due to molecular shape effects. Performing calibration and analysis under denaturing conditions may be desirable in this circumstance. Addition of a denaturant such as 4 to 6 M guanidinium hydrochloride, 4 to 6 M urea, or 0.1 to 1% sodium dodecylsulfate (SDS) to the mobile phase can be used to convert calibrants and analytes to random coil conformations. However, denaturants can reduce the effective porosity of the column. Also, surfactants such as SDS may bind strongly to the column and be difficult to remove. It is advisable to dedicate the column to such an application. High concentrations of chaotropic salts such as urea and guanidinium HCl can compromise pump and injector seals and should never be left standing in the HPLC system following use.

Use of Probes to Characterize Column Performance

When installing a new SEC column, the values of V_0 and V_M should be determined using appropriate probes. The value of V_0 can be measured using a large biopolymer outside the exclusion limit of the column; high-MW DNA (e.g., calf thymus DNA) is often used. The blue dextran used for measuring V_0 on soft-gel columns may give erroneous values on some high-performance SEC columns due to hydrophobic binding. The value of V_M is determined using a very hydrophilic small molecule that can be detected in the UV. Popular choices are cyanocobalamin (vitamin B_{12}) and glycyl tyrosine.[6]

Nonideal interactions can be also be characterized using small-molecule probes.[7] Ideally, these probes should elute at V_M. Cation-exchange interactions can be detected by excessive retention of arginine or lysine.[8] Ion-exclusion effects can be characterized by early elution of citrate or glutamic acid. Hydrophobic interactions can be detected by late elution of phenylethyl alcohol or benzyl alcohol.[8]

Exploiting Nonideal Interactions

While nonideal interactions can prevent accurate estimates of molecular weight, they may be exploited to achieve a separation if SEC is being used for preparative isolation or to profile a sample. The same techniques used to suppress nonideal interactions (modulating salt levels, adding organic modifiers, and manipulating pH) can be used to enhance these interactions.

Coupling SEC Columns

SEC columns may be coupled in series either to increase efficiency or to increase the fractionation range. Because resolution is increased by the square root of column efficiency, operating two columns in tandem to double the number of theoretical plates will increase the resolution by 2½, or 40%. As there are very few options in optimizing an SEC separation, coupling columns is often the only road to achieving the desired separation. In cases where the distribution of analyte sizes is larger than can be accommodated with a single pore-size column, coupling columns of different pore sizes can increase the MW range of the separation. The order of coupling is not important. In either application of coupled columns, the columns should be closely matched in terms of efficiency.

ADVANTAGES AND LIMITATIONS OF SEC

SEC offers several advantages that make it a desirable technique for both pre-parative and analytical applications. First, separations are rapid: with an 8×300 mm analytical column operated at 1 ml/min, all analytes elute in about 10 min. Second, because the stationary phase is designed to eliminate interactions with the sample, SEC columns exhibit excellent recovery of mass and biological activity. Third, because all separations are performed under isocratic conditions, peak area and retention time precision are high.

There are four limitations to SEC. First, the resolving power is quite modest compared to interactive chromatographic modes. The peak capacity (i.e., the maximum number of baseline-resolved peaks in a separation) is approximately 5 to 10. For an SEC column with a fractionation range from 10 to 500 kDa, this implies that two proteins can be resolved if they differ in molecular weight by a factor of two. This means that SEC is useful as an analytical tool only for samples containing a limited number of components. The second limitation of SEC is the low volume and mass loading capacity. As a consequence of these two limitations, SEC is more likely to be used as a later step in a purification scheme. A third limitation of SEC is modest column lifetime, particularly for silica-based SEC columns. When operated with aqueous buffers at neutral pH, SEC-column lifetime is typically shorter than that of a silica-based reversed-phase column operated with aqueous–organic solvent systems. A final limitation of SEC is the accuracy of molecular weight estimates. Although SDS-PAGE and mass spectrometry provide more accurate values, the first technique is laborious and the second very expensive. SEC may be preferred when a quick and rough estimate of molecular weight is satisfactory.

APPLICATIONS

SEC can be used to accomplish a class separation in which one component of the sample elutes in either excluded volume or permeation volume; we term this application as *group fractionation*. Alternatively, SEC can be used to resolve two or more species within the included volume (e.g., between V_0 and V_P). We term this application simply as *fractionation*.

Although resolution in SEC is relatively low as compared with other tech-niques (e.g., SDS-PAGE), it allows analysis of the native protein. Thus, we can obtain a glimpse into the tertiary or quaternary structure of the molecule. Indeed, proteins such as bovine serum albumin (BSA) are usually resolved into monomer, dimer, tetramer, etc., by SEC. Caution should be exercised, as aggregates can go undetected if sheer forces disrupt them. The Stokes radius increases for partially unfolded or denatured proteins, and SEC detects such changes as an increase in apparent MW.

Group Fractionation

SEC can be used for group fractionation when the molecule of interest has a significant MW difference relative to contaminants. The most common uses are desalting, buffer exchange, and the removal of excess reactants during protein modification. Buffer exchanges and desalting are performed when the sample is at a pH or has a composition that is not suitable for our purposes, e.g., when the sample is in a UV-absorbing buffer and we want to measure the UV absorbance of the protein. Protein treatment (e.g., reduction of S–S bonds) and modifications (e.g., acetylation) result in an excess of reactants. Often the reactants are much smaller than the modified protein, and they can be removed by group fractionation. Some molecules are purchased with specific protecting groups that need to be removed before they can be used for synthesis. In this regard, Figure 7.4 shows a typical chromatogram obtained when deprotecting oligonucleotides containing a modifiable chemical group. The oligonucleotide has a MW of approximately 7500 Da, whereas the protecting group and excess reagents have MWs below 1000 Da. The chromatographic packing is selected to exclude globular proteins with MW higher than 5000 Da. However, because oligonucleotides are usually linear (unless they have complementary sequences), their retention is lower and apparent MW in SEC is higher (more than 2 times) than that of a globular protein. In Figure 7.4A, the UV signal at 260 nm is totally saturated, and any resolution between the oligonucleotide and the low-MW contaminants is obscured. In such a case, the injection and collected volumes can be predetermined. As a rule of thumb, the injection volume should not exceed 20% of the column volume, and the collected volume should be approximately 1.5 times the injection volume. Monitoring at a wavelength that is not saturated during the procedure (i.e., a wavelength with less sensitivity) can be helpful in visualizing more detail of the chromatogram as shown in Figure 7.4B, in which the detector signal is collected at 300 nm. Of course, better resolution is obtained when the MW difference between the sample and the contaminants is greater (we routinely obtain baseline resolution when desalting proteins with MW higher than 40,000 Da). When desalting will be used as a routine technique, we find it useful to inject a smaller volume (2 to 3% of the column volume) during purification development to obtain better resolution and then evaluate the peak shape for tailing, which can severely affect sample recovery. If the peak tailing is excessive, other packings are tried or, if the application permits, the buffer composition is changed. It is necessary to keep in mind that in SEC, column length and mobile-phase flow can also be used to manipulate resolution.

There are several other techniques that can be used to perform group fractionation. These include dialysis, ultrafiltration, ultracentrifugation, tangential

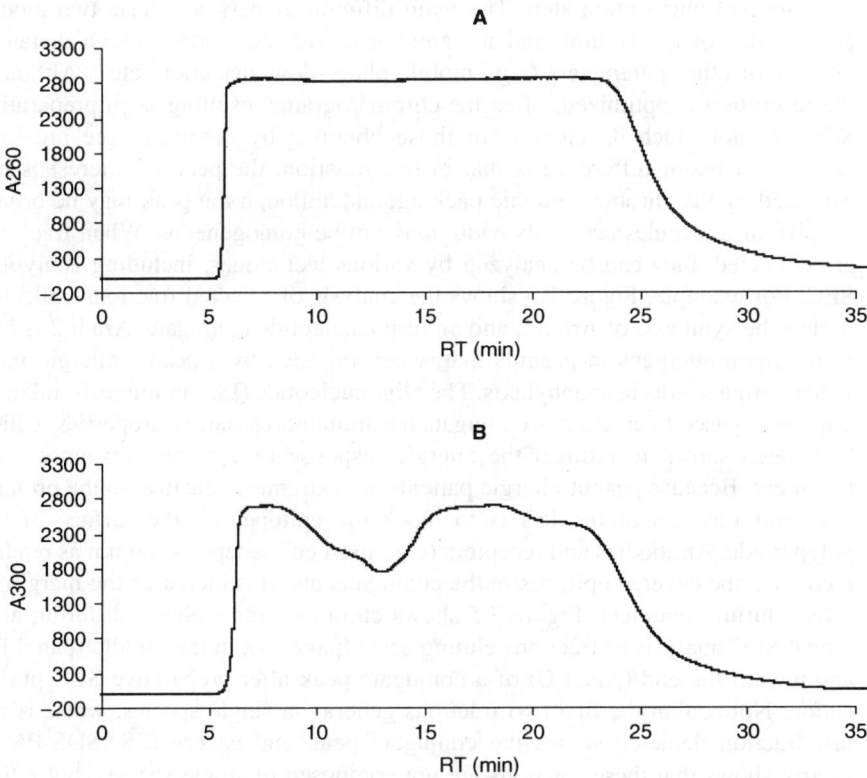

Figure 7.4 Preparative chromatogram showing SEC group separation of a 22mer phosphorothiolated oligonucleotide and its protecting group. Because of the high volume and concentration loaded, the UV signal at 260 nm is totally saturated (panel A), and no resolution is observed between the oligonucleotide and the low-MW contaminants. A second wavelength removed from the absorption maxima of the oligonucleotide is shown in panel B (300 nm).

flow filtration, etc. Selection of the appropriate method depends on variables such as sample volume (ultrafiltration requires larger volumes), sample stability (dialysis is a slow process), differences in MW (the chromatographic packings offer a wider selection of pore sizes), and cost. Ultrafiltration is more amenable and cost-effective for scaled-up processes (tens to thousands of milliliters).

Fractionation

During the development of sample-fractionation methods, the main goal is to achieve the highest resolution possible between the molecule of interest and the contaminants. Fractionation by SEC is used both as an analytical tool and as a (preparative) purification step. The main differences between these two modes are the size of the column and the amount of sample loaded, which dictate a number of other parameters (e.g., mobile-phase flow, detection, etc.). Although the resolution is optimized, often the chromatograms resulting from preparative SEC are not much different from those obtained by desalting (see previous section). A major difference is that in fractionation, the peak of interest is not excluded by the chromatographic packing, and although the peak may be broad, the size of molecules across its width may not be homogeneous. When fractions are collected, they can be analyzed by various techniques, including analytical SEC. For example, Figure 7.5 shows the analysis of selected fractions collected during the synthesis of Ara h 2 and an oligonucleotide conjugate. Ara h 2 is one of the main allergens in peanuts that, when ingested by a peanut-allergic individual, often results in anaphylaxis. The oligonucleotide (ISS, immunostimulatory sequences) used to create this conjugate has immunoregulatory properties, which have been shown to redirect the allergic response to a normal response after treatment. Because peanut-allergic patients are extremely sensitive to the protein, a second function of the ISS is to block the epitopes on the surface of the polypeptide. Antibodies and receptors (e.g., mast cell receptors) do not as readily recognize the covered epitopes in the conjugate, and this increases the margin of safety during treatment. Figure 7.5 shows chromatograms obtained during analytical SEC analysis of fractions eluting early (panel A), in the middle (panel B), and toward the end (panel C) of a conjugate peak after preparative SEC purification. Notice that the first two fractions generate a single species, whereas the last fraction depicted shows the conjugate peak and excess ISS. SDS-PAGE clearly shows that these fractions are not composed of single species, but rather of protein containing discrete amounts of ISS; nevertheless, analytical SEC provides the data required for pooling of the final product for this and other conjugates. Notice that the mobile phase for the analytical SEC contains a small amount (10%) of methanol. The addition of the organic solvent was done with two purposes: reduced sample tailing and prevention of microbial growth in the HPLC. Before adding any additive to the mobile phase, it is important to show that the sample and the chromatography are not affected in undesirable ways. Because in many laboratories HPLC instruments are used in the "micro-preparative" mode (lab bench-scale purification), system sanitation must be performed routinely and microbial growth prevented to avoid interference in biological assays (*in vitro* and *in vivo*), which usually are performed with the HPLC-purified material. One of the most persistent effects of microbial growth is sample contamination with endotoxin.

Amb a 1 is another allergenic protein that we have conjugated to ISS. Amb a 1 is the main allergen of ragweed pollen, and when conjugated, also generates a family of molecules containing protein and various amounts of oligonucleotides (Amb a 1 immunostimulatory complexes, or AIC). Preparative SEC is used to purify the main product from excess reactants, and analytical SEC is used to evaluate the purity (and relative concentration by the deflection of the detector

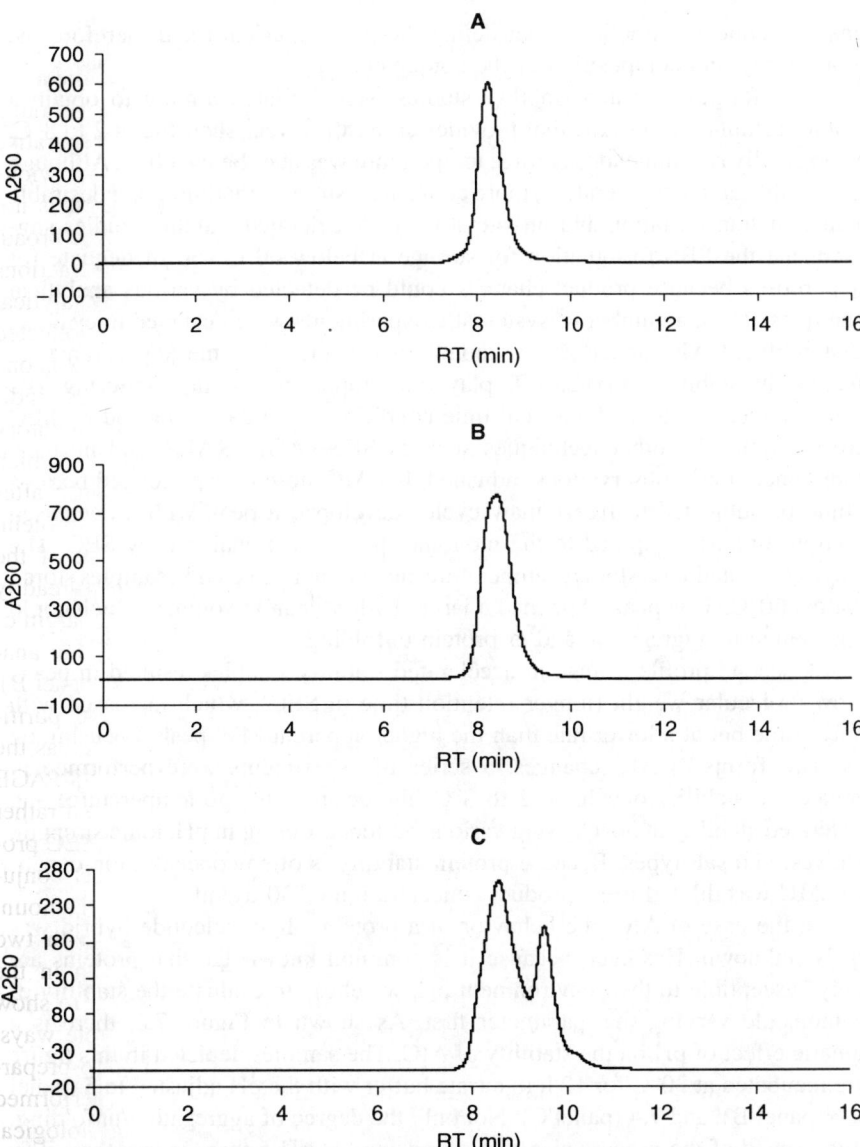

Figure 7.5 Analytical SEC of selected fractions collected during the synthesis of Ara h 2 and an oligonucleotide conjugate. The fractions were chosen to represent early eluting species (panel A, higher MW), middle of the peak (panel B), and late eluting (panel C, lower MW) of the conjugate peak. Notice that the first two fractions generate a single species, whereas the last fraction depicted shows the conjugate peak and excess ISS. Size exclusion chromatography was performed using a Biosep 3000 column (Phenomenex, Torrance, CA) 4.6 cm × 30 cm, using PBS + 10% methanol as mobile phase flowing at 0.35 ml/min. The injection volume was set to 20 µl, and detection was performed by UV absorption at 260 nm.

signal) of collected fractions, that define the pooling criteria and therefore the overall purity and composition of the conjugate.

AIC formulation and stability studies were initiated aiming to obtain a suitable formulation for AIC that provides at least a 1-year shelf life at 2 to 8°C. The originally recommended storage temperature was at or below 60°C. Although AIC is stable at that temperature, more convenient storage conditions are desirable because of transportation and on-site storage. Accelerated stability studies suggested that the PBS formulation for storage at below 60°C was inadequate for our purposes because product changes could be detected by various analytical techniques. Thus, a number of systematic experiments were designed to evaluate the stability of AIC under various conditions to elucidate the key parameters affecting the stability of AIC. SEC played an important role in monitoring AIC stability under accelerated and real-time conditions. The data collected by SEC were supported by other techniques such as SDS-PAGE, RALS, and intrinsic fluorescence. Early observations indicated that AIC stored for prolonged periods of time or subjected to freeze/thaw cycles developed a peak with a decreased retention time as compared to the monomer peak when analyzed by SEC. The change correlated with storage temperature and was not detected in samples stored at below 60°C. This peak, thus, had a larger hydrodynamic volume, which could be attributed to aggregation and/or protein unfolding.

A second profile change in accelerated stability samples resulted in peaks of low molecular weight (longer retention time in SEC), which increased with storage time but at a lower rate than the higher apparent MW peak. Focusing on these two forms of AIC changes, a series of experiments were performed to evaluate the stability of AIC at 2 to 8°C (the desired storage temperature) and accelerated stability at 30°C, using various buffers differing in pH, ionic strength, additives, and salt types. Because protein stability is often concentration-dependent, AIC was diluted to its product concentration of 30 µg/ml.

In the case of AIC, the behavior of a protein–oligonucleotide hybrid was largely unknown. However, because it is common knowledge that proteins are highly susceptible to their environment pH, we chose to evaluate the stability of the molecule varying this parameter first. As shown in Figure 7.6, there is a dramatic effect of pH on the stability of AIC. The samples depicted in this figure were incubated at 30°C for 12 h in citrate buffer with the pH adjusted to 5 (panel A), 6 (panel B), and 7.4 (panel C). Not only the degree of aggregation/unfolding, but the speed of the process also was surprising. At pH 5 there is less than 50% monomer remaining after incubation; at pH 6 the process is slower than at pH 5, but it is still significant. Notice that aggregation/unfolding practically stops at pH 7.4. The peak with a retention time >10 min is due to a buffer component. SDS-PAGE analysis indicated that the fronting peak was composed of aggregated and unfolded AIC. This was concluded because the relative area of the high-MW peak in SEC was much higher than the aggregate band in SDS-PAGE. Because proteins are denatured (unfolded) during SDS-PAGE, only chemically bound

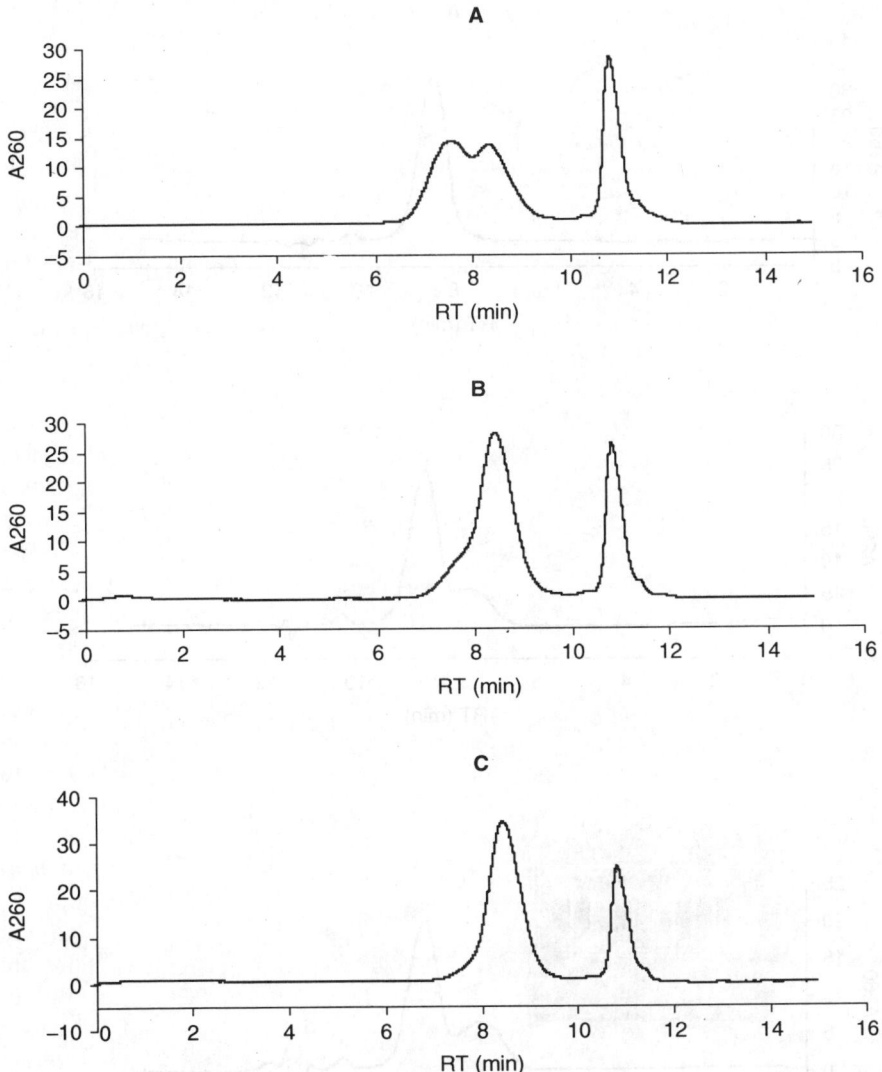

Figure 7.6 Effect of pH on the stability of AIC after incubation at 30°C for 12 h in citrate buffer with the pH adjusted to 5 (panel A), 6 (panel B), and 7.4 (panel C). The peak with a retention time >10 min is due to a buffer component. Panel C consist mainly of monomer, and the front peak observed in panels A and B is due to aggregation/unfolding of the conjugate. Analysis conditions as described in Figure 7.5.

aggregates remained. We conducted SEC in the presence of urea to unfold the protein and disrupt physical aggregates, and then a more consistent relative area between the two techniques was obtained for the aggregate.

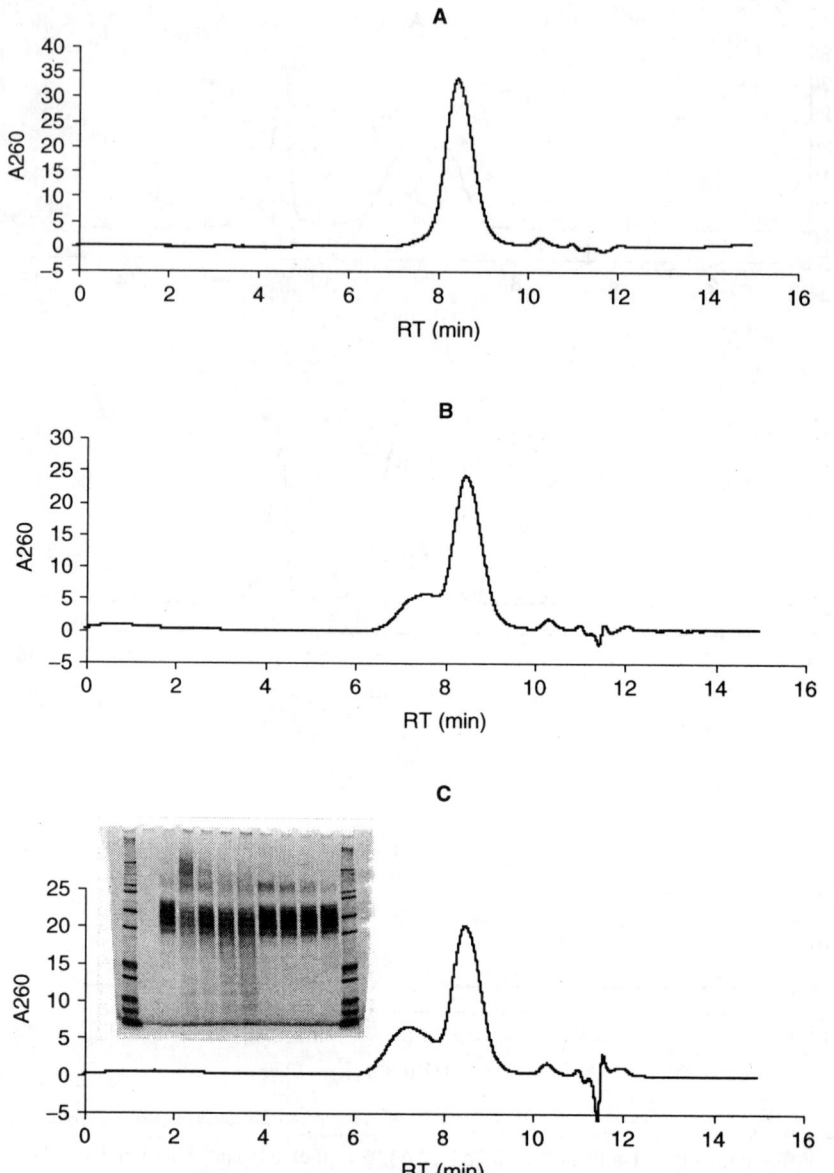

Figure 7.7 Effect of salt concentration on the stability of AIC after incubation for 24 h at 30°C in phosphate buffer pH 7.2 in the absence of salt (panel A), 0.5 *M* NaCl (panel B); and 1 *M* NaCl (panel C). AIC exposed to high concentration of NaCl shows a fronting shoulder. Analysis conditions as described in Figure 7.5. The gel shown as an inset in panel C also depicts AIC. From left to right, Lanes (1) MW standards, (2) blank, (3) reference AIC, (4) AIC incubated in the presence of 1 *M* NaCl, (5) AIC incubated in the presence of 0.5 *M* NaCl, (6–7) AIC incubated in the absence NaCl, (8–11) reference AIC, and (12) MW standards.

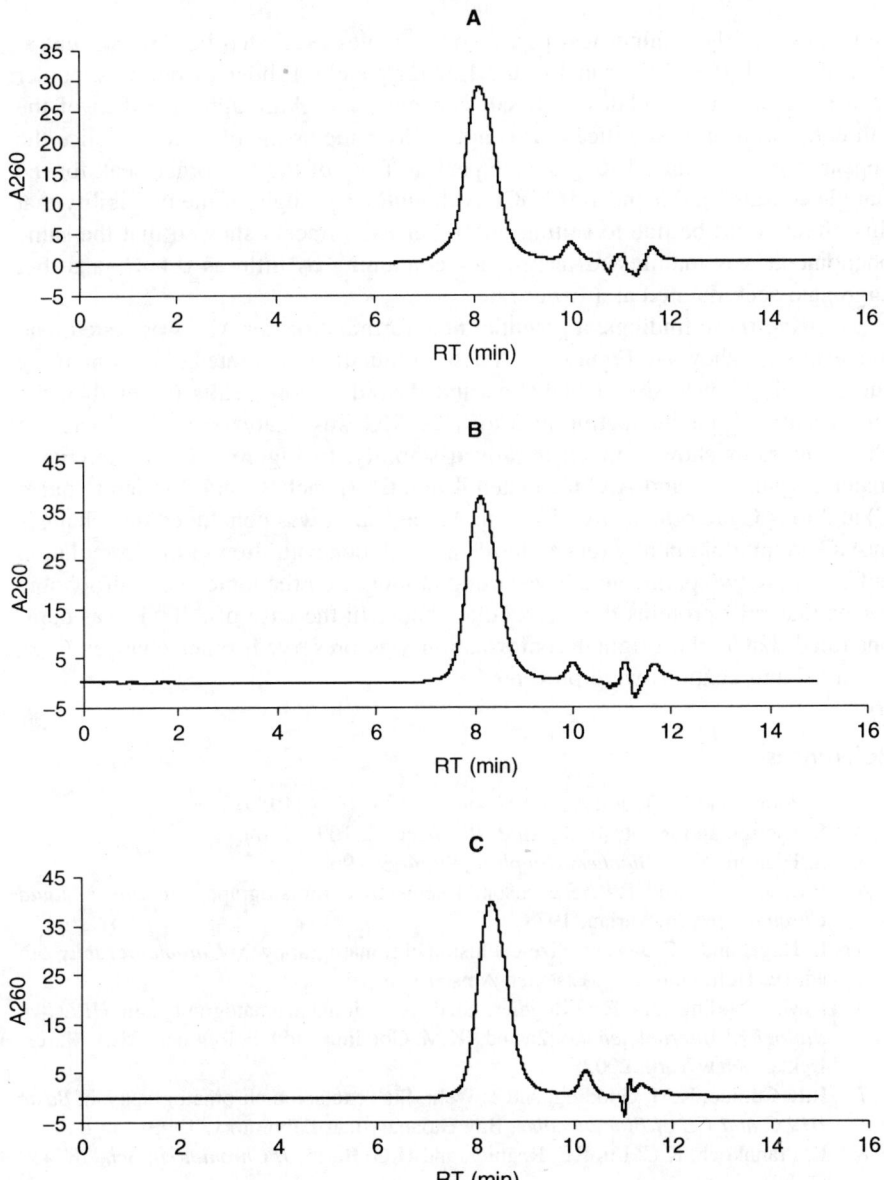

Figure 7.8 Stability of AIC in borate buffer containing sucrose (pH 8) after storage at 2 to 8°C for various time periods as analyzed by SEC. AIC reference material (panel A), AIC incubated 4 months (panel B), and 7 months (panel C) at 2 to 8°C are practically identical. Analysis conditions as described in Figure 7.5.

Next, the effect of salt concentration was evaluated. Figure 7.7 shows SEC analysis of AIC incubated for 24 h at 30°C in phosphate buffer pH 7.2 containing various amounts of NaCl. AIC incubated in the absence of NaCl (panel A) generates a fairly symmetrical peak. Both samples incubated in high concentrations of NaCl (0.5 *M* for panel B and 1 *M* for panel C) show a fronting shoulder with an area that correlates with salt concentration. Although the effect of the salt concentration is significant, it is not as dramatic as the pH effect. Notice the appearance of an unfolded/aggregate peak in front of the monomer peak for the sample containing 0.5 and 1 *M* NaCl. Although we thought of the possibility that this effect could be due to salting-out, other experiments showed that the same phenomena was manifested in samples containing as little as 0.1 *M* salts, but aggregate accumulated at a lower rate.

Using these findings, a potential new formulation for AIC was tested, and the results are shown in Figure 7.8. A new formulation (a borate buffer containing sucrose, pH 8) was slightly alkaline and devoid of ionic salts (other than the buffer salts). After incubation at 2 to 8°C, AIC was analyzed by SEC and the chromatograms showed much improved stability. In Figure 7.8, AIC reference material (panel A) and AIC incubated 4 months (panel B) and 7 months (panel C) at 2 to 8°C are practically identical. At the end, it was concluded that changes in AIC are predominantly related to pH and salt concentration in the formulation buffer. These two parameters have strong influence on the ionic and hydrophobic forces that give proteins their particular shape. In the case of AIC, it was demonstrated that if the original conformation was preserved, other changes (e.g., chemical aggregation) were prevented.

References

1. J. Porath and P. Flodin, *Nature (London)*, 183, 1657 (1959).
2. S. Hjerten and R. Mosbach, *Anal. Biochem.*, 3, 109 (1962).
3. S. Hjerten, *Arch. Biochem. Biophys.*, 99, 466 (1962).
4. E.L. Johnson and R.L. Stevenson, Exclusion chromatography, in *Basic Liquid Chromatography*, Varian, 1978.
5. L. Hagel and J.C. Janson, Size-exclusion chromatography, in *Chromatography*, 5th ed. (E. Heftmann, Ed.), Elsevier, Amsterdam, 1992.
6. K.M. Gooding and F.E. Regnier, Size exclusion chromatography, in *HPLC of Biological Macromolecules*, 2nd ed. (K.M. Gooding and F.E. Regnier, Eds.), Marcel Dekker, New York, 2002.
7. R.L. Cunico, K.M. Gooding, and T. Wehr, Size exclusion chromatography, in *Basic HPLC and CE of Biomolecules*, Bay Bioanalytical Laboratory, 1999.
8. E. Pfannkoch, K.C. Lu, F.E. Regnier, and H.G. Barth, *J. Chromatogr. Sci.*, 18, 430 (1980).

8

Slab Gel Electrophoresis
for Protein Analysis

David E. Garfin

INTRODUCTION

Gel electrophoresis holds a special place among the methods for protein analysis. It is fair to state that it is the most widely used procedure in protein studies. At some point in their work on proteins, researchers are likely to use gel electrophoresis or at least contemplate using it. The technique is familiar to everyone working with biomolecules and to the general public as well through accounts of its use in genomics, proteomics, diagnostics, and forensics. The widespread utilization of gel electrophoresis is undoubtedly because of its high resolution and its ease of use.

Of all the forms of electrophoresis, sodium dodecyl sulfate–polyacrylamide gel electrophoresis (SDS-PAGE) is by far the most commonly practiced method. It has become an everyday procedure in protein laboratories. Many people do not realize that there is any kind of gel electrophoresis other than SDS-PAGE. It provides an easy way to assess the complexity of a sample or the purity of a preparation. SDS-PAGE is particularly useful for monitoring the fractions obtained during protein purification by other techniques such as chromatography. It also allows samples from different sources to be compared for protein content. One of the more important features of SDS-PAGE is that it is a simple, reliable method with which to estimate the molecular weights of proteins.

SDS-PAGE is a denaturing technique in which proteins are broken down to their constituent polypeptide chains. Nondenaturing procedures are also available for use when it is desirable to maintain biological activity or antigenicity. However, it is more difficult to extract easily interpretable information from nondenaturing gels than from SDS-PAGE.

Although nondenaturing systems can give information about the charge isomers of proteins, this information is best obtained by isoelectric focusing (IEF). An IEF run can show charge heterogeneity in proteins not apparent with other types of electrophoresis. Proteins thought to be a single species by SDS-PAGE analysis are sometimes found by IEF to consist of multiple species. A more thorough determination of the composition of a protein preparation is obtained upon combining IEF with SDS-PAGE in two-dimensional polyacrylamide gel electrophoresis (2-D PAGE). 2-D PAGE is the primary separation tool in the field of proteomics research because it is capable of separating thousands of proteins in a single gel. When desired, protein identifications are obtainable by immunoblotting, which combines antibody specificity with the separation power of 1- or 2-D gel electrophoresis.

GEL ELECTROPHORESIS

Despite some refinements in the methods, the basic principles and protocols of gel electrophoresis have not changed appreciably since their introduction. Proteins are introduced into a gel matrix and separated by the combined effects of an electrical field, buffer ions, and the gel itself, which acts as a protein sieve. At the completion of the electrophoresis run, separated proteins in the gel are stained to make them visible, then analyzed qualitatively or quantitatively. The topic has been covered in numerous texts, methods articles, and reviews.[1-11] In addition, apparatus and reagents for analytical and preparative gel electrophoresis are available from several suppliers.

Proteins are charged molecules and migrate under the influence of electric fields. For the purposes of gel electrophoresis, the two most important physical properties of proteins are their electrophoretic mobilities and their isoelectric points (pIs). The surrounding medium influences both the electrophoretic mobilities and pIs of proteins, and this is exploited in various ways. In gel electrophoresis, the largest influence on protein migration comes from the sieving properties of the gels. Factors such as pH and the amounts and kinds of ions and denaturants in the system also influence migration.

The rate of migration of a protein per unit of field strength [velocity ÷ (magnitude of electric field)] is called its *electrophoretic mobility*. It is relatively easy to show[1,6,12] that the electrophoretic mobility of a particle is given by the ratio of its charge to its friction coefficient [charge ÷ (friction coefficient)]. Attempts to derive physical properties from the electrophoretic mobilities of proteins have been generally unsuccessful because of the complexity and size of protein molecules. Thus, electrophoretic mobility is used as a descriptive concept rather than as an analytical tool. It is relatively unimportant except in discussions of how the properties of gels and the compositions of buffer systems influence migration rates during electrophoresis.

The pH of the electrophoresis buffer determines the charges on the proteins in the sample being run. This means that the directions of motion of proteins in electrical fields depend on the pH of the electrophoresis buffer. Proteins are amphoteric molecules with net charges that vary with the pH of their local environment. For every protein there is a specific pH at which its net charge is zero. This pH is the so-called *isoelectric point* or pI of the protein. A protein is positively charged in solutions at pH values below its pI and negatively charged when the pH is above its pI. Thus, proteins are cationic when the pH is below their pIs and anionic when pH is above their pIs. On the other hand, when ionic detergents are employed in electrophoresis, the charges on the detergents determine the directions of migration and the pH of the buffer becomes relatively unimportant.

Apparatus for Gel Electrophoresis

The equipment and reagents for gel electrophoresis are readily available and familiar to laboratory workers. Particularly noteworthy is the steady increase in the popularity of precast polyacrylamide gels since their introduction in the early 1990s (see Section 8.2.4). Precast gels provide researchers with off-the-shelf convenience and reproducibility and help to make gel electrophoresis and IEF commonplace laboratory practices.

Cells for gel electrophoresis are relatively simple. Most electrophoresis cells are variants of a standard design (Figure 8.1). Gels are formed in glass or plastic cassettes that are suspended vertically between anode and cathode buffer compartments. Electrodes of platinum wire and jacks for making electrical contact with the electrodes connect the gels and buffers to the power source. Samples are introduced into wells formed in the gels at the time of casting. The rectangular slab format allows multiple samples to be run and compared in the same gel. Manufacturers provide thorough instructions for using their electrophoresis cells.

Conventional gels are of the order of 20 cm wide × 20 cm long. The so-called mini-cells and midi-cells allow rapid analysis and are adequate for relatively uncomplicated samples. The design of the small cells allows runs to be completed in as little as 35 to 45 min as compared to 6 to 7 h for the larger gels. Mini-gels are about 7 cm long × 8 cm wide while midi-gels are of the order of 15 × 10 cm. Gel thickness is varied by means of spacers inserted into the cassettes. Standard gel thickness is 0.75 or 1 mm, which allows for adequate loads, high sensitivity staining, and good heat dissipation. Some gels can hold as many as 26 samples. All three size categories of gels can be used for nearly all purposes. However, the separation between bands is greater with longer gels. Closely spaced bands are easier to distinguish from one another, their bandwidths are easier to measure, and they are more easily cut out from a large gel than from a small one.

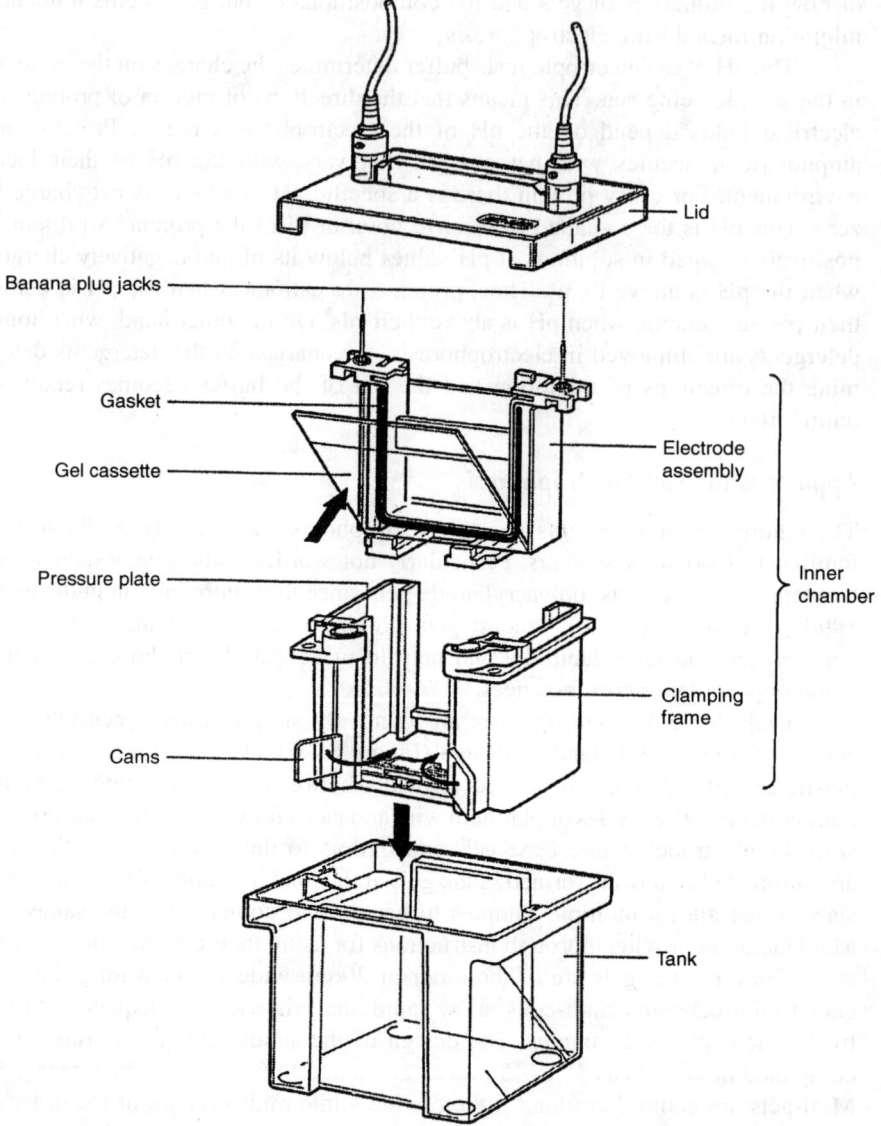

Banana plug jacks

Gasket

Gel cassette

Pressure plate

Cams

Lid

Electrode
assembly

Inner
chamber

Clamping
frame

Tank

Figure 8.1 Exploded view of an electrophoresis cell. The components of the Bio-Rad Mini-PROTEAN 3 are shown. The inner chamber can hold one or two gels. It contains an electrode assembly and a clamping frame. The interior of the inner assembly constitutes the upper buffer compartment (usually the cathode compartment). The chamber is placed in the tank to which buffer is added. This constitutes the lower (anode) buffer compartment. Electrical contact is made through the lid.

It is good practice to use small gels while the conditions for sample preparation are being optimized or when the optimum nondenaturing system is being determined. This is because small gels provide rapid results to encourage thorough optimization studies. Large gels should be used for comprehensive analyses.

Regulated direct current (DC) power supplies designed for electrophoresis allow control of every electrophoretic mode. Constant voltage, constant current, or constant power conditions can be selected. Many power supplies have timers and some have integrators allowing runs to be automatically terminated after a set time or number of volt-hours (important in IEF). All modes of operation can produce satisfactory results, but for best results and good reproducibility some form of electrical control is important. The choice of which electrical parameter to control is almost a matter of preference. The major limitation is the ability of the chamber to dissipate the heat generated by the electrical current.

Gels

No other separation medium can resolve the components of complex mixtures of proteins as well and as easily as a gel electrophoresis system. Gels and the associated buffer systems establish the migration rates of proteins as appropriate for the intended separation. Gels also hold proteins in place at the ends of runs until they can be stained for visualization. Polyacrylamide gels are the principal media for protein electrophoresis. Agarose gels are used in some applications such as for the separation of proteins larger than about 500 kDa and for immunoelectrophoresis,[1,10] but they are not as useful or versatile as polyacrylamide gels.

Polyacrylamide gels are particularly well suited for protein electrophoresis for several reasons. (1) The pores of polyacrylamide gels are roughly the same size as proteins and function as three-dimensional sieves. (2) Pore size is determined by the conditions of polymerization and can be easily altered. (3) Gel formation is easy and reproducible and gels can be cast in different sizes and shapes. (4) Polyacrylamide gels are hydrophilic and electrically neutral. (5) Polyacrylamide does not bind proteins. (6) Polyacrylamide gels are transparent to light at wavelengths above nearly 250 nm (important for visualization of stained gels).

The standard gel-forming reaction is shown in Figure 8.2. Acrylamide and the cross-linker *N,N*-methylenebisacrylamide (*bis*) are mixed in aqueous solution and then copolymerized by means of a vinyl addition reaction initiated by free radicals.[13–17] Gel formation occurs as acrylamide monomer polymerizes into long chains cross-linked by bis molecules. The resultant interconnected meshwork of fiberlike structures has both solid and liquid components. It can be thought of as a mass of relatively rigid fibers that create a network of open spaces (the pores) all immersed in liquid (the buffer). The liquid in a gel maintains the gel's three-dimensional shape. Without the liquid, the gel would dry to a thin film. At the same time, the gel fibers retain the liquid and prevent it from flowing away.

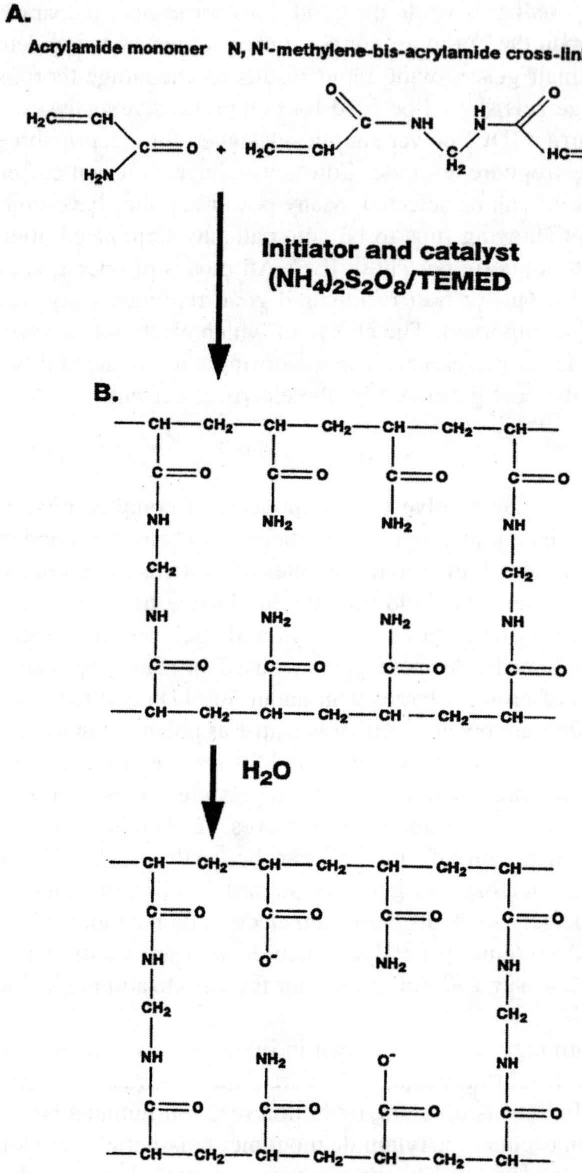

A.

Acrylamide monomer   N, N'-methylene-bis-acrylamide cross-linker, "bis"

Initiator and catalyst
$(NH_4)_2S_2O_8$/TEMED

B.

H_2O

Figure 8.2 Polyacrylamide gel formation and hydrolysis of acrylamide to acrylate. (A) Acrylamide and *N,N*-methylenebisacrylamide (bis) are copolymerized in a reaction catalyzed by ammonium persulfate $[(NH_4)_2S_2O_8]$ and TEMED. (B) A very short stretch of cross-linked polyacrylamide is represented. Cross-linking between similar structures leads to the formation of ropelike bundles of polyacrylamide that are themselves cross-linked together forming the gel matrix. In the lower portion of (B) is shown how pendant, neutral carboxamide groups can become hydrolyzed to charged carboxyls.

During electrophoresis proteins move between the pores of the gels so that pore size has a great influence on the separation. However, pore size is difficult to measure directly. It is operationally defined by the size limit of proteins that can be forced through a gel. The collective experience of many years of gel electrophoresis has established gel compositions suitable for proteins of nearly every size range. From a macroscopic point of view, migrating proteins segregate into discrete regions or zones corresponding to their individual gel-mediated mobilities. When the electric field is turned off, the proteins stop moving. The gel matrix constrains the proteins at their final positions long enough for them to be stained to make them visible. An example of a one-dimensional separation of proteins is shown in Figure 8.3. In this configuration, the protein pattern is one of multiple bands with each band containing one protein or a limited number of proteins with similar molecular weights.

By convention, polyacrylamide gels are characterized by a pair of values, %T and %C. In this convention, %T is the weight percentage of total monomer (acrylamide + *bis*) in g/100 ml, and %C is the proportion of *bis* as a percentage of total monomer. The effective pore size of a polyacrylamide gel is an inverse function of the total monomer concentration (%T) and a biphasic function of %C. When %T is increased at a fixed %C, the number of polymer fiber chains increases and the pore size decreases. On the other hand, when %T is held constant and %C is increased from low values, pore size decreases to a minimum at about 5%C. With further increases in %C from the minimum, pore size increases, presumably because of the formation of shorter, thicker bundles of polymer fiber chains. For most protein separations, 2.6%C has been found appropriate (37.5 parts acrylamide and 1 part bis). Gels with low %T (e.g., 7.5%T) are used to separate large proteins, whereas gels with high %T (e.g., 15%T) are used for small proteins.

An example of the effect of pore size on the separation of a set of native proteins is shown in Figure 8.4. The 4%T, 2.67%C gel shown on the left is essentially nonsieving. Proteins in the artificial sample migrate in the gel more or less on the basis of their free mobility. The 8%T, 2.67%C gel on the right sieves the proteins shown and demonstrates the combined effects of charge and size on protein separation. The relative positions of some proteins are shifted in the sieving gel as compared to the nonsieving one.

Acrylamide gels can be cast so that they consist of gradients of pore size.[1,11,16] Pore-gradient gels are mostly used in SDS-PAGE of samples containing both large and small proteins. Acrylamide concentrations in gradient gels increase linearly from top to bottom so that the pores get smaller with the distance into the gels. As proteins move through gradient gels from regions of relatively large pores to regions of relatively small pores, they encounter greater and greater resistance to migration. Small proteins remain in gradient gels longer than in single percentage gels so that both large and small proteins can be retained in the same gel. Thus, gradient gels are popular for analyses of complex mixtures of proteins spanning wide molecular-mass ranges. However, gradient gels cannot match the resolution obtainable with properly chosen single-concentration gels.

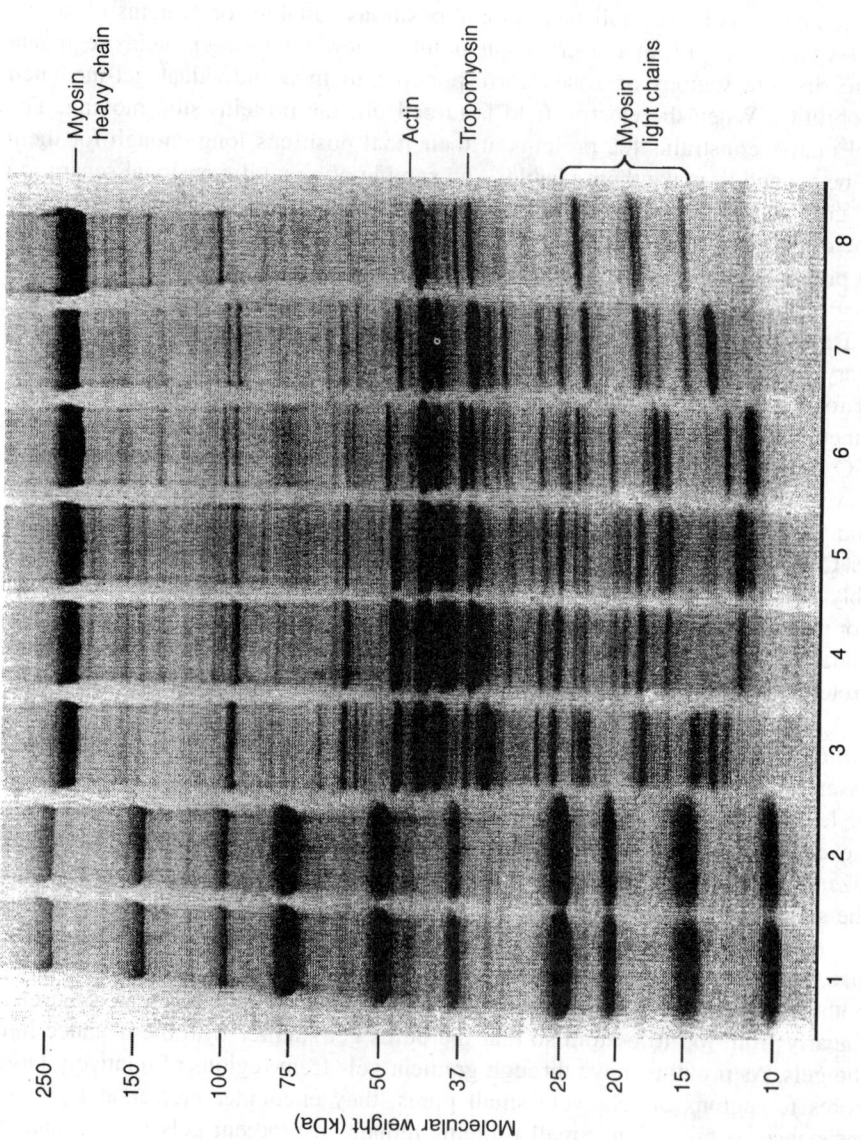

Figure 8.3 A typical analytical SDS-PAGE gel. Extracts of muscle proteins from the meat of five different fish varieties were separated by SDS-PAGE in a precast mid-size gel (Bio-Rad Criterion Gel), consisting of a 4 to 20%T polyacrylamide gel gradient. Separated proteins were visualized with colloidal Coomassie Brilliant Blue G-250. The lanes contain proteins from the following sources: lanes 1 and 2, marker proteins; lane 3, shark; lane 4, salmon; lane 5, trout; lane 6, catfish; lane 7, sturgeon; lane 8, mixture of rabbit actin and myosin. The molecular weights of the marker proteins are shown at the left and serve to calibrate the gel. Rabbit muscle proteins are identified on the right. The salmon and trout patterns (lanes 4 and 5) are very similar as expected, given the close evolutionary relationship of the two species. All of the fish samples appear to contain muscle proteins similar to those of rabbit.

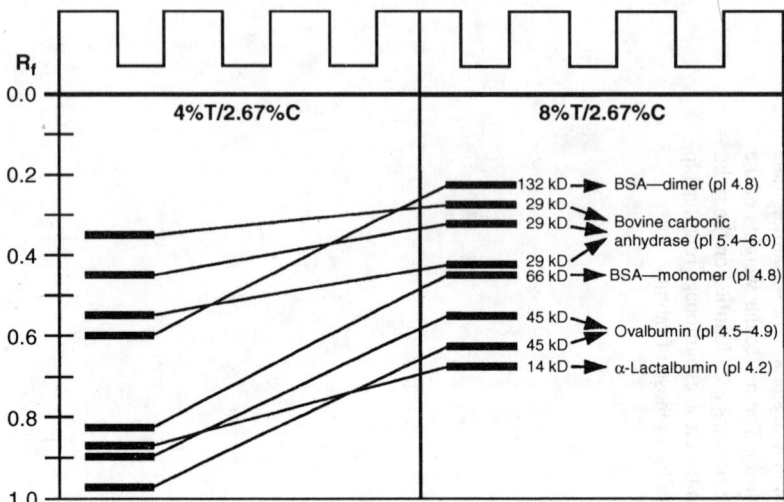

Figure 8.4 Effect of pore size on protein migration. The diagram shows the migration patterns of a set of proteins run in gels of differing %T with the Ornstein–Davis native, discontinuous buffer system. The diagram on the left was generated from 4%T, 2.67%C gels, whereas the pattern on the right was obtained from gels with 8%T, 2.67%C. The diagrams are drawn to scale. The slanted lines connect the bands representing the same proteins in the two diagrams. Note the large mobility shifts of BSA dimer and α-lactalbumin between the two gel types.

Buffer Systems

The electrical current in an electrophoresis cell is carried largely by the ions supplied by buffer compounds. Proteins constitute only a small proportion of the current-carrying ions in an electrophoresis cell. Buffer systems for electrophoresis are classified as either continuous or discontinuous, depending on whether one or more buffers are used. They are further classified as native or denaturing, depending on whether their compositions maintain or destroy protein structure and activity.

Continuous Buffer Systems

Continuous systems use the same buffer, at constant pH, in the gel, sample, and electrode reservoirs. With continuous systems, the sample is loaded directly on the gel in which separation will occur. The sample application buffer is the same as the gel and electrode buffer, but at about half the concentration. The localized voltage drop that results from decreased conductivity in the sample solution helps drive sample proteins into the gel and sharpens protein bands. Once inside a gel, proteins are separated on the basis of their individual (gel-mediated) mobility differences. Bandwidths are highly dependent on the height of the applied sample volume, which should be kept as small as possible, thus restricting continuous systems to high-concentration samples for best resolution.

Almost any buffer can be used for electrophoresis in a continuous system. Solutions of relatively low ionic strength are best suited for electrophoresis because these keep heat generation at a minimum. On the other hand, protein aggregation may occur if the ionic strength is too low. The choice of buffer will depend on the proteins being studied, but in general the concentrations of electrophoresis buffers are in the range of 0.01 to 0.1 *M*.

Discontinuous Buffer Systems

Discontinuous buffer systems (often called multiphasic buffer systems) employ different ions in the gel and electrode solutions. These systems are designed to sharpen starting zones for high-resolution separations, even with dilute samples. The sharpening of sample starting zones is called *stacking*. It is an electrochemical phenomenon based on mobility differences between proteins and carefully chosen leading and trailing buffer ions.[2,6,13] Samples are applied in dilute gel buffer and sandwiched between the gel and the electrode buffer. When the electric field is applied, *leading ions* from the gel (e.g., Cl) move ahead of the sample proteins while *trailing ions* from the electrode buffer (e.g., glycinate) migrate behind the proteins. The proteins in the sample become aligned between the leading and trailing ion fronts in the order of decreasing mobility. Proteins are said to be *stacked* between the two buffer ion fronts. The analogy is that of a stack of coins. The width of the stack is no more than a few hundred micrometers with possible protein concentrations there approaching 100 mg/ml.[18] Electrophoretic stacking concentrates proteins into regions narrower than can be achieved by mechanical means. This has the effect of minimizing overall bandwidths. Dilute samples require discontinuous buffers for best results. With high-concentration samples above nearly 1 mg/ml, continuous systems provide adequate results.[19]

In order to allow the stack to develop, the gels used with discontinuous systems are divided into two distinct segments. The smaller, upper portion is called the *stacking gel*. It is cast with appreciably larger pores than the lower *resolving gel* (or separating gel) and serves mainly as an anticonvective medium during the stacking process. Separation takes place in the resolving gel, which has pores of roughly the same size as the proteins of interest. Once proteins enter the resolving gel, their migration rates are slowed by the sieving effect of the small pores. In the resolving gel, the trailing ions pass the proteins, and electrophoresis continues in the environment supplied by the electrode buffer. The proteins are said to become unstacked in the resolving gel. They separate there on the basis of size and charge.

The runs are monitored and timed by means of the buffer front. The migration of the buffer front as it moves through the gel can be followed by the change in the index of refraction between the regions containing the leading and trailing ions. It is usual to add tracking dye to the sample solution. Tracking dye moves with the buffer front and aids in visualization of its motion.

Table 8.1   Continuous Buffers for Electrophoresis of Native Proteins

Buffer pH[a]	Basic Component	Amount for 5X Solution	Acidic Component	Amount for 5X Solution
3.8	β-Alanine 1X = 30 mM	13.36 g/l	Lactic acid 1X = 20 mM	7.45 ml/l[b]
4.4	β-Alanine 1X = 80 mM	35.64 g/l	Acetic acid 1X = 40 mM	11.5 ml/l
4.8	Gaba 1X = 80 mM	41.24 g/l	Acetic acid 1X = 20 mM	5.75 ml/l
6.1	Histidine 1X = 30 mM	23.28 g/l	Mes 1X = 30 mM	29.28 g/l
6.6	Histidine 1X = 25 mM	19.4 g/l	Mops 1X = 30 mM	31.40 g/l
7.4	Imidazole 1X = 43 mM	14.64 g/l	Hepes 1X = 35 mM	41.71 g/l
8.1	Tris 1X = 32 mM	19.38 g/l	Epps 1X = 30 mM	37.85 g/l
8.7	Tris 1X = 50 mM	30.29 g/l	Boric acid 1X = 25 mM	7.73 g/l
9.4	Tris 1X = 60 mM	36.34 g/l	Caps 1X = 40 mM	44.26 g/l
10.2	Ammonia 1X = 37 mM	12.5 ml/l	Caps 1X = 20 mM	22.13 g/l

[a] Listed buffer pH is ±0.1 unit. Do not adjust the pH with acid or base. Remake buffers outside the given range.
[b] Lactic acid from an 85% solution.
Source: Adapted from McLellan, T., *Anal. Biochem.*, 126: 94 (1982).

For detailed descriptions of the electrochemical processes that operate with discontinuous buffer systems, consult References 1, 2, 4–7, 13, and 20. Mathematically inclined readers might want to follow the development of multiphasic buffer theory as presented in References 21 to 23.

Native Buffer Systems

The choice of native electrophoresis system depends on the particular proteins of interest. There is no universal buffer system ideal for the electrophoresis of all native proteins. Both protein stability and resolution are important considerations in buffer selection. Recommended choices are the Ornstein–Davis discontinuous system[21,24] and McLellan's continuous buffers.[25]

The set of buffers compiled by McLellan provides the simplest way to carry out the electrophoresis of proteins in their native state.[25] McLellan's buffers range from pH 3.8 to pH 10.2, all with relatively low conductivity (Table 8.1). By using different buffers from the set it is possible to compare the effect of pH changes on protein mobility while maintaining similar electrical conditions. This is demonstrated

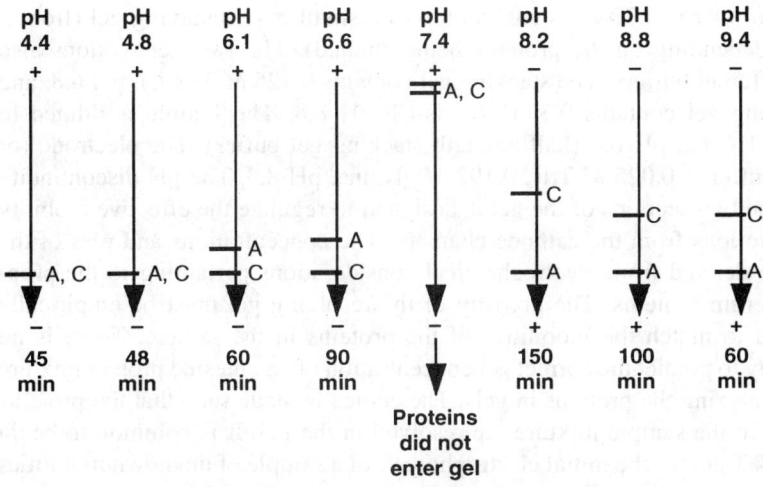

Figure 8.5 Effect of pH on protein mobility. Hemoglobin A (pI 7.1) and Hemoglobin C (pI 7.4) were electrophoresed in eight of the McLellan native, continuous buffer systems (Table 8.1). The diagram is drawn to scale. Migration is from top to bottom as shown by the vertical arrows. Bands marked A or C indicate the positions of the two hemoglobin variants in each gel representation. The polarities of the voltages applied to the electrophoresis cell are indicated by + and − signs above and below the vertical arrows. Run times are shown below the arrows. Note the polarity change between the gel at pH 7.4 and the one at pH 8.2. This reflects the pIs of the two proteins (and was accomplished by reversing the leads of the electrophoresis cell at the power supply).

in Figure 8.5. The illustration is a line-drawing representation, drawn to scale, of the relative positions of two hemoglobin variants, A and C, run under comparable electrical conditions in different McLellan buffers. HbA has a lower isoelectric point (pI 7.1) than HbC (pI 7.4). At pH 7.4, neither protein carries enough charge to move into the gel. At the acidic pHs tested, HbC is more highly charged and moves farther through the gel than HbA. The situation is reversed at the basic pHs tested. Note the differences in polarity and run times of the various runs.

Other buffers that have been used for continuous, native electrophoresis are Tris-glycine (pH range 8.3 to 9.5),[19] Tris-borate (pH range 8.3 to 9.3),[26] and Tris-acetate (pH range 7.2 to 8.5).[27] Borate ions[26] can form complexes with some sugars and can therefore influence resolution of some glycoproteins.

Ornstein[21] and Davis[24] developed the first high-resolution PAGE system for native proteins. Their popular system is still in widespread use. It was designed for the analysis of serum proteins, but works well for a broad range of protein types. The Ornstein–Davis buffers should be the first discontinuous system tried when working with a new, native sample.

Gels for the Ornstein–Davis method are cast in two sections. A large-pore stacking gel (4%T, 2.7%C) is cast on top of a small-pore resolving gel (from 5 to 30%T depending on the proteins being studied). The two gel sections also contain different buffers. The stacking gel contains 0.125 M Tris-Cl, pH 6.8, and the resolving gel contains 0.375 M Tris-Cl, pH 8.8. The sample is diluted in 0.0625 M Tris-Cl, pH 6.8 (half-strength stacking gel buffer). The electrode (or running) buffer is 0.025 M Tris, 0.192 M glycine, pH 8.3. The pH discontinuity between the two sections of the gel is designed to regulate the effective mobility of glycinate ions from the cathode chamber. The concentrations and pHs of the buffers are derived from electrochemical considerations pertaining to the properties of serum proteins. The porosity of the resolving gel must be empirically determined to match the mobilities of the proteins in the sample. There is no reliable way to predict the correct gel concentration of an untested protein mixture without analyzing the proteins in gels. The choice is made such that the proteins of interest in the sample mixture are resolved in the gel. It is common to begin with a 7.5%T gel for the initial electrophoresis of a sample of unknown mobilities and then to try higher-concentration gels (and sometimes lower-concentration gels such as 5%T).

Tris-sulfate/Tris-borate, Tris-formate/Tris-borate, and Tris-citrate/Tris-borate have been advocated as electrophoresis buffers.[5] For basic proteins, a low-pH alanine–acetate system[28] is often used.

Denaturing Buffer Systems

Because it is not yet possible to calculate the physical properties of proteins from mobility data alone, researchers turn to SDS-PAGE denaturing systems in order to estimate protein molecular weights. Sample treatment for SDS-PAGE includes reduction of disulfide bonds and heating proteins in the presence of the surfactant. Breakage of inter- and intramolecular bonds in this process dissociates proteins into their polypeptide subunits and converts the subunits to forms that can be separated on the basis of their molecular weights. Moreover, SDS solubilizes most proteins, so SDS-PAGE is applicable to a wide range of sample types. The electrophoretic band patterns obtained by SDS-PAGE are appreciably easier to interpret than those from native PAGE.

The most popular electrophoresis system is the discontinuous buffer system devised by Laemmli.[29] Laemmli added SDS to the standard Ornstein–Davis buffers and developed a simple denaturing treatment. Sample preparation for SDS-PAGE is quite easy.[3,11,16] Nevertheless, it is as important to the system as the electrophoresis buffers. Proteins are simply brought to near-boiling in sample buffer (0.0625 M Tris-Cl, pH 6.8) containing 5% (v/v) 2-mercaptoethanol (a thiol reducing agent) and 2% (w/v) SDS. The treatment simultaneously breaks disulfide bonds and dissociates proteins into their constituent polypeptide subunits. SDS monomer binds to the polypeptides and causes a change in their conformations. For most proteins, 1.4 g of SDS binds per gram of polypeptide (approximately one SDS molecule per two amino acids).[30] The properties of the detergent overwhelm the properties of the polypeptides. In particular, the charge densities of SDS–polypeptides are independent of the pH in the range from 7 to 10.[5,31] Negatively charged micelles of SDS coat polypeptides in a more or less regular manner. SDS–polypeptides assume similar shapes. However, it is not entirely clear as to what exactly this shape is. It was for a long time thought that SDS–polypeptide complexes were rodlike particles.[30,32] There is some evidence, however, that most SDS–polypeptide complexes adopt a structure like beads on a string in which spherical SDS micelles are distributed along the unfolded polypeptide chain.[33] Regardless of their exact shapes, the collective experience of many years is that the electrophoretic mobilities of SDS–polypeptides are so nearly identical that they can be compared on the basis of size by means of gel electrophoresis and that the sizes of the complexes are proportional to the molecular weights of the polypeptides.

Laemmli's buffers as usually described are more elaborate than strictly necessary. Most presentations of this method utilize the two different gel buffers of the Ornstein–Davis system with SDS added to them. Because SDS so dominates the electrophoresis system, the buffer in the stacking gel can be the same as the buffer in the resolving gel. Results are the same whether the stacking gel is cast at pH 6.8 or at pH 8.8. Also, gels do not need to be cast with SDS in them. The SDS in the sample buffer is sufficient to saturate the proteins with the detergent. The SDS in the cathode buffer overtakes the proteins in the sample and at 0.1% is sufficient for maintaining saturation during electrophoresis. This distinction is important for the commercial manufacturing of gels for SDS-PAGE (Subsection 8.2.4). When following standard protocols for SDS-PAGE, it is acceptable to use resolving gel buffer in both the stacking gel and the resolving gel and to omit SDS from the polymerization mixture.[11]

Other systems for SDS-PAGE have been developed. Weber and Osborn's continuous, denaturing SDS-PAGE system uses pH 7 sodium phosphate.[34] This system helped establish the utility of SDS in electrophoresis as a means for estimating the molecular weights of proteins. The Weber–Osborn system is a popular one, but the lack of stacking limits its use to high-concentration samples for best resolution. A Tris-sulfate/Tris-borate buffer system has been shown to fractionate SDS-saturated proteins in the 2- to 300-kDa range with very sharp bands.[35] Replacement of Tris in the Laemmli SDS-PAGE system with its analog ammediol (2-amino-2-methyl-1,3-propanediol) resolves polypeptides in the 1- to 10-kDa size range, but the bands are less sharp than with either the Laemmli or sulfate/borate systems.[31] The sulfate/borate and ammediol systems receive attention in discussions of SDS-PAGE, but neither is very popular. Systems that substitute taurine for glycine in the running buffer[36,37] can give improved resolution of the smaller polypeptides in the sample. In addition, the buffer systems that do not employ glycine in the running buffer are preferred when proteins are to be extracted from the gels for amino acid analysis.

An SDS-PAGE system based on the use of Tricine instead of glycine in the electrode buffer provides excellent separation of small polypeptides.[38] Peptides as small as 1 kDa are resolvable in Tricine-SDS gels. In particular, 16.5%T, 3%C separating gels are used for separations in the range from 1 to 70 kDa. Stacking gels in this system are 4%T, 3%C. Resolution is sometimes enhanced by inclusion of a 10%T, 3%C spacer gel between the resolving and stacking gels. Tricine-SDS resolving gels contain 1 M Tris-Cl, pH 8.45, and 13% glycerol. It is not necessary to include SDS in the gel buffer, but the glycerol is important to impart a viscosity that seems necessary for resolving small peptides. Electrode buffer is 0.1 M Tris, 0.1 M Tricine, 0.1% (w/v) SDS, pH 8.25. Sample buffer is 0.1 M Tris-Cl, pH 8.45, 1% (w/v) SDS, 2% (v/v) 2-mercaptoethanol, 20% (w/v) glycerol, and 0.04% Coomassie Brilliant Blue (CBB) G-250. Sample buffer should contain no more than 1% SDS for best resolution of small polypeptides (1 to 5 kDa). Proteins of very low molecular mass are not completely fixed and may diffuse from the gels during staining. This system is quite popular for polypeptide analysis.

The cationic detergent cetyltrimethylammonium bromide (CTAB) has been used as an alternative to SDS for gel electrophoresis of proteins. The most successful application of CTAB employs a discontinuous buffer system with sodium (from NaOH) as the leading ion and arginine as the trailing ion with Tricine as the counter ion and buffer.[39] This method uses no reducing agent in the sample buffer, and protein solutions are not boiled prior to electrophoresis. As a result, many enzymes retain their activities. Nonetheless, CTAB coats proteins thoroughly enough that it can be used for molecular weight determinations in analogy with SDS.

Very basic proteins such as histones and ribosomal proteins are separated in acetic acid–urea gels without SDS.[40,41]

Precast Gels

The biggest change in gel electrophoresis since the advent of polyacrylamide gels in the 1960s is the commercial availability of precast gels. Since the early 1990s, several companies have made a wide variety of precast polyacrylamide gels available to the research community. Most of the precast gels offered are Laemmli SDS-PAGE gels of differing %T and numbers of wells. Because the Laemmli SDS-PAGE gel is so overwhelmingly popular, alternative types of electrophoresis gels have tended to be ignored by researchers, and the companies have focused on this bias.

It took more than 20 years for manufacturers to devise production and distribution networks for delivering consistently high-quality gels to customers. The problems that the companies faced stem from the limited shelf life of polyacrylamide gels. Because polyacrylamide gels hydrolyze over time as shown in Figure 8.2, they are inherently unstable. At basic pH, the pendant, neutral carboxamide groups ($-CO-NH_2$) of the acrylamide monomers hydrolyze to ionized carboxyl groups (-COO) that can interact with some proteins. In addition, counter ions from the buffer neutralize the carboxy groups. The waters of hydration associated with the counter ions disrupt the integrity of the pores. Over extended periods of storage, band sharpness and resolution slowly deteriorate. The shelf life of a gel cast in the Laemmli gel buffer (pH 8.8) is about 3 to 4 months at 4°C. After this time, band patterns begin to deteriorate noticeably.

It is not possible to cast large volumes of Laemmli-type gels and to hold them in a warehouse for long periods of time. Consequently, a great deal of planning goes into the decisions of how many gels of each different type are to be cast at any particular time. Manufacturing and distribution issues have now been largely addressed, and customers can now be guaranteed that they will receive gels that can be stored for several weeks before they are used. A limited number of precast gel products are available that are cast with neutral pH buffers.[42] These gels have longer shelf lives than gels made according to the Laemmli formulation. The band patterns obtained with neutral-pH gels are different from those obtained with Laemmli gels. Nevertheless, they are becoming popular because of the convenience of extended shelf life.

People were initially drawn to precast gels because of difficulties in casting gradient gels. The convenience of being able to buy gradient gels that are already made rather than casting them is very appealing. The same holds true for single-percentage gel types as well. For all but the most demanding situations there is little reason to cast gels by hand. The gel types most in demand are 7.5%T, 10%T, 12%T, 4 to 15%T, and 4 to 20%T.

Precast gels differ from hand-cast gels in three ways: (1) they are cast with a single buffer throughout, (2) they are cast without SDS, and (3) they are cast without a sharp demarcation between the stacking and resolving gels. As pointed out previously, because SDS dominates the system, using different buffers in the stacking and resolving gels as in the original Laemmli formulation has no practical value. The two different buffers would mingle together on storage without elaborate means to keep them separate. In addition, during electrophoresis, SDS from the cathode buffer sweeps past the proteins in the resolving gel and keeps them saturated with SDS even when there is initially no SDS in the gel. Precast gels are thus made without SDS. This is beneficial to both the manufacturer and the user. SDS tends to form bubbles in the pumping systems used to deliver monomer solutions to gel cassettes, causing problems with monomer delivery. In addition, SDS micelles can trap acrylamide monomer and lead to heterogeneity of the gel structure.

When gels are cast by hand, it is customary to allow the resolving gel to harden before the stacking gel is placed on top of it. This practice is acceptable because hand-cast gels are usually used within a short period of time. On the other hand, when gels are cast this way and stored, the stacking gel eventually begins to pull away from the resolving gel. The gap that forms between the two gels leads to lateral spreading of the stacked proteins and destruction of the stack as it leaves the upper gel. For this reason, precast gels are cast in a continuous manner with the stacking-gel monomer solution added on top of the resolving-gel monomer solution before gelation. This means that the separation between the two gel segments is a gradual one rather than a sharp one. Even though the distance between the two gels is short, the transition between gel segments exists as a short gradient of %T. Proteins unstack gradually rather than abruptly. Because of this, the bands obtained with precast gels are not quite as sharp as those obtained with hand-cast gels.

Choice of System

Different projects and protein samples have different requirements. The decision as to which gel electrophoresis system to use depends on the needs of each particular project.

Native Proteins

Continuous buffer systems are preferred for native work because of their simplicity. Furthermore, some native proteins have a tendency to aggregate and precipitate at the very high protein concentrations reached during the stacking process employed in discontinuous electrophoresis. Consequently, aggregated proteins might not enter the resolving gel or they might cause streaking as accumulated protein slowly dissolves during a run. If the proteins of interest behave in this manner, it is best to use some form of continuous buffer system.

The pH of the electrophoresis buffer must be in the range over which the proteins of interest are stable or where they retain their biological activity. The pH should also be properly chosen with respect to the pI. The pH of the gel buffer should be far enough away from the pIs of the proteins of interest that they carry enough net charge to migrate through the gel in a reasonable time (at least one-half pH unit). On the other hand, separation of two proteins is best near one of their pIs because the isoelectric protein will barely move in that pH range. (Figure 8.5 shows how the buffer choice determines migration rates.) The choice of pH is often a compromise between considerations of resolution and protein stability. For best results with continuous systems, the concentrations of the proteins of interest should be at least 1 mg/ml to keep sample volume at a minimum. The sample should be loaded in gel buffer diluted to one-fifth to one-half strength (some form of buffer exchange may be required). The decreased ionic strength of diluted buffer causes a voltage to develop across the sample, which assists in driving the proteins into the gel.

The choice of proper gel concentration (%T) is, of course, critical to the success of the separation because it heavily influences separation. Too high %T can lead to exclusion of proteins from the gel, and too low %T can decrease sieving (see Figure 8.4). One approach, useful with the McLellan continuous buffers (Table 8.1), is to use relatively large-pore gels (6%T or 7%T) and to alter mobilities with pH. An approach for discontinuous systems is to start with a

7.5%T gel, and then, if that is not satisfactory, to try a number of gel concentrations between 5%T and 15%T.

Denatured Proteins

It is easier to choose suitable gel concentrations (%T) for SDS-PAGE than for native protein gels because the separation of SDS–polypeptides is dependent mainly on chain length. Laemmli gels with 7.5%T resolve proteins in the 40- to 200-kDa range, those with 10%T resolve 20- to 200-kDa proteins, 12%T gels separate proteins in the 15- to 100-kDa range, and 15% gels separate 6- to 90-kDa proteins (Figure 8.6).

Sample Preparation

Samples for SDS-PAGE by the Laemmli procedure are prepared in diluted gel buffer containing SDS, 2-mercaptoethanol, glycerol, and bromophenol blue tracking dye. It is most efficient to prepare a stock solution of sample buffer (0.0625 M Tris-Cl, pH 6.8, 2% SDS, 25% glycerol, and 0.01% bromophenol blue) containing everything but 2-mercaptoethanol and to add this reagent (to 5%) just before use. In some situations, it can be instructive to omit 2-mercaptoethanol and leave disulfide bonds intact.[3] The glycerol provides density for applying the sample on the stacking gel under the electrode buffer. The tracking dye allows both sample application and the electrophoretic run to be monitored (it migrates with the ion front). There is sufficient SDS present in the sample buffer to ensure saturation of most protein mixtures.[30] Except in the rare instances when the sample is in a very high-ionic-strength solution (>0.2 M salts), it can be dissolved 1:1 (v/v) in stock sample buffer. It is much better, though, to dilute the sample at least 1:4 (v/v) with the sample buffer stock. The amount of sample protein to load on a gel depends on the detection method to be used (Subsection 8.2.8). Enough of the protein of interest must be loaded on the gel for it to be subsequently located. Detection in gels requires on the order of 1 μg of total protein for easy visibility of bands stained with anionic dyes such as CBB R-250 or 0.1 μg of total protein with silver staining. Complete dissociation of most proteins is achieved by heating diluted samples to 95 to 100°C for 2 to 5 min.

For native, discontinuous gels, upper gel buffer diluted twofold to fivefold for sample application is commonly used. Tracking dye and glycerol are added to these samples also, and protein concentrations should fall within the same limits as for SDS-PAGE. With discontinuous systems, the volume of sample is not very important as long as the height of the stacking gel is at least twice the height of the sample volume loaded on the gel. Continuous systems require minimal sample volumes for best resolution.

Careful sample handling is important when sensitive detection methods are employed. Silver-stained SDS-PAGE gels sometimes show artifact bands in the 50- to 70-kDa molecular mass region and irregular but distinctive vertical streaking parallel to the direction of migration. The appearance of these artifacts has been

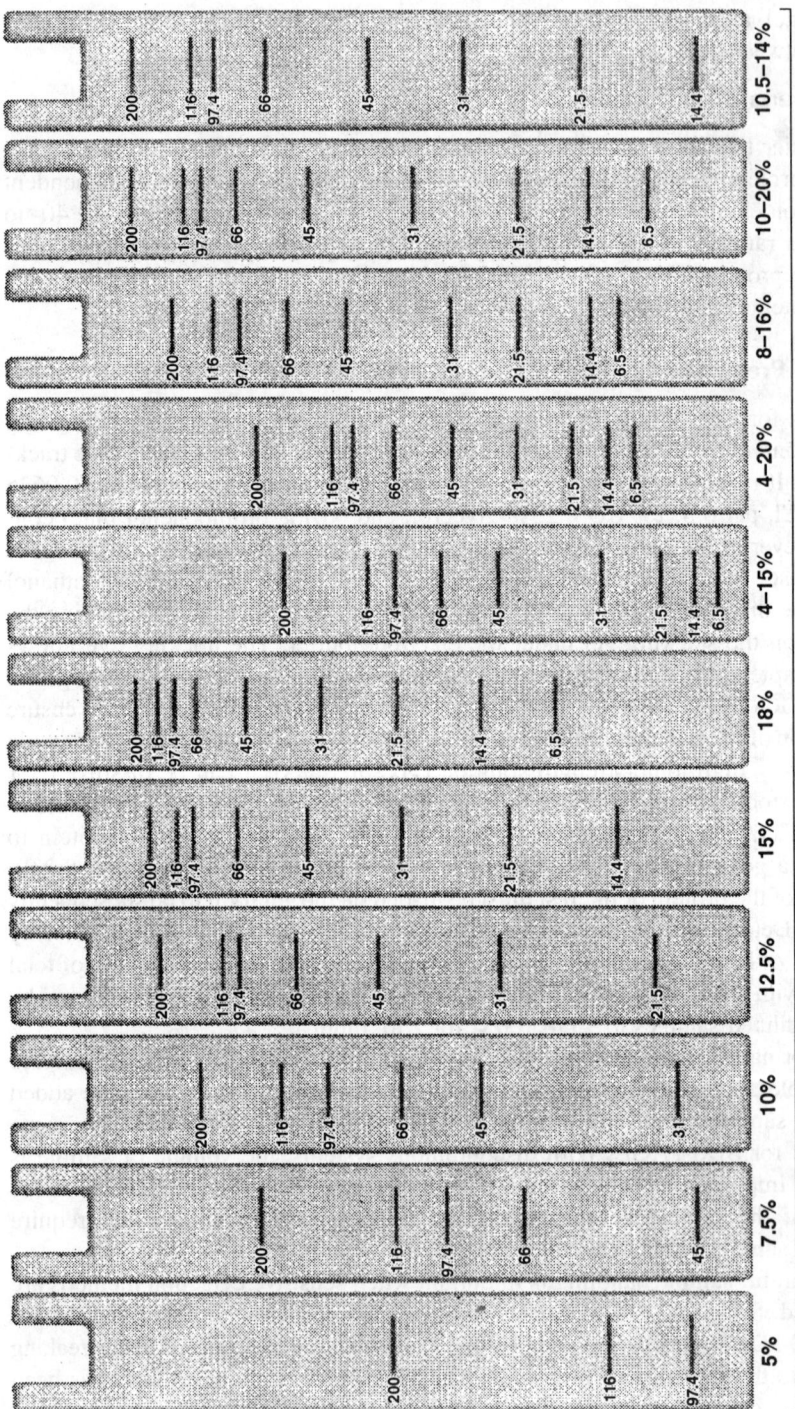

Figure 8.6 Protein migration charts. The relative positions of SDS-PAGE standards (Bio-Rad, broad range standards) are shown for several single percentage and gradient gels of the Laemmli type. The chart is useful as an aid in the selection of the appropriate gel types to match the molecular weight ranges of different protein samples.

attributed to the reduction of contaminant skin keratin inadvertently introduced into the samples.[43] The best remedy for the keratin artifact is to avoid introducing it into the sample in the first place. Monomer solution, stock sample buffer, gel buffers, and upper electrode buffer should be filtered through nitrocellulose and stored in well-cleaned containers. It also helps to clean the gel apparatus thoroughly with detergent and to wear gloves while assembling the equipment.

Prepared samples are placed in sample wells of gels with microliter syringes or micropipettes. Both types of liquid-handling device provide good control of sample volume. Syringes must be thoroughly rinsed between applications to avoid cross-contamination of different samples. Standard pipette tips are too wide to fit into narrow sample wells, but several thin tips, specifically designed for sample application, are available. The choice of sample-loading device is one of personal preference.

Electrical Considerations

During an electrophoresis run, electrical energy is converted into heat, called Joule heat. This heat can have many deleterious consequences, such as band distortion, increased diffusion, enzyme inactivation, and protein denaturation. All good electrophoresis chambers are designed to transfer the heat generated in the gel to the outside environment. In general, electrophoresis should be carried out at voltage and current settings at which the run proceeds as rapidly as allowed by the ability of the chamber to draw off heat. That is, the run should be as fast as possible without exceeding desired resolution and distortion limits, and these can only be determined empirically for any given system. Each experiment will impose its own criteria on cooling efficiency. Nearly all electrophoresis runs can be carried out on the laboratory bench, but some delicate proteins may require that the runs be conducted in the cold room or with circulated coolant.

It is important to bear in mind that an electrophoresis gel is an element in an electrical circuit and as such obeys the fundamental laws of electricity. Each gel has an intrinsic resistance, R, determined by the ionic strength of its buffer (R changes with time in discontinuous systems). When a voltage V is impressed across the gel, a current I flows through the gel and the external circuitry. Ohm's law relates these three quantities: $V = IR$, where V is expressed in volts, I in amperes, and R in ohms. In addition, power P, in watts, is given by $P = IV$. The generation of Joule heat, H, is related to power by the mechanical equivalent of heat, 4.18 J/cal, so that $H = (P/4.18)$ cal/sec.

With continuous buffer systems, the resistance of the gel is essentially constant, although it will decrease a bit during a run as the buffer warms. With the discontinuous Ornstein–Davis or Laemmli buffers, R increases during the course of a run as the chloride ions are exchanged by glycinate. For runs at constant current in Laemmli gels, the voltage (IR), power (I^2R), and consequently the heat generated in the gel chamber increase during the run. Under constant voltage conditions, current (V/R), power (V^2/R), and heat generation decrease during electrophoresis as R increases. Thus, runs carried out under constant current conditions are faster but hotter than runs done at constant voltage. Voltage and current should be set to keep H below the heat dissipation limit of the electrophoresis chamber. Follow the recommendations of the manufacturer for the proper electrical settings to use with any particular cell. Vertical cells are usually run at electric field strengths of 10 to 20 V/cm or currents in the range of 15 to 25 mA/mm of gel thickness.

The voltage applied to an electrophoresis cell is divided across three distinct resistance regions (Figure 8.7). The buffer paths from the open ends of the gel to the electrode wires form two of these regions. These two resistance regions are usually ignored, but they should be kept in mind for electrical analysis when experimenting with electrode buffers having very high or very low conductivity. The gel buffer is the third resistance region. It is the most important component in the system both electrically and electrochemically. With the Laemmli SDS system, the buffers create two different resistive sections. The low-resistance leading Cl ion forms a resistance segment that runs ahead of the higher-resistance trailing glycinate ion segment. Taken together, the two gel-segment resistors act as a voltage divider. The voltage across either one of the gel segments is proportional to the resistance of that segment (Figure 8.7B). The voltage across the chloride section provides the force that pulls the ion front through the gel, whereas the voltage across the glycinate section pulls the proteins through the gel. This proportioning of the applied voltage can cause two gels of the same %T to run differently if their gel buffers are different. For example, the final band pattern in a 12%T Laemmli gel with a gel buffer at pH 8.6 looks like the band pattern of a 10%T Laemmli gel with a gel buffer at pH 8.8. The gel at pH 8.6 also takes about 20% longer to run than the gel at pH 8.8. The differences in the properties of the two gel types are due to the increased conductivity of the pH 8.6 gel relative to the pH 8.8 gel. The extra chloride needed to drop the pH of 0.375 M Tris from 8.8 to 8.6 brings about the increased conductivity (0.12 M Cl vs. 0.19 M Cl). A subtle electrochemical process related to the ionization of glycine accentuates this effect, which can sometimes be used to advantage.[44] On the other hand, the electrode buffer is electrically less of a factor in the run than is the gel buffer. Its function is to provide a source of both glycinate and Tris ions to respectively replace the chloride ions that migrate from the gel at the anode and the Tris ions that migrate from the gel at the cathode. Concentration differences of as much as fivefold up or down from the prescribed formulation of electrode buffer are tolerated, but they do influence run time.

Detection of Proteins in Gels

Gels are run for either analytical or preparative purposes. The intended use of the gel imposes restrictions on the amount of protein to be loaded and the means of detection. At one time it was popular to use radioactive labeling of proteins, with detection of proteins done by autoradiography. It is now more common to

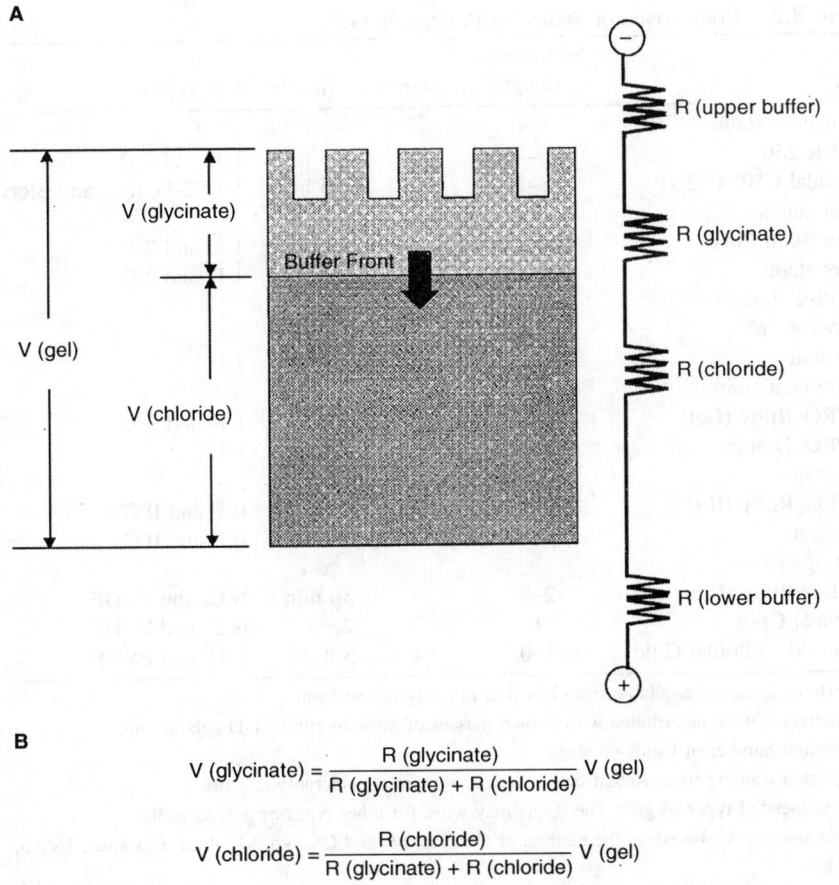

Figure 8.7 The electrophoresis gel as an element of an electrical circuit. With discontinuous buffer systems, the gel acts as a voltage divider as illustrated for the Ornstein–Davis or Laemmli system. (A) The voltage across the gel is divided between two resistors denoted by R (glycinate) and R (chloride). The resistors indicating the upper and lower buffer reservoirs are roughly equivalent and can be ignored. (B) The magnitudes of the voltages across the glycinate and chloride regions of a gel are shown in the two equations. They are proportional to the voltage across the gel and the relative resistances of the corresponding regions. The voltage across the lower portion of the gel, V (chloride), is responsible for pulling the buffer front through the gel, whereas the upper voltage, V (glycinate), acts to move the proteins in the samples. Migration rates and separations are influenced by the conductivities of the gel and running buffers.

make proteins in gels visible by staining them with dyes or metals.[45–48] Each type of protein stain has its own characteristics and limitations with regard to the sensitivity of detection and the types of proteins that best bind to the stain (Table 8.2).

Table 8.2 Comparison of Stains for Proteins in Gels

Stain[a]	Sensitivity, in ng[b]	Steps[c]	Time[d]	Gel Types[e]
Coomassie stains				
CBB R-250	36–47	2	2.5 h	1-D and 2-D
Colloidal CBB G-250	8–28	3	2.5 h	1-D, 2-D, IPG, and Blots
Silver stains				
Silver Stain Plus[f]	0.6–1.2	3	90 min	1-D and 2-D
Silver stain[g]	0.6–1.2	7	2 h	1-D and 2-D
Negative stains				
Copper stain[h]	6–12	3	10 min	1-D
Zinc stain[i]	6–12	3	15 min	1-D
Fluorescent stains				
SYPRO[j] Ruby (Gel)	1–10	2	3 h	1-D and 2-D
SYPRO[j] Orange	4–8	1	45 min	1-D
IEF stains				
SYPRO[j] Ruby (IEF)	2–8	2	2 h[k]	IEF and IPG
IEF stain	40–50	2	3 h	IEF and IPG
Blotting stains				
SYPRO[j] Ruby (Blot)	2–8	3	50 min	N.C. and PVDF[l]
Colloidal Gold	1	3	2 h	N.C. and PVDF
Enhanced Colloidal Gold	0.01–0.1	4	3 h	N.C. and PVDF

[a] All stains listed are available from Bio-Rad in ready-to-use form.
[b] Sensitivities were determined with known masses of proteins run in 1-D gels or blots.
[c] Minimum number of hands-on steps.
[d] Estimated staining time. Actual manipulations require considerably less time.
[e] Recommended types of gels. The stains may work for other types of gels as well.
[f] Silver staining kit based on the method of Gottlieb, M. and Chavko, M., *Anal. Biochem.*, 165: 33 (1987).
[g] Silver staining kit based on the method of Merril, C.R., Goldman, D., Sedman, S.A., and Ebert, M.H., *Science*, 211: 1437 (1981).
[h] Lee, C., Levin, A., and Branton, D., *Anal. Biochem.*, 166: 308 (1987).
[i] Fernandez-Patron, C., Castellanos-Serra, L., and Rodriguez, P., *BioTechniques*, 12: 564 (1992).
[j] SYPRO is a trademark of Molecular Probes, Inc., Eugene, OR.
[k] SYPRO Ruby IEF stain requires an overnight incubation.
[l] N.C.: nitrocellulose membrane. PVDF: polyvinylidene fluoride membrane.

If the purpose of gel electrophoresis is to identify low-abundance proteins (e.g., low-copy-number proteins in a cell extract or contaminants in a purification scheme), then a high protein load (0.1 to 1 mg/ml) and a high-sensitivity stain such as silver or fluorescence should be used. When the intention is to obtain enough protein for use as an antigen or for sequence analysis, then a high protein load should be applied to the gel and the proteins visualized with a staining procedure that does not fix the proteins in the gel, e.g., colloidal CBB G-250 (Subsection 8.2.8.1). Furthermore, for purposes of quantitative comparisons, stains with broad linear ranges of detection response should be used.

The sensitivity that is achievable in staining is determined by (1) the amount of stain that binds to the proteins, (2) the intensity of the coloration, and (3) the difference in coloration between stained proteins and the residual, background coloration in the body of the gel (signal-to-noise ratio). With all of the common stains, unbound stain molecules can be washed out of the bodies of the gels without removing much stain from the proteins.

Staining of all types of gels is done in the same way. Gels are removed from the cassettes, then washed, fixed, and incubated with staining solution in any convenient container such as a glass casserole or a photography tray. Staining is most commonly done at room temperature with gentle agitation (e.g., on an orbital shaker platform). Gloves should always be worn when staining gels because fingerprints (and fingers) will stain. Permanent records of stained gels can be obtained by photographing them, drying them with the appropriate apparatus on filter paper or between sheets of cellophane, or by capturing electronic images of them.

All stains interact differently with different proteins. No stain is universal in that it will stain all proteins in a gel proportionally to their quantities. The only observation that seems to hold for most of the positive stains is that they interact best with basic amino acids. For critical analyses, replicate gels should be stained with two or more different kinds of positive stain. Of all the stains available, colloidal CBB appears to stain the broadest spectrum of proteins. It is instructive, especially with 2-D PAGE gels, to follow a colloidal CBB-stained gel with silver staining[49-51] or to follow a fluorescence stain with colloidal CBB or silver. Very often, this double staining procedure will show a few differences in the two protein patterns. With CBB and silver stains, the order in which they are used often does not seem to be important.[52,53] With the fluorescent SYPRO stains (Subsection 8.2.8.4), CBB staining should always follow SYPRO staining rather than the reverse order. This is because the two dye molecules have such strong affinity for each other that proteins stained by the first dye are preferentially stained by the second dye.

Dye Staining

CBB R-250 is the standard stain for protein detection in polyacrylamide gels. It and the G-250 variety (CBB G-250) are wool dyes that have been adapted to the staining of proteins in gels. The R and G designations signify red and green hues, respectively. Easy visibility requires on the order of 0.1 to 1 μg of protein per band. The staining solution consists of 0.1% CBB R-250 (w/v) in 40% methanol (v/v), 10% acetic acid (v/v), which also fixes most proteins in gels. Absolute sensitivity and staining linearity depend on the proteins being stained.

CBB G-250 is less soluble than the R-250 variety. In acidic solutions it forms colloidal particles that are too large to penetrate surface gel pores and can be formulated into a staining solution that requires little or no destaining. It can also be formulated to be environmentally benign. This stain is somewhat more sensitive than CBB R-250, in part because of increased signal-to-noise ratios because the bulk of the gel matrix does not pick up excess stain. Staining with colloidal CBB G-250 can be linear over two orders of magnitude of protein concentration for some proteins. Gels containing low-molecular weight polypeptides ($\approx$1000 Da) can be stained with CBB G-250 with minimum fixation time (and minimal potential loss of material).

Silver Staining

There are a number of different silver staining methods. Some are available in kit form from various manufacturers. Others do not lend themselves to commercial kits. For a discussion of the mechanisms of silver staining, see Reference 54. Silver staining can be as much as 100 times more sensitive than CBB dye staining. All silver staining procedures have many manual steps and the decision as to when to terminate color development is quite subjective.

Copper and Zinc Staining

Proteins in SDS-PAGE gels can be stained negatively with copper or zinc. For negative staining with copper, SDS-PAGE gels are incubated for a short time in a copper chloride solution, then washed with water.[55] Blue-green precipitates of copper hydroxide form in the bodies of the gels except where there are high concentrations of SDS, such as those bound to the proteins. Clear protein bands can be easily seen against the blue-green backgrounds and photographed with the gels on black surfaces. Proteins are not permanently fixed by this method and can be quantitatively eluted after chelating the copper.

A method using the combination of zinc and imidazole produces similar, negatively stained SDS-PAGE gels.[56] Zinc imidazolate forms precipitates in gels except at the sites where precipitation is inhibited by SDS–protein bands.[57] The resultant gels are opaque white with clear regions at the sites of the protein bands. They too are best viewed with the gel on a black surface.

With both the copper and zinc methods, the resultant negatively stained images of the electrophoresis patterns are intermediate in sensitivity between the CBB dyes and silver staining. Neither of the negative staining methods is recommended for 2-D PAGE gels because neither gives good quantification or discrimination of closely clustered spots.

Fluorescent Stains

The rare earth chelate stains have desirable features that make them popular in high-throughput laboratories.[58] They are end point stains with little background staining (high signal-to-noise characteristics), and they are sensitive and easy to use. The rare earth chelate compounds possess three distinct domains: (1) one domain binds a rare earth ion such as ruthenium, (2) a chromophoric domain is responsible for detection of the rare earth ion, and (3) a third domain reversibly binds to proteins. Because these compounds are fluorescent, they require an imaging device capable of providing high-intensity illumination at the excitation wavelength, band pass filters for excitation and emission wavelengths, and a detector such as a photographic or CCD camera (Section 8.6). Sensitivity varies from protein to protein, but can exceed that of silver stain. Linearity can extend to three orders of magnitude. The most popular types of fluorescent protein stains are the SYPRO™ class of compounds. SYPRO Orange is recommended for 1-D SDS-PAGE with SYPRO Ruby used for 1-D and 2-D SDS-PAGE and for native gels. (SYPRO is a trademark of Molecular Probes, Inc., Eugene, OR.)

Separation and Resolution

Stained proteins in gels appear visually as bands of variable width, each having a uniform cross section. This is because the eye tends to sharpen the boundaries between the colored bands and the clear background (ignoring the negative stains). In reality, proteins are distributed with Gaussian profiles with peaks at the centers of the bands and widths dependent on the amount of protein present. Image-acquisition instruments and analysis software (Section 8.6) display the true distributions and base quantifications on the true band shape. For much of the work in electrophoresis, measurements of bandwidths and interband spacing can be done with a ruler. Often mere qualitative comparisons of the band patterns of different protein samples are sufficient. For very accurate quantitative analyses of protein mixtures, digital image analysis is required.

Discussions of electrophoretic data handling usually include mention of *separation* and *resolution*. Although the two terms are not synonymous, they are often treated as such. In the terminology of separation science, separation refers to the distance between two adjacent band centers. Because bands are seen as being sharply defined with clearly evident blank spaces between adjacent bands, for practical purposes, separation is often taken to be the distance between the top of the faster running of two adjacent bands and the bottom of the slower one. It is the distance between the top of the bottom band and the bottom of the top band. This definition seems preferable to the rigorous one in electrophoresis.

Resolution, on the other hand, is a more technical term. It refers to the distance between adjacent bands relative to their bandwidths and acknowledges the fact that proteins are distributed in Gaussian profiles with overlapping distributions. The numerical expression for resolution is obtained by dividing the distance between the centers of adjacent bands by some measure of their average bandwidths. It expresses the distance between band centers in units of bandwidth and gives a measure of the overlap between two adjacent bands. For preparative applications, when maximal purity is desired, two proteins to be isolated should be separated by at least a bandwidth. In many applications it is sufficient to be able to simply discern that two bands are distinct. In this case bands can be less than a bandwidth apart.

Anomalous Migration in SDS-PAGE

Detergents disrupt protein–lipid and protein–protein interactions and play a large role in gel electrophoresis.[59,60] SDS is the most common detergent used in PAGE analysis. Most proteins are readily soluble in SDS, making SDS-PAGE a generally applicable method. In SDS-PAGE, the quality of the SDS is of prime importance. The effects of impurities in SDS are unpredictable. Of the contaminants, the worst offenders are probably the alkyl sulfates other than dodecyl sulfate (C_{12}), especially decyl sulfate (C_{10}), tetradecyl sulfate (C_{14}), and hexadecyl sulfate (C_{16}).[61,62] These bind to proteins with different affinities, thereby affecting mobilities. Lipophilic contaminants in SDS preparations, including dodecanol, can be trapped in SDS–protein complexes and SDS micelles leading to loss of resolution. Only purified SDS should be used for electrophoresis, but even with pure SDS, various glycoproteins, lipoproteins, and nucleoproteins can bind the detergent irregularly. The resultant SDS–polypeptides then migrate anomalously with respect to their molecular masses. Nevertheless, anomalous behavior in SDS-PAGE can serve as a diagnostic for certain posttranslational modifications.

Several types of proteins do not behave as expected during SDS-PAGE.[1,16] Incomplete reduction, which leaves some intra- or intermolecular disulfide bonds intact, makes some SDS-binding domains unavailable to the detergent so that the proteins are not saturated with SDS. Glycoproteins and lipoproteins also migrate abnormally in SDS-PAGE because their nonproteinaceous components do not bind the detergent uniformly. Proteins with unusual amino acid sequences, especially those with high lysine or proline content, very basic proteins, and very acidic proteins behave anomalously in SDS-PAGE, presumably because the charge-to-mass ratios of the SDS–polypeptide complexes are different from those that would be expected from size alone. Very large SDS–proteins, with molecular masses in the several hundred-kilodalton range, often do not migrate as expected. Polypeptides smaller than about 12,000 Da are not resolved well in most SDS-PAGE systems. In most cases, they do not separate from the band of SDS micelles that forms behind the leading ion front. The Tricine buffer system was devised to overcome this difficulty.

Molecular Weight Estimation

One of the reasons that SDS-PAGE became the most popular method of gel electrophoresis is that it can be used to estimate molecular masses.[1,4,10,16,63] To a first approximation, migration rates of SDS–polypeptides are inversely proportional to the logarithms of their molecular weights. The larger the polypeptide, the slower it migrates in a gel. Molecular weights are determined in SDS-PAGE by comparing the mobilities of test proteins to the mobilities of known protein markers. At one time, when samples for SDS-PAGE were run in individual tubes, it was necessary to normalize migration rates to a common parameter so that the different tube gels could be compared. This was because tube gels differ in length. The normalizing parameter that is still used is the relative mobility, R_f, defined as the mobility of a protein divided by the mobility of the ion front. In practice, when all gels are run for the same length of time, R_f is calculated as the quotient of the distance traveled by a protein from the top of the resolving gel divided by the distance migrated by the ion front. The distance to the ion front is usually taken as the distance to the tracking dye (measured or marked in some way before staining). With slab gels, this normalization is less important provided that a lane of standards is run in the same gel as the samples whose masses are to be determined. It is sufficient to compare migration distances of samples and standards. Plots of the logarithms of protein molecular weights (log M_r) vs. their migration distances fit reasonably straight lines.

In each gel, a lane of standard proteins of known molecular masses is run in parallel with the test proteins. After staining the gel to make the protein bands visible (Subsection 8.2.8), the migration distances are measured from the top of the resolving gel. The gel is calibrated with a plot of log M_r vs. migration distances for the standards. The migration distances of the test proteins are compared with those of the standards. Interpolation of the migration distances of test proteins into the standard curve gives the approximate molecular masses of the test proteins.

Pore-gradient SDS-PAGE gels can also be used to estimate molecular masses. In this case, log M_r is proportional to log (%T). With linear gradients, %T is proportional to distance migrated, so that the data can be plotted as log M_r vs. log (migration distance).

Standard curves are actually sigmoid in shape (Figure 8.8). The apparent linearity of a standard curve may not cover the full range of molecular weights for a given protein mixture in a particular gel. However, the mathematical function log M_r varies sufficiently slowly with changes in its argument (M_r) that fairly accurate molecular weight estimates can be made by interpolation, and even extrapolation, over relatively wide ranges. The approximate useful ranges of single-percentage SDS-PAGE gels for molecular-mass estimations is as follows: 40,000 to 200,000 Da, 7.5%T; 30,000 to 100,000 Da, 10%T; 15,000 to 90,000 Da, 12%T; 10,000 to 70,000 Da, 15%T. Mixtures of standard proteins with known molecular weights are available commercially for calibrating electrophoresis gels.

The semilogarithmic plots used for molecular weight determination are holdovers from the days when people had only graph paper and straight edges for curve fitting. (From a mathematical point of view, log M_r should be the independent variable [x-axis] and migration distance should be the dependent variable [y-axis], and not as usually drawn.) Several computer programs allow for standard curves to be fit with mathematical functions other than the semilogarithmic model. Some other types of curves actually fit the data better than the semilog function. Nevertheless, the semilogarithmic model for standard curves is the accepted form. It is important to bear in mind that the molecular weights obtained using Laemmli SDS-PAGE are those of the polypeptide subunits and not of native, oligomeric proteins. Moreover, proteins that are incompletely saturated with SDS, very small polypeptides, very large proteins, and proteins conjugated with sugars or lipids behave anomalously in SDS-PAGE, as mentioned in the preceding text. Nevertheless, SDS-PAGE provides reasonable molecular-mass estimates for most proteins.

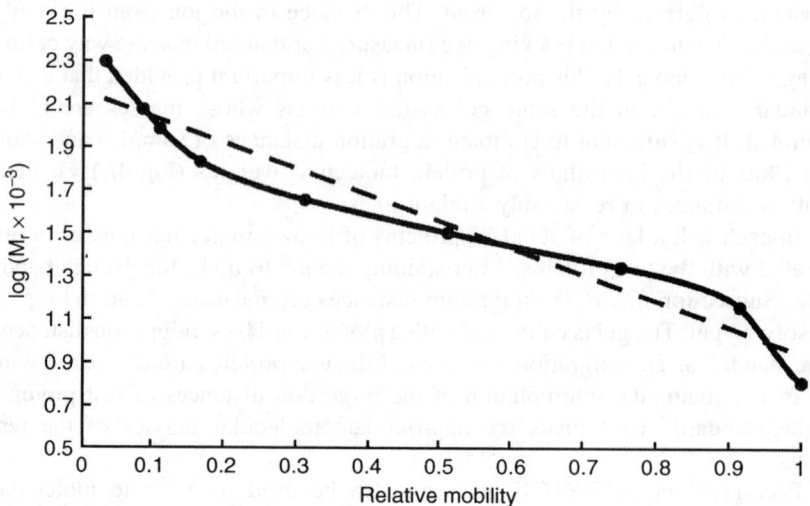

Figure 8.8 A representative calibration curve for molecular weight estimation. In the run that is plotted in this figure (solid line), Bio-Rad SDS-PAGE standards (broad range) with molecular weights of 200, 116.2, 97.4, 66.2, 45, 31, 21.5, 14.4, and 6.5 kDa (top to bottom, closed circles) were separated in a 15%T SDS-PAGE gel. The plot of $\log_{10} (M_r \times 10^3)$ vs. Relative mobility (R_f) shows the inherent nonlinearity of such curves. The straight-line segment in the middle of the plot is the most accurate range for molecular weight estimations. Larger polypeptides experience greater sieving than do those in the middle range of the plot, so that the upper portion of the curve has a different slope than does the middle. Small polypeptides experience less sieving than the others and also deviate from the straight line. It is customary to estimate molecular weights from a "best fit" straight line (dashed line). This is sufficient for many purposes and is acceptable because the mathematical logarithm function changes slowly with its argument.

ISOELECTRIC FOCUSING

IEF is an electrophoretic method in which proteins are separated on the basis of their pIs.[1,2,4–6,10,64–69] It makes use of the property of proteins that their net charges are determined by the pH of their local environments.

Proteins carry positive, negative, or zero net electrical charge, depending on the pH of their surroundings. The net charge of any particular protein is the (signed) sum of all of its positive and negative charges. The ionizable acidic and basic side chains of the constituent amino acids and prosthetic groups of the protein determine the net charge. If the number of acidic groups in a protein exceeds the number of basic groups, the pI of that protein will be at a low pH value and the protein is classified as being *acidic*. When the basic groups outnumber the acidic groups in a protein, the pI will be high and the protein is classified as being *basic*. Proteins show considerable variation in pIs, but pI values

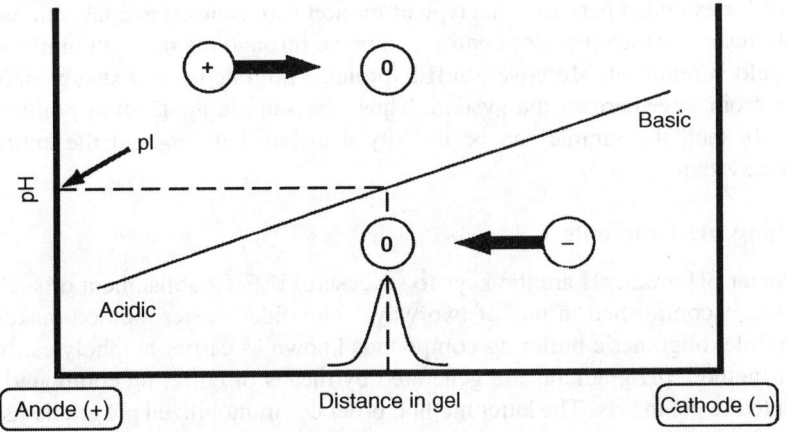

Figure 8.9 Isoelectric focusing. The motion of a protein undergoing isoelectric focusing is depicted (circles). The protein is shown near its pI in a pH gradient. Both the pH gradient and the motion of the protein are governed by an applied electric field. At pH values lower than the pI, the protein is positively charged (+) and it is driven toward the cathode as shown by the arrow. Above its pI, the protein is negatively charged (–) and it moves toward the anode. There is no net electrical force on the protein at its pI (0). The protein focuses in a Gaussian distribution centered at the pI.

usually fall in the range of pH 3 to 12 with a great many having pIs between pH 4 and pH 7.[70–72]

Proteins are positively charged in solutions at pH values below their pI and negatively charged above their pIs. Thus, at pH values below the pI of a particular protein, it will migrate toward the cathode during electrophoresis. At pH values above its pI, a protein will move toward the anode. A protein at its pI will not move in an electric field.

When a protein is placed in a medium with a linear pH gradient and subjected to an electric field, it will initially move toward the electrode with the opposite charge (Figure 8.9). During migration through the pH gradient, the protein will either pick up or lose protons. As it does so, its net charge and mobility will decrease and the protein will slow down. Eventually, the protein will arrive at the point in the pH gradient equaling its pI. There, being uncharged, it will stop migrating. If a protein at its pI should happen to diffuse to a region of lower pH, it will become protonated and be forced toward the cathode by the electric field. If, on the other hand, it diffuses into a pH higher than its pI, the protein will become negatively charged and will be driven toward the anode. In this way, proteins condense, or focus, into sharp bands in the pH gradient at their individual, characteristic pI values.

Focusing is a steady-state mechanism with regard to pH. Proteins approach their respective pI values at differing rates, but remain relatively fixed at those pH values for extended periods. This type of motion is in contrast to conventional electrophoresis in which proteins continue to move through the medium until the electric field is removed. Moreover, in IEF, proteins migrate to their steady-state positions from anywhere in the system. Thus, the sample application point is arbitrary. In fact, the sample can be initially distributed throughout the entire separation system.

Establishing pH Gradients

Stable, linear pH gradients are the keys to successful IEF. Establishment of such gradients is accomplished in one of two ways. The older, easier method makes use of mobile, oligomeric buffering compounds known as carrier ampholytes. In the other method, pH gradients are generated by means of buffering compounds grafted into the gel matrix. The latter method produces immobilized pH gradients.

Carrier ampholytes (amphoteric electrolytes) are mixtures of molecules containing multiple aliphatic amino and carboxylate groups. Some varieties contain sulfonic acid and phosphoric acid residues. Carrier ampholytes are small, multicharged organic buffer molecules about 300 to 1000 Da in size. They have closely spaced pI values and initially high conductivity. Ampholytes are included directly in IEF gel solutions at the time of casting. Under the influence of an electric field, carrier ampholytes partition into a smooth pH gradient that increases linearly from the anode (acidic) to the cathode (basic). The slope of a pH gradient is determined by the pH interval covered by the carrier ampholyte mixture and the distance between the electrodes. The use of carrier ampholytes is the most common and simplest means for forming pH gradients.

Immobilized pH gradients (IPGs) are formed by incorporating (acrylamido) buffers into polyacrylamide gels. Acrylamido buffers are derivatives of acrylamide. Each one contains both a reactive double bond and a buffering group. The general structure is $CH_2 = CH\text{-}CO\text{-}NH\text{-}R$, where R contains either a carboxyl [-COOH] or a tertiary amino group [e.g., $-N(CH_3)_2$]. Acrylamido buffers are covalently incorporated into polyacrylamide gels at the time of casting. In any given gradient, some of the acrylamido compounds act as buffers, whereas others serve as titrants. Published formulations and methods are available for casting the most common gradients.[66,68] Because the buffering compounds are fixed in place in the separation medium, gradients are stable over extended runs. This can prove important with many proteins that require long focusing times. IPGs are, however, more difficult and expensive to cast than carrier ampholyte gels. IPGs are commercially available in sheet form in a few pH ranges. A greater variety of pH ranges are available in IPGs that have been cut into strips for the IEF first dimension of 2-D PAGE.

IEF is a high-resolution technique that can routinely resolve proteins differing in pI by less than 0.05 pH unit. Under nondenaturing conditions, antibodies, antigens, and enzymes can retain their activities during IEF. The proper choice of ampholyte or IPG range is very important to the success of a fractionation.

Ideally, the pH range covered by an IEF gel should be centered on the pI of the proteins of interest. This ensures that the proteins of interest focus in the linear part of the gradient with many extraneous proteins excluded from the separation zone.

With carrier ampholytes, concentrations of about 2% (w/v) are best. Ampholyte concentrations below 1% (w/v) often result in unstable pH gradients. At concentrations above 3% (w/v), ampholytes are difficult to remove from gels and can interfere with protein staining.

Gels for Isoelectric Focusing

IEF is carried out in large-pore polyacrylamide gels that serve mainly as anti-convective matrices. A common composition of gels for IEF is 5%T, 3.3%C (29 parts acrylamide and 1 part bis). The best configuration for analytical IEF is the horizontal polyacrylamide slab gel. This is because the horizontal configuration allows use of ultrathin gels. Ultrathin gels (<0.5 mm) allow the use of very high field strengths and, therefore, very high resolution. Gels are cast with one exposed face on glass plates or specially treated plastic sheets. They are placed on cooling platforms and run with the exposed face upward or downward, depending on the particular setup. Electrolyte strips, saturated with 0.1 to 1 M phosphoric acid at the anode and 0.1 to 1 M sodium hydroxide at the cathode, are often placed directly on the exposed surface of the IEF gel. Electrodes of platinum wire maintain contact between the electrical power supply and the electrolyte strips. In another possible configuration, the gel and its backing plate are suspended between two carbon rod electrodes.

Precast IEF gels are available for carrying out carrier-ampholyte electrofocusing. A selection of IPG sheets is also available for horizontal IEF. Vertical IEF gels have the advantages that the electrophoresis equipment for running them is available in most laboratories and they can hold relatively large sample volumes. Because vertical electrophoresis cells cannot tolerate very high voltages, this orientation is not capable of the ultrahigh resolution of horizontal cells. To protect the proteins in the sample and the materials of the electrophoresis cells (mainly the gaskets) from caustic electrolytes, alternative catholyte and anolyte solutions are substituted in vertical IEF runs. As catholyte, 20 mM arginine, 20 mM lysine are recommended in vertical slab systems. The recommended anolyte is 70 mM H_3PO_4, but it can be substituted with 20 mM aspartic acid, 20 mM glutamic acid.

Electrofocusing can also be done in tubes, and this configuration once constituted the first dimension of 2-D PAGE.[73] Because of difficulties in handling and reproducibility with tube gels, IPG strips have largely replaced them.

Sample Preparation and Loading for IEF

A fundamental problem with IEF is that some proteins tend to precipitate at their pI values. Carrier ampholytes sometimes help overcome pI precipitation, and they are usually included in the sample solutions for IPG strips. In addition, nonionic detergents or urea are often included in IEF runs to minimize protein precipitation.

Urea is a common solubilizing agent in gel electrophoresis. It is particularly useful in IEF, especially for moderately soluble proteins. Urea disrupts hydrogen bonds and is used in situations in which hydrogen bonding can cause unwanted aggregation or formation of secondary structures that affect mobilities. Dissociation of hydrogen bonds requires high urea concentrations (7 to 8 M). If complete denaturation of proteins is sought, samples must be treated with a thiol-reducing agent to break disulfide bridges. Urea must be present in the gels during electrophoresis, but, unlike SDS, urea does not affect the intrinsic charge of the sample polypeptides. Urea solutions should be used soon after they are made, or they should be treated with a mixed-bed ion-exchange resin to avoid protein carbamylation by cyanate in old urea. Protein solutions in urea should never be left standing for extended periods of time or heated above 30°C.

Some proteins, especially membrane proteins, require detergent solubilization during isolation. Ionic detergents, such as SDS, are not compatible with IEF, although nonionic detergents, such as octylglucoside, and zwitterionic detergents, such as 2-[(3-cholamidopropyl)dimethylammonio]-1-propane-sulfonate (CHAPS) and its hydroxyl analog CHAPSO, can be used. NP-40 and Triton X-100 often perform satisfactorily, but some preparations may contain charged contaminants.

Concentrations of CHAPS and CHAPSO, or octylglucoside of 1 to 2% in the gel, are recommended. Some proteins may require as high as 4% detergent for solubility. Even in the presence of detergents, some samples may have stringent salt requirements. Salt should be present in a sample only if it is an absolute requirement. Carrier ampholytes contribute to the ionic strength of the solution and can help to counteract a lack of salts in a sample. When urea or detergents are not needed, samples (1 to 10 µl) can be loaded in typical biochemical buffers, but better results can be obtained with solutions in deionized water, 1% glycerol, 2% ampholytes, or 1% glycine. Suitable samples can be prepared by dialysis or gel filtration.

Good visualization of individual bands generally requires a minimum of 0.5 µg of protein per band with dye staining or 50 ng of protein per band with silver staining (Subsection 8.2.8). One of the simplest methods for applying samples to thin polyacrylamide gels is to place filter paper strips impregnated with sample directly on the gel surface. Up to 25 µl of sample solution can be conveniently applied after absorption into 1-cm squares of filter paper. A convenient size for applicator papers is 0.2 × 1 cm, holding 5 µl of sample solution. Alternatively, 1- to 2-µl samples can be placed directly on the surface of the gel. In most cases, IPG strips for 2-D PAGE (which are provided in dehydrated form) are rehydrated in sample-containing solution prior to electrophoresis.[74] Rehydration loading allows higher protein loads to be applied to gels than do other methods. It is particularly popular because of its simplicity.

There are few rules regarding the positioning of the sample on the IEF gel. In general, samples should not be applied to areas where they are expected to focus. To protect the proteins from exposure to extreme pH, the samples should not be applied closer than 1 cm from either electrode. Forming the pH gradients in carrier ampholyte IEF gels before sample application also limits the exposure of proteins to pH extremes. When acidic, narrow-range IPGs are used (e.g., pH 3 to 6), the sample is best applied at the cathode end of the strip. With basic, narrow-range IPGs (e.g., pH 7 to 10), the sample should be loaded at the anode side of the strip.

Power Conditions and Resolution in Isoelectric Focusing

The pH gradient and the applied electric field determine the resolution of an IEF run. According to both theory and experiment,[1,64,69] the difference in pI between two resolved adjacent protein IEF bands (ΔpI) is directly proportional to the square root of the pH gradient and inversely proportional to the square root of the voltage gradient (field strength) at the position of the bands: ΔpI $\propto$ [(pH gradient)/(voltage gradient)]$^{\frac{1}{2}}$. Thus, to minimize ΔpI, a narrow pH range and high applied voltage give high resolution (small ΔpI) in IEF.

In addition to the effect on resolution, high electric fields also result in shortened run times. However, high voltages in electrophoresis are accompanied by large amounts of generated heat at the beginning of a run. Thus, there are limitations on the magnitudes of the electric fields that can be applied depending on the ionic strengths of the solutions used in IEF. Voltages are usually ramped up to the desired final values to give salts in the samples time to clear from the gel before high-resolution focusing is begun. Because of their higher surface-to-volume ratio, thin gels are better able to dissipate heat than thick ones and are therefore capable of higher resolution (high voltage). Electric fields used in IEF are generally of the order of 100 V/cm. At focusing, currents drop to nearly zero because the current carriers have stopped moving by then.

Detection of Proteins in Isoelectric Focusing Gels

IEF gels differ from those for gel electrophoresis in that they have relatively large pores and they contain relatively large amounts of carrier ampholytes. It is possible for proteins to diffuse in the large-pore gels during some staining procedures, and some stains will interact with carrier ampholytes. The standard staining solution for proteins in IEF gels uses a combination of CBB R-250 and Crocein Scarlet in an ethanol–acetic acid solution containing cupric sulfate. The Crocein Scarlet binds rapidly to proteins and helps fix them in the large-pore IEF gels.[75] The cupric sulfate enhances stain intensity.[76] The procedures of staining and destaining are similar to those for CBB R-250 used alone but yield better signal-to-noise ratios. IEF gels can also be silver stained for increased detection sensitivity. However, some silver stains will turn the plastic backing sheets of

IPGs into mirrors. An easy way to stain IPGs is to immerse them for 1 h in colloidal CBB G-250 followed by two 10-min water washes. There is also a version of SYPRO Ruby stain specifically formulated for use with both carrier ampholyte and IPG-IEF gels.

TWO-DIMENSIONAL GEL ELECTROPHORESIS

2-D PAGE is the highest resolution method available for separating proteins.[73,77–82] The technique combines a first dimension of IEF, with a second separation by SDS-PAGE in a perpendicular direction. The technique is a true orthogonal procedure in that the two separation mechanisms are based on different physical principles (they are orthogonal in that sense) and the two separations are done at right angles to one another (they are geometrically orthogonal). The resolution obtained in the first dimension separation is retained when the IEF gel is mated to the second, SDS, gel.[82] It is this feature that gives 2-D PAGE exceptional resolution and distinguishes it from other separation methods. Thousands of polypeptides can be resolved in a single 2-D PAGE slab gel. The technique works best with soluble proteins such as those from serum or cytoplasm. It is relatively labor intensive for an electrophoresis technique, requiring a relatively high skill level for best results.

The best approach for 2-D PAGE is to run the IEF first dimension using IPG strips, and the best approach to obtaining IPG strips is to purchase them already made. Common practice is to denature proteins for 2-D PAGE to their constituent polypeptide chains so that polypeptide sequences can be matched to their corresponding gene sequences. This means that the IEF dimension is carried out in the presence of urea, CHAPS, carrier ampholytes, and a disulfide reducing agent such as dithiothreitol. Following IEF, an IPG strip is first treated with SDS and reducing and alkylating agents, then inserted into the gel cassette on top of the SDS-PAGE slab gel. The SDS-PAGE gel is run and stained as with one-dimensional electrophoresis. The difference between 1-D PAGE and 2-D PAGE gels is that the protein patterns in 2-D PAGE are spots rather than bands (Figure 8.10).

A large part of the success of a 2-D PAGE run is determined by careful sample preparation. This topic and several of the nuances of 2-D PAGE are outside the scope of this chapter. Those interested should consult References 77 to 85. See also www.expasy.ch and www.proteomeworkssystem.com and associated links.

IMMUNOBLOTTING

Proteins bound to the surfaces of synthetic membranes retain their antigenicity and are accessible to antibody probes. The most common membrane-based immunoassay technique is called *immunoblotting* or, more popularly, *Western blotting*. In Western blotting, proteins are transferred from an electrophoresis gel to a

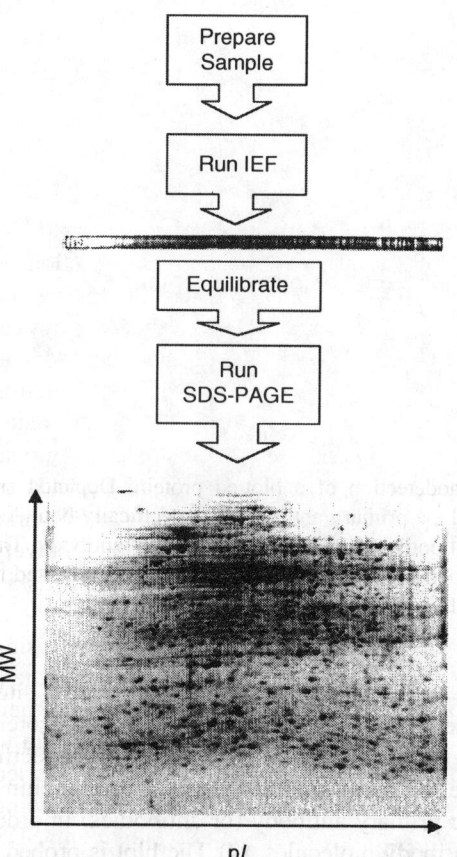

Figure 8.10 2-D polyacrylamide gel electrophoresis. Proteins from a lysate of *Escherichia coli* were subjected first to IEF in a 17-cm IPG strip spanning the pH range of 4 to 7. The strip containing focused proteins was prepared for SDS-PAGE, transferred to an 18 × 20 cm, 8 to 16%T gel, and electrophoresed in the Laemmli buffer system. At the end of the electrophoresis run, the separated proteins in the gel were stained with SYPRO Ruby Gel Stain, and the image shown was captured with a laser-based instrument. A second IPG strip was run in parallel and stained with colloidal CBB G-250. Its image is shown above the 1-D PAGE gel.

support membrane and then probed with antibodies. The method combines the resolution of PAGE (1-D or 2-D) with the specificity of immunoassays and enables the definitive identification of individual proteins in complex mixtures.[86–89]

Western blotting consists of the following steps (Figure 8.11): (1) Proteins are electrophoretically transferred from a gel to a membrane surface. The transferred proteins bind to the surface of the membrane and are immobilized in a

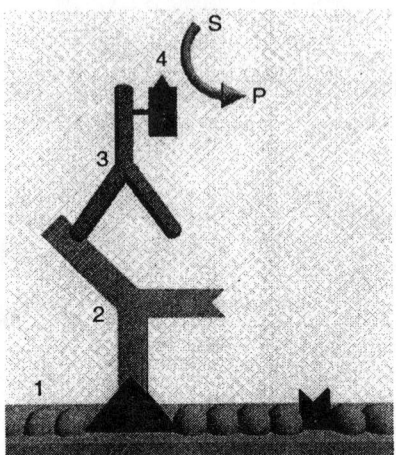

Figure 8.11 Specific enzymatic immunodetection of a blotted protein. Depicted are blocked binding sites on the membrane (1), a primary antibody (2) specifically bound to an antigenic protein, and a secondary antibody (3) bound to the primary antibody. The secondary antibody is conjugated to a reporter enzyme (4). Substrate (S) is converted to insoluble product (P) at the site of the antigen.

pattern that is an exact replica of the gel. (2) Unoccupied protein-binding sites on the membrane are saturated with detergent and (or) some kind of inert protein that will not interact with the antibody probes. This is done to prevent nonspecific binding of antibodies to the membrane and is called either blocking or quenching. (3) The blot is probed with a specific primary antibody (or antibodies) in order to tag the proteins of interest with antibody molecules. (4) The blot is probed a second time with an antibody that recognizes the species of the primary antibody. Secondary antibodies are conjugated to some kind of reporter group. In Western blotting the reporter group is a detectable enzyme. The most common enzymes used in Western blotting are alkaline phosphatase and horseradish peroxidase. The site of the protein of interest is thus tagged with an enzyme through the intermediaries of the primary and secondary antibodies. (5) Enzyme substrates are incubated with the blot. They are converted into insoluble, detectable (visible) products leaving a colored trace at the site of the band or spot representing the protein of interest.

Apparatus for Blotting

Electrotransfer from a gel to a membrane is done by directing an electric field across the thickness of the gel to drive proteins out of the gel and on to the membrane. There are two types of apparatus for electrotransfer: (1) buffer-filled tanks and (2) "semidry" transfer devices (Figure 8.12).

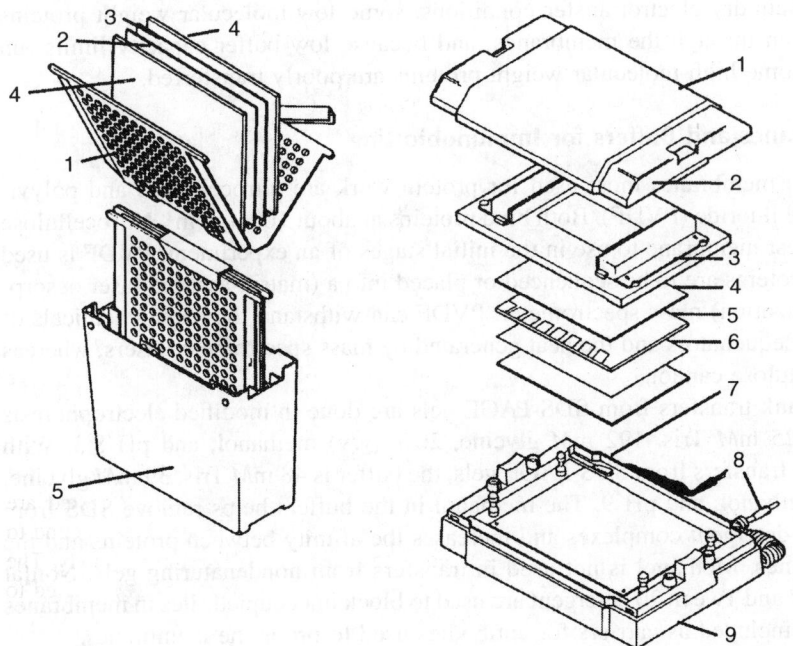

Figure 8.12 Two types of electrotransfer apparatus. At the left a tank transfer cell is shown in an exploded view. The cassette (1) holds the gel (2) and transfer membrane (3) between buffer-saturated filter paper pads (4). The cassette is inserted vertically into the buffer-filled tank (5) between positive and negative electrodes (not shown). A lid with connectors and leads for applying electrical power is not shown. On the right side of the figure is shown an exploded view of a semidry transfer unit. The gel (5) and membrane (6) are sandwiched between buffer-saturated stacks of filter paper (4) and placed between the cathode assembly (3) and anode plate (7). A safety lid (1) attaches to the base (9). Power is applied through cables (8).

    Transfer tanks are made of plastic with two electrodes mounted near opposing tank walls. A nonconductive cassette holds the membrane in close contact with the gel. The cassette assembly is placed vertically into the tank parallel to the electrodes and submerged in electrophoresis buffer. A large volume of buffer in the tank dissipates the heat generated during the transfer.

    In semidry blotting, the gel and membrane are sandwiched horizontally between two stacks of buffer-wetted filter papers in direct contact with two closely spaced solid-plate electrodes. The close spacing of the semidry apparatus provides for high field strengths. The term semidry refers to the limited amount of buffer that is used in the stacks of filter paper.

    Tanks rather than semidry apparatus should be used for most routine work. With tanks, transfers are somewhat more efficient than with semidry devices.

Under semidry electrotransfer conditions, some low-molecular weight proteins are driven through the membranes, and because low buffer capacity limits run times, some high-molecular weight proteins are poorly transferred.

Membranes and Buffers for Immunoblotting

The two membranes most used for protein work are nitrocellulose and polyvinylidene fluoride (PVDF). Both bind proteins at about 100 µg/cm². Nitrocellulose is the best membrane to use in the initial stages of an experiment. PVDF is used when proteins are to be sequenced or placed into a (matrix-assisted laser desorption ionization) mass spectrometer. PVDF can withstand the harsh chemicals of protein sequenators and the heat generated by mass spectrometer lasers, whereas nitrocellulose cannot.

Tank transfers from SDS-PAGE gels are done in modified electrophoresis buffer, 25 mM Tris, 192 mM glycine, 20% (v/v) methanol, and pH 8.3. With semidry transfers from SDS-PAGE gels, the buffer is 48 mM Tris, 39 mM glycine, 20% methanol, and pH 9. The methanol in the buffers helps remove SDS from protein–detergent complexes and increases the affinity between proteins and the membranes. Methanol is not used in transfers from nondenaturing gels. Nonfat dry milk and Tween 20 detergent are used to block unoccupied sites in membranes and are included as carriers for antibodies used to probe the membranes.

Immunodetection

Appropriate primary antibodies can be produced in any convenient animal, such as rabbits or mice. Antibodies to many important proteins can be purchased from a number of commercial vendors. Secondary antibodies (e.g., goat antirabbit immunoglobulin) conjugated to alkaline phosphatase or horseradish peroxidase are also commercially available. The preferred substrate for alkaline phosphatase is the mixture of 5-bromo-4-chloro-3-indolyl phosphate (BCIP) and nitroblue tetrazolium (NBT). The substrate BCIP is dephosphorylated by the enzyme and then oxidized in a reaction coupled to reduction of NBT. The resultant highly visible product is a purple-colored formazan. Good substrates for horseradish peroxidase are 4-(chloro-1-naphthol) or diaminobenzidine (and hydrogen peroxide). Chemiluminescent substrates for horseradish peroxidase are based on oxidation of luminol. The luminol substrate provides the most sensitive signal of the blotting substrates, but requires photographic exposures or specially configured imaging devices. For procedures and more detail than can be provided here, consult References 86 to 89.

Total Protein Detection

For proper identification of the proteins of interest in a blot, immunodetected proteins must be compared to the total protein pattern of the gel. This requires the indiscriminate staining of all the proteins in the blot. Colloidal gold stain is a very sensitive reagent for total protein staining. It consists of a stabilized sol of colloidal gold particles. The gold particles bind to proteins on the surfaces of membranes. Detection limits are in the low hundreds of picogram range and can be enhanced by an order of magnitude by subsequent treatment with silver.

CBB G-250 is another popular total protein stain. Researchers blotting 2-D PAGE gels particularly favor it because it is compatible with mass spectrometry. Stained blots provide good media for archiving 2-D PAGE separations. A version of SYPRO Ruby, formulated for blots, is a very sensitive total protein stain.

IMAGE ACQUISITION AND ANALYSIS

Several types of imaging systems and associated software are commercially available for analyzing gels stained with just about any kind of stain.[90–92] These instruments greatly simplify data acquisition and analysis and the archiving of gel patterns.

The three categories of image-acquisition devices used with electrophoresis gels are (1) document scanners, (2) charge-coupled device (CCD) cameras, and (3) laser-based detectors. Document scanners as configured for densitometry are for measurements on gels stained with one of the colored materials: CBB, silver, copper, or zinc. They operate in visible light illumination, 400 to 750 nm, with dynamic ranges extending to 3 O.D. The linear-array CCD detectors used with the better densitometers can distinguish adjacent features that are separated by 50 μm or greater (spatial resolution), which is more than adequate for most gel applications.

The better CCD camera instruments are cooled to increase their signal-to-noise ratios. They operate with illumination provided by either light boxes (UV or visible) for transmittance measurements or overhead lamps for epi-illumination. Filters are used to view fluorescent signals. CCD camera instruments are very versatile, and they can acquire images from gels stained with colored or fluorescent compounds. The epi-illumination feature allows CCD cameras to capture images of blots on opaque membranes. The spatial resolution obtainable with the cameras is entirely dependent on the properties of the lenses used and the area being imaged, but is generally in the 100- to 200-μm range. Their dynamic ranges for quantification often exceed four orders of magnitude.

Laser devices are the most sophisticated image-acquisition tools. They are particularly useful for gels labeled with fluorescent dyes because the lasers can be matched to the excitation wavelengths of the fluorophores. Detection is generally with photomultiplier tubes. Some instruments incorporate storage phosphor screens for detection of radiolabeled and chemiluminescent compounds (not discussed in this chapter). Resolution depends on the scanning speed of the illumination module and can be as low as 10 μm.

An imager is the most significant investment of all electrophoresis apparatus. As with all significant purchases, comparison shopping among the available products is highly recommended. In practice, researchers access the data in their gels through the analysis software, and the software should be a primary consideration in any imaging system. Good software will be able to use data from most imaging devices. However, dedicated software designed for use with particular instruments provides the desirable feature of controlling the imagers with the software.

Software for 1-D gel analysis (Figure 8.13) defines lanes and bands, quantifies bands, constructs standard curves, and determines molecular weights. Images can be adjusted for contrast, processed in various ways, annotated, and exported to other files for publication or document control. 2-D analysis software (Figure 8.14) defines and quantifies spots in 2-D PAGE gels. Those programs that use Gaussian spot modeling are better able to quantify proteins in overlapping spots than the programs that define spots by contours. Programs for 2-D analysis include statistical software designed for quantitative comparisons of large numbers of gels. The programs are also set up for analysis of spot patterns derived from differentially expressed proteins, and some can query databases to assist in protein identifications. They can also be used for image adjustments, annotation, and export in a variety of file formats.

ACKNOWLEDGMENTS

I thank my colleagues at Bio-Rad Laboratories for assistance with the illustrations. Lee Olech provided Figure 8.2, Laurie Usinger provided the image used in Figure 8.3 and Figure 8.13, Adriana Harbers provided Figure 8.4, Figure 8.5, and Figure 8.8, and Mingde Zhu provided the images for Figure 8.10.

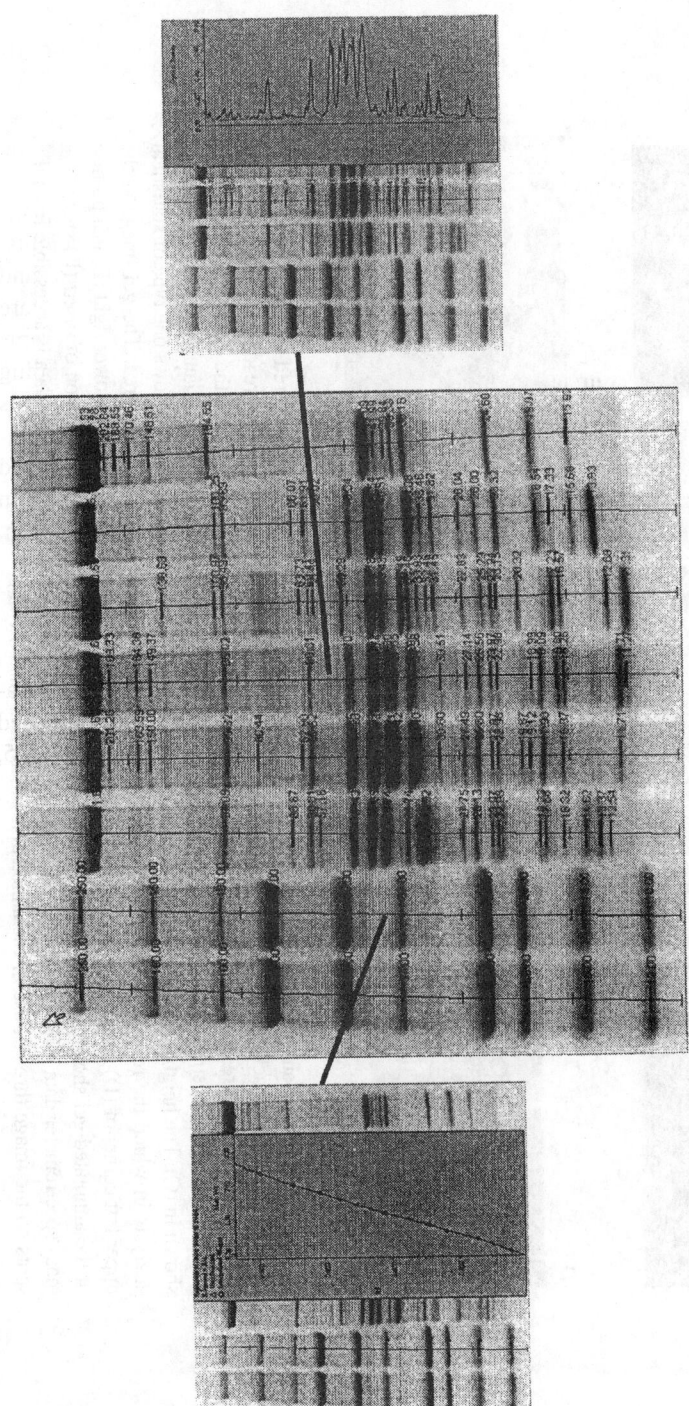

Figure 8.13 Image analysis software for 1-D gels. The molecular weights of the proteins in the gel of Figure 8.3 were digitally determined. The software first identified the individual lanes (vertical lines) and protein bands (horizontal bars). A standard curve was generated from the molecular weights of the marker proteins in lane 2 as shown in the insert at the left of the figure. The software automatically calculated the molecular weights of all the other protein bands in the gel and displayed them numerically on the image. The insert at the right of the gel image shows a densitometric scan of lane 5.

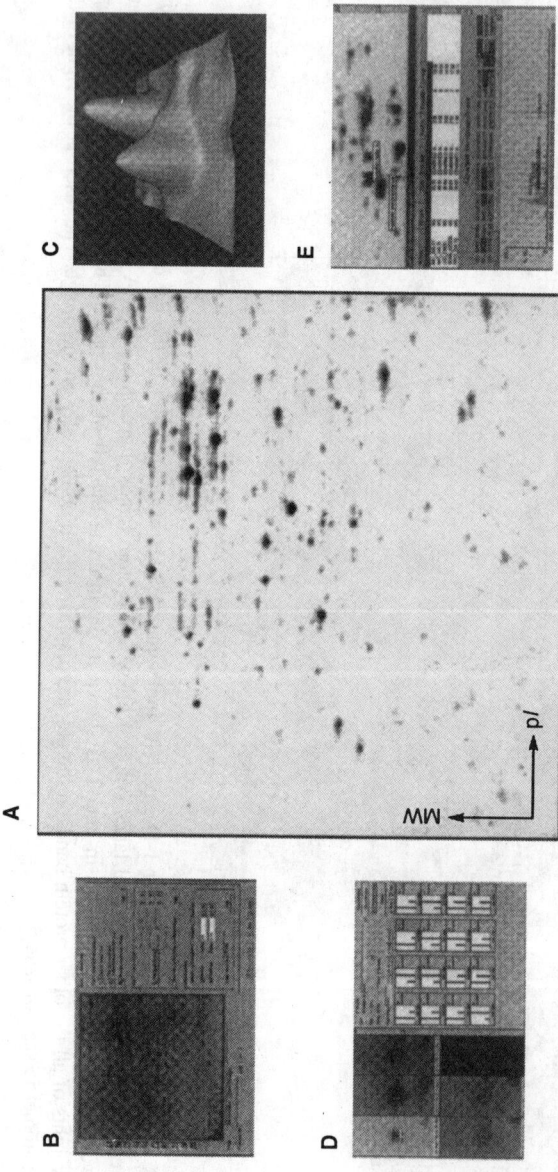

Figure 8.14 Image analysis software for 2-D gels. The image of a 2-D PAGE gel was filtered to remove streaks and other blemishes and background "noise" was subtracted. Protein spots on the filtered image were identified and converted to their Gaussian representations (A). Insert (B) shows the software "wizard" used to set the parameters for image filtering, background subtraction, and spot detection. A 3-D representation of two adjacent Gaussian-modeled spots is shown in (C). The heights of the cones represent the intensities of the two spots. Insert (D) shows one type of statistical analysis in which the relative intensities of identical spots in four different gels are compared. The gel image at the upper left corner of (D) is from a master image made from the four experimental gels. At the lower right (E) is a portion of the annotated gel showing two proteins identified by mass spectrometry. The bottom portion of insert (E) shows the mass spectrometer data used to identify one of the proteins. This type of mass spectrometer data is accessible from the spots on the image through use of the annotation function of the software.

References

1. Andrews, A.T., *Electrophoresis: Theory, Techniques, and Biochemical and Clinical Applications,* 2nd ed., Oxford University Press, Oxford, 1986.
2. Låås, T., in *Protein Purification: Principles, High Resolution Methods, and Applications* (Eds. J.-C. Janson and L. Ryden), p. 349, VCH Press, Weinheim, 1989.
3. Garfin, D.E., in *Methods in Enzymology* (Ed., M.P. Deutscher), Vol. 182, p. 425, Academic Press, San Diego, CA, 1990.
4. Dunn, M.J., *Gel Electrophoresis: Proteins,* BIOS Scientific, Oxford, 1993.
5. Allen, R.C. and Budowle, B., *Gel Electrophoresis of Proteins and Nucleic Acids,* De Gruyter, Berlin, 1994.
6. Garfin, D.E., in *Introduction to Biophysical Methods for Protein and Nucleic Acid Research* (Eds. J.A. Glasel and M.P. Deutscher), p. 53, Academic Press, San Diego, 1995.
7. Hames, B.D. (Ed.), *Gel Electrophoresis of Proteins: a Practical Approach,* 3rd ed., Oxford University Press, Oxford, 1998.
8. Makowski, G.S. and Ramsby, M.L., in *Protein Structure: a Practical Approach,* 2nd ed. (Ed. T.E. Creighton), p. 1, Oxford University Press, Oxford, 1997.
9. Goldenberg, D.P., in *Protein Structure: a Practical Approach,* 2nd ed. (Ed. T.E. Creighton), p. 187, Oxford University Press, Oxford, 1997.
10. Westermeier, R., *Electrophoresis in Practice: a Guide to Methods and Applications of DNA and Protein Separations,* 3rd ed., Wiley-VCH Press, Weinheim, 2001.
10a. Booz, M.L., in *Molecular Biology Problem Solver: a Laboratory Guide* (Ed. A.S. Gerstein), p. 331, Wiley-Liss, New York, 2001.
11. Garfin, D.E., in *Essential Cell Biology: a Practical Approach* (Eds. J. Davey and M. Lord), p. 197, Oxford University Press, Oxford, 2003.
12. Cantor, C.R. and Schimmel, P.R., *Biophysical Chemistry, Part 2: Techniques for the Study of Biological Structure and Function,* p. 676, W.H. Freeman, San Francisco, 1980.
13. Chrambach, A., *The Practice of Quantitative Gel Electrophoresis,* VCH Press, Weinheim, 1985.
14. Righetti, P.G., *J. Biochem. Biophys. Methods, 19*: 1 (1989).
15. Boschetti, E., *J. Biochem. Biophys. Methods, 19*: 21 (1989).
16. Shi, Q. and Jackowski, G., in *Gel Electrophoresis of Proteins: a Practical Approach,* 3rd ed. (Ed. B.D. Hames), p. 7, Oxford University Press, Oxford, 1998.
17. Bio-Rad Laboratories, *Bulletin No. 1156,* Bio-Rad Laboratories, Hercules, CA, (2001).
18. Chen, B., Rodbard, D., and Chrambach, A., *Anal. Biochem., 89*: 596 (1978).
19. Chen, B., Griffith, A., Catsimpoolas, N., Chrambach, A., and Rodbard, D., *Anal. Biochem., 89*: 609 (1978).
20. Chrambach, A. and Jovin, T.M., *Electrophoresis, 4*: 190 (1983).
21. Ornstein, L., *Ann. N.Y. Acad. Sci., 121*: 321 (1964).
22. Jovin, T.M., *Biochemistry, 12*: 871; 879; 890 (1973).
23. Kleparnik, K. and Bocek, P., *J. Chromatogr., 569*: 3 (1991).
24. Davis, B., *Ann. N.Y. Acad. Sci., 121*: 404 (1964).
25. McLellan, T., *Anal. Biochem., 126*: 94 (1982).
26. Margolis, J. and Kenrick, K.G., *Anal. Biochem., 25*: 347 (1968).
27. Fairbanks, G., Steck, T.L., and Wallach, D.F.H., *Biochemistry, 10*: 2606 (1971).

28. Reisfeld, R.A., Lewis, U.J., and Williams, D.E., *Nature, 195*: 281 (1962).
29. Laemmli, U.K., *Nature, 227*: 680 (1970).
30. Nielsen, T.B. and Reynolds, J., in *Methods in Enzymology* (Eds. C.H.W. Hirs and S.N. Timasheff), Vol. 48, p. 3, Academic Press, New York, 1978.
31. Wyckoff, M., Rodbard, D., and Chrambach, A., *Anal. Biochem., 78*: 459 (1977).
32. Reynolds, J.A. and Tanford, C., *Proc. Natl. Acad. Sci. U.S.A., 66*: 1002 (1970).
33. Westerhuis, W.H.J., Sturgis, J.N., and Niederman, R.A., *Anal. Biochem., 284*: 143 (2000).
34. Weber, K. and Osborn, M., *J. Biol. Chem., 244*: 4406 (1969).
35. Neville, D.M., Jr., *J. Biol. Chem., 246*: 6328 (1971).
36. Shafer-Nielsen, C., *Electrophoresis, 8*: 20 (1987).
37. Rabilloud, T., Valette, C., and Lawrence, J.J., *Electrophoresis, 15*: 1552 (1994).
38. Schägger, H. and von Jagow, G., *Anal. Biochem., 166*: 368 (1987).
39. Akins, R.E., Levin, P.M., and Tuan, R.S., *Anal. Biochem., 202*: 172 (1992).
40. Panyim, S. and Chalkley, R., *Arch. Biochem. Biophys., 130*: 337 (1969).
41. Rabilloud, T., Girardot, V., and Lawrence, J.J., *Electrophoresis, 17*: 67 (1996).
42. Engelhorn, S. and Updyke, T.V., U.S. Patent 5,578,180 (1996).
43. Ochs, D., *Anal. Biochem., 135*: 470 (1983).
44. Makowski, G.S. and Ramsby, M.L., *Anal. Biochem., 212*: 283 (1993).
45. Wirth, P.J. and Romano, A., *J. Chromatogr. A, 698*: 123 (1995).
46. Merril, C.R. and Washart, K.M., in *Gel Electrophoresis of Proteins: a Practical Approach,* 3rd ed. (Ed. B.D. Hames), p. 53, Oxford University Press, Oxford, 1998.
47. Allen, R.C. and Budowle, B., *Protein Staining and Identification Techniques*, Bio-Techniques Books, Natick, MA, 1999.
48. Rabilloud, T., *Anal. Biochem., 72*: 48A (2000).
49. Irie, S., Sezaki, M., and Kato, Y., *Anal. Biochem., 126*: 350 (1982).
50. DeMoreno, M.R., Smith, J.F., and Smith, R.V., *Anal. Biochem., 151*: 466 (1985).
51. Ross, M. and Peters, L., *BioTechniques, 9*: 532 (1990).
52. Dzandu, J.K., Deh, M.E., Barratt, D.L., and Wise, G.E., *Proc. Natl. Acad. Sci. U.S.A., 81*: 1733 (1984).
53. Vediyappan, G., Bikandi, J., Braley, R., and Chaffin, W.L., *Electrophoresis, 21*: 956 (2000).
54. Rabilloud, T., *Electrophoresis, 11*: 785 (1990).
55. Lee, C., Levin, A., and Branton, D., *Anal. Biochem., 166*: 308 (1987).
56. Fernandez-Patron, C., Castellanos-Serra, L., and Rodriguez, P., *BioTechniques, 12*: 564 (1992).
57. Fernandez-Patron, C., Castellano-Serra, L., Hardy, E., Guerra, M., Estevez, E., Mehl, E., and Frank, R.W., *Electrophoresis, 19*: 2398 (1998).
58. Patton, W.F., *Electrophoresis, 21*: 1123 (2000).
59. Neugebauer, J.M., in *Methods in Enzymology* (Ed. M.P. Deutscher), Vol. 182, p.239, Academic Press, San Diego, CA, 1990.
60. Hjelmeland, L.M. and Chrambach, A., *Electrophoresis, 2*: 1 (1981).
61. Brown, E.G., *Anal. Biochem., 174*: 337 (1988).
62. Lopez, M.F., Patton, W.F., Utterback, B.L., Chung-Welch, N., Barry, P., Skea, W.M., and Cambria, R.P., *Anal. Biochem., 199*: 35 (1991).
63. Hames, B.D., in *Gel Electrophoresis of Proteins: a Practical Approach,* 2nd ed. (Eds. B.D. Hames and D. Rickwood), p. 1, IRL Press, Oxford, 1990.

64. Righetti, P.G., *Isoelectric Focusing: Theory, Methodology, and Applications*, Elsevier, Amsterdam, 1983.
65. Låås, T., in *Protein Purification: Principles, High Resolution Methods, and Applications* (Eds. J.-C. Janson and L. Ryden), p. 376, VCH Press, Weinheim, 1989.
66. Righetti, P.G., *Immobilized pH Gradients: Theory and Methodology*, Elsevier, Amsterdam, 1990.
67. Garfin, D.E., in *Methods in Enzymology* (Ed. M.P. Deutscher), Vol. 182, p. 459, Academic Press, San Diego, 1990.
68. Righetti, P.G., Bossi, A., and Gelfi, C., in *Gel Electrophoresis of Proteins: a Practical Approach*, 3rd ed. (Ed. B.D. Hames), p. 127, Oxford University Press, Oxford, 1998.
69. Garfin, D.E., in *Handbook of Bioseparations* (Ed. S. Ahuja), p. 263, Academic Press, San Diego, 2000.
70. Wilkins, M.R. and Gooley, A.A., in *Proteome Research: New Frontiers in Functional Genomics* (Eds. M.R. Wilkins, K.L. Williams, R.D. Appel, and D.F. Hochstrasser), p. 35, Springer, Berlin, 1997.
71. Langen, H., Röder, D., Juranville, J.-F., and Fountoulakis, M., *Electrophoresis*, *18*: 2085 (1997).
72. Schwartz, R., Ting, C.S., and King, J., *Genome Res.*, *11*: 703 (2001).
73. Harrington, M.G., Gudeman, D., Zewert, T., Yun, M., and Hood, L., in *Methods: a Companion to Methods in Enzymology* (Ed. M.G. Harrington), Vol. 3, No. 2, p. 98, Academic Press, San Diego, 1991.
74. Sanchez, J.-C., Rouge, V., Pisteur, M., Ravier, F., Tonella, L., Moosmayer, M., Wilkins, M.R., and Hochstrasser, D.F., *Electrophoresis*, *18*: 324 (1997).
75. Crowle, A.J. and Cline, L.J., *J. Immunol. Methods*, *17*: 379 (1977).
76. Righetti, P.G. and Drysdale, J.W., *J. Chromatogr.*, *98*: 271 (1974).
77. Herbert, B.R., Sanchez, J.-C., and Bini, L., in *Proteome Research: New Frontiers in Functional Genomics* (Eds. M.R. Wilkins, K.L. Williams, R.D. Appel, and D.F. Hochstrasser), p. 13, Springer, Berlin, 1997.
78. Hanash, S.M., in *Gel Electrophoresis of Proteins: A Practical Approach*, 3rd ed. (Ed. B.D. Hames), p. 189, Oxford University Press, Oxford, 1998.
79. Link, A.J. (Ed.), *2-D Proteome Analysis Protocols*, Human Press, Totowa, NJ, 1999.
80. Rabilloud, T. (Ed.), *Proteome Research: Two-Dimensional Gel Electrophoresis and Identification Methods*, Springer, Berlin, 2000.
81. Bio-Rad Laboratories, *Bulletin No. 2651*, Bio-Rad Laboratories, Hercules, CA (2001).
82. Anderson, N.G., Matheson, A., and Anderson, N.L., *Proteomics*, *1*: 3 (2001).
83. Rabilloud, T., *Electrophoresis*, *17*: 813 (1996).
84. Molloy, M.P., *Anal. Biochem.*, *280*: 1 (2000).
85. Garfin, D.E., *Trends Anal. Chem.*, *22*: 263 (2003).
86. Bjerrum, O.J. and Heegard, N.H. (Ed.), *Handbook of Immunoblotting of Proteins*, Vols. 1 and 2, CRC Press, Boca Raton, FL, 1988.
87. Baldo, B.A. and Tovey, E.R. (Eds.), *Protein Blotting: Methodology, Research and Diagnostic Applications*, Karger, Basel, 1989.
88. Dunbar, B.S. (Ed.), *Protein Blotting: A Practical Approach*, Oxford University Press, Oxford, 1994.

89. Ledue, T.B. and Garfin, D.E., in *Manual of Clinical Laboratory Immunology,* 5th ed. (Eds. N.R. Rose, E. Conway de Macario, J.D. Folds, H.C. Lane, and R.M. Nakamura), p. 54, ASM Press, Washington, D.C., 1997.
90. Patton, W.F., *J. Chromatogr. A,* 698: 55 (1995).
91. Patton, W.F., *BioTechniques,* 28: 944 (2000).
92. Miller, M.D., Jr., Acey, R.A., Lee, L.Y.-T., and Edwards, A.J., *Electrophoresis, 22*: 791 (2001).

9

Capillary Electrophoresis of Biopharmaceutical Proteins

Roberto Rodriguez-Diaz, Stephen Tuck, Rowena Ng,
Fiona Haycock, Tim Wehr, and Mingde Zhu

INTRODUCTION

In the development of new biopharmaceutical molecules, there is a constant need for analytical methods that provide critical information in areas that range from early characterization to routine analysis of approved products. Past experience indicates there are few projects in drug development that can be addressed by "standard" analytical procedures. Even well-established techniques often have to be modified to better suit the analysis of new samples. For this reason, a broad range of techniques is already an integral part of laboratories in the biopharmaceutical industry.

Methods development starts with a relatively high number of techniques to characterize and test samples. The number of protocols is often reduced once the critical parameters and the methods that identify them have been defined. The analyst must evaluate the initial techniques with respect to their purposes. If the goal is to generate research data, the practicality of the method and its limitations are not of primary concern; if the goal is to use the technique as part of a test procedure, it has to be evaluated in terms of its potential to meet full validation. Critical procedures (e.g., release testing) that cannot be validated will bring a project to an expensive halt. For these reasons, this chapter provides basic principles as well as limitations of capillary electrophoresis (CE) as applied to the analysis of real biopharmaceutical molecules.

When a pharmaceutical is in the early stages of drug development (e.g., preclinical phases), the amount of material is limited and little is known about the characteristics of the molecule. Thus, CE is an attractive technique for developmental work due to its low sample consumption. Multiple injections can be made from a few microliters of sample, thus allowing optimization and early qualification of methods.

CE provides analysis based on orthogonal separation principles compared to other techniques as well as high resolving power. Like slab gel electrophoresis, CE is a family of techniques that resolve sample components by differences in intrinsic molecular characteristics such as size, mass, charge, differential interaction, and isoelectric point (pI).

As an analytical chemist whose task is to make measurements, knowing how experimental parameters affect measurements results in increased troubleshooting ability and expertise. For these reasons, basic theoretical principles and their applications will be presented.

Critical Factors for CE Methods Development

When developing a CE method for routine analysis of samples in our laboratory, the first three parameters to be checked are reproducibility, sensitivity, and throughput. There is little use for any analytical method that does not meet these three requirements.

Reproducibility in CE is affected by many parameters, and a detailed description of these factors is beyond the scope of this chapter. Some of the most common problems associated with poor reproducibility include sample composition, sample–capillary wall interactions, and sample instability. Sample composition may affect injection of the analyte and cause fluctuations in the current. Instability of sample components, including excipients, under analysis conditions can hinder the separation and reproducibility of the method. Undesirable sample–capillary wall interactions and sample instability have a profound effect on peak pattern, migration time, and resolution. Peak pattern irreproducibility and loss of resolution are unpredictable and weaken the value of the data obtained. Migration-time variations can be addressed by using internal standards or reference standards. In addition, coinjections of reference and sample materials are useful for identifying peaks. The main approaches to minimizing sample interactions with the column wall are capillary coatings and buffer additives. In some cases it may be necessary to use both. Although there are some logical guidelines to follow in the selection of the capillary and additives, the final choice relies heavily on empirical experimentation.

CE is based on the use of narrow-bore capillaries with internal diameters typically betwen 20 and 150 μm. Because most commercial instruments equipped with ultraviolet/visible (UV-Vis) absorption detectors use a segment of the same capillary as the detection cell, the path length in CE is much less compared to those in HPLC or spectrometry. Therefore, the most commonly used CE detectors can present problems with sensitivity, especially when detecting impurities at or below the 1% level if the sample is not concentrated enough. Because UV-Vis absorption is so simple and universal, a number of strategies have been developed to extend the sensitivity of CE. One of these approaches is the use of extended path-length cells (e.g., bubble cells or Z-cells). Although the improvement in detection can be significant, alternative capillaries like these should be carefully evaluated because resolution may be compromised, and the available configurations and internal wall modifications are limited. Detectors that employ fluorescence with continuum sources do not greatly improve sensitivity, but fluorescence detection with laser-based sources dramatically increases sensitivity. The advantages and drawbacks of the most common detectors will be addressed later in this chapter.

In CE, a single sample is injected at the inlet of the capillary and multiple samples are analyzed in series. This contrasts with conventional electrophoresis in which multiple samples are run in parallel as lanes on the same gel. When sample throughput is a problem in slab gel electrophoresis, it is easy to order another set of gel boxes and a power supply. Due to the cost, ordering another CE system may not be a viable solution. This limitation in sample throughput is somewhat compensated by the ability to process samples automatically using an autosampler and by the speed of the analysis. CE is highly suitable in applications with a small number of samples requiring short analysis time, e.g., during process monitoring or in a QC laboratory with low sample analysis requirements for that particular application. If high throughput is essential, multiple capillary systems are commercially available.

By evaluating the limitations of CE, one is left to wonder if there are any real applications that can be developed using this technique. The fact is that researchers have found clever ways to overcome the problems described in the preceding text. Their efforts resulted in procedures that provide greater convenience or information hard to obtain with other technologies.

PRINCIPLES AND PRACTICE OF CAPILLARY ELECTROPHORESIS

This section provides a brief discussion of the basic theoretical concepts of CE (including separation mechanisms), a description of CE instrumentation, and some guidelines in selecting conditions for a CE separation. Readers interested in more detailed presentations of CE theory and practice may consult References 1 to 8. Several general reviews of CE have been published,[9-11] as well as specific reviews of protein analysis by CE.[12-16]

CE is frequently compared to two of the mainstream techniques in the protein laboratory: HPLC and gel electrophoresis. The comparison to HPLC focuses on the instrumentation format and modes of detection, whereas gel electrophoresis shares its separation principles with CE. CE exploits the same molecular differences of the sample components to achieve separation as slab gel electrophoresis. The basis of separations in CE is differential migration of proteins in an applied electric field due to intrinsic characteristics of the sample (e.g., mass-to-charge ratio, pI) or to active participation of the separation matrix (e.g., sieving separations based on size). Despite the similarities, CE also provides capabilities that are not possible with either of the two other techniques.

Gel electrophoresis provides a simple method for separating complex protein mixtures. Because proteins are visualized using stains that may not be linearly incorporated in the gel, the intensity of the stained bands may be poorly correlated with the amount of protein. For this reason, gel electrophoresis is at best a semiquantitative technique capable of generating relative purity results. In CE, separations are commonly performed in free solution, i.e., in the absence of any support such as gel matrices. This allows the replacement of the capillary's content in between analyses and therefore the automation of the process. The use of UV-transparent fused-silica capillaries enables direct on-line optical detection of focused protein zones, eliminating the requirement for sample staining. The detection systems available to CE provide true quantitative capabilities.

Basic Theory of Capillary Electrophoresis

CE is a technique with a very high power of resolution. This is attributed to low diffusion and high plate numbers obtained from the absence of band-broadening factors (e.g., eddy diffusion, equilibrium dynamics, etc.) other than diffusion, which is also minimized by short analysis time.

CE is a family of techniques similar to those found in conventional electrophoresis: zone electrophoresis, displacement electrophoresis, isoelectric focusing (IEF), and sieving separations. Other modes of operation unique to CE include micellar electrokinetic chromatography (MEKC) and capillary electrochromatography (CEC).

As the name implies, CE separates sample components within the lumen of a narrow-bore capillary (20 to 150 μm) filled with a buffered electrolyte. High electric fields (hundreds of volts/centimeters in practice, but sometimes in excess of 1000 V/cm) can be used in CE because the capillary contains a small volume of electrolyte and a relatively large surface area to dissipate the heat generated by the electric current (Joule heat). High-voltage applications result in reduced analysis time and therefore less diffusion.

A schematic representation of a CE system is presented in Figure 9.1. In this diagram, the CE components have obvious counterparts to those found in slab gel electrophoresis. Instead of buffer tanks there are two small buffer reservoirs, and the capillary takes the place of the gel (or more accurately, a gel lane). The capillary is immersed in the electrolyte-filled reservoirs, which also make contact with the electrodes connected to a high-voltage power supply. A new feature to the conventional gel electrophoresis format is the presence of an on-line detection system.

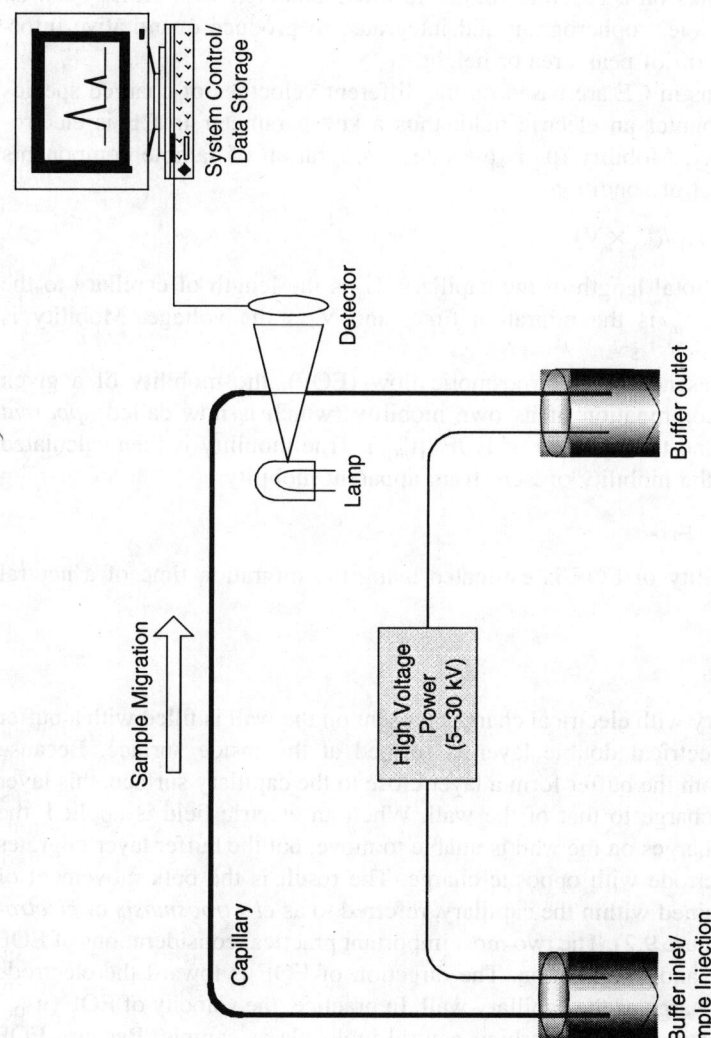

Figure 9.1 Schematic representation of a basic capillary electrophoresis system. The main components include a capillary (commonly contained within a housing that allows for temperature control), a power supply, and a detector. Automation is achieved through the use of computer-controlled setting of solutions and samples, displacement forces (to replace capillary contents and for hydrodynamic injection), and automatic data collection. (Courtesy of Agilent Technologies.)

A typical analysis starts by filling the capillary with fresh electrolyte, then the sample is introduced at one end of the capillary (the inlet), and analytes are separated as they migrate through the capillary toward the outlet end. As separated components migrate through a section at the far end of the capillary, they are sensed by a detector and an electronic signal is sent to a recording device. The data output (peaks on a baseline similar to those obtained with HPLC) can be displayed as an electropherogram and integrated to produce quantitative information in the form of peak area or height.

Separations in CE are based on the different velocities of charged species when they encounter an electric field; thus a key parameter in CE is electrophoretic mobility. Mobility (μ) is the rate of migration of sample components under a given set of conditions:

$$\mu = (L_t \times L_d)/(t_m \times V)$$

where L_t is the total length of the capillary, L_d is the length of capillary to the detection point, t_m is the migration time, and V is the voltage. Mobility is expressed in $cm^2\ V^{-1}\ s^{-1}$.

In the presence of electroosmotic flow (EOF), the mobility of a given molecule is a combination of its own mobility (which is now called *apparent mobility*, μ_{app}) and the mobility of EOF (μ_{EOF}). True mobility is then calculated by subtracting the mobility of EOF from apparent mobility:

$$\mu = \mu_{app} - \mu_{EOF}$$

The mobility of EOF is estimated using the migration time of a neutral marker.

Electroosmosis

When a capillary with electrical charges present on the wall is filled with a buffer solution, an electrical double layer is formed at the inside surface. Because counter-ions from the buffer form a layer close to the capillary surface, this layer is of opposite charge to that of the wall. When an electric field is applied, the fixed layer of charges on the wall is unable to move, but the buffer layer migrates toward the electrode with opposite charge. The result is the bulk movement of the liquid contained within the capillary, referred to as *electroosmosis* or *electro-endosmosis* (Figure 9.2). The two most important practical considerations of EOF are its velocity and its direction. The direction of EOF is toward the electrode with the same charge as the capillary wall. In practice, the velocity of EOF (v_{EOF}) is determined using a UV-absorbing neutral molecule as sample. Because EOF is easily affected by many parameters including protein interaction with silica surfaces, an internal marker, usually the same neutral molecule employed to determine EOF, is used as an internal standard. Under these conditions, proteins with net positive charge will migrate faster than the rate of EOF. (EOF velocity is higher than the mobility of most proteins, and thus the polarity of the power

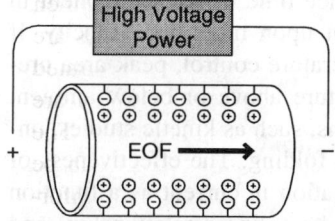

Figure 9.2 Formation of an electrical double layer responsible for electroendosmotic flow in an uncoated fused-silica capillary. The negative charges on the surface of the capillary are "neutralized" by positive charges of cations present in the buffer, which form an electrical layer near the surface of the capillary. When the electric field is applied, the positive charges migrate toward the negative electrode, generating a bulk flow of the solution contained within the column. Electroosmosis exhibits a flat profile, in contrast to hydraulic flow, which is parabolic.

supply is set so that EOF flows toward the detector side of the capillary.) Anionic proteins will migrate toward the cathode at a rate that is the difference between their electrophoretic velocity and v_{EOF}. A variety of capillary surface treatments and buffer additives developed for reducing protein adsorption and controlling EOF are described in detail in "Minimization of Nonspecific Protein–Wall Interactions."

Capillary Electrophoresis Instrumentation

The core components of a CE instrument are a power supply, a detector, and devices that allow for temperature control of the capillary and sample compartment. A wide variety of commercial CE instruments are available, from simple modular systems to fully integrated automated systems under computer control.

Power Supply

Most systems are equipped with power supplies that can be operated in constant voltage, constant current, and constant power modes. Common limits are 30 kV and 300 μA for constant voltage and constant current, respectively. Software-controlled polarity switching and programmable gradients are conveniences.

Detectors

The main detectors used in CE are briefly described in "Detection of Proteins." The most common detectors include absorbance, fluorescence, and on-line coupling with mass spectrometry (MS).

Capillary Temperature Control

Temperature control of the capillary environment is essential for attaining satisfactory reproducibility. Inadequate temperature control results in variable migration times. In CE, peak area depends upon the residence time of the component in the detector light path and therefore is dependent upon migration velocity. If migration times vary because of inadequate temperature control, peak area precision will be poor. Control of capillary temperature above or below ambient temperature may be desirable in special applications, such as kinetic studies, on-column enzyme assays, or in the study of protein folding. The effectiveness of capillary thermostating can be determined by variation in current as a function of voltage. According to Ohm's law, this should be a linear relationship, and deviation from linearity in an Ohm's law plot is indicative of poor efficiency in heat dissipation by the capillary temperature control system. For the use of CE as a research tool, programmable temperature gradients are useful for some applications.

Other devices or capabilities may be necessary for the performance of techniques such as electrochromatography (high pressure and pressure on during the run), sieving using polymer solutions (pressure higher than 50 psi), and capillary isoelectric focusing (CIEF) with pressure mobilization (programmable pressure).

Preparative Capillary Electrophoresis

Although the total amount of sample that is loaded into the capillary is extremely small, the use of CE as a "preparative" technique is highly desirable. A preparative technique is one in which samples are collected and then used as product or starting material for other analytical assays. Modern characterization methods require minute amounts of sample to provide such information as amino acid content, sequence, and activity. The use of larger-bore capillaries (e.g., 100 to 200 μm) is recommended for preparative analysis because more sample volume is injected for these capillaries. For short columns with large internal diameters, the use of viscosity-enhancing agents may be necessary to avoid siphoning. In addition, it is important that the additive does not interfere with the technique used for further analysis. Multiple analyses to enrich sample concentration are often necessary.

Capillary Electrophoresis Separation Modes

One of the major advantages of CE as a separation technique is the wide variety of separation modes available. Analytes can be separated on the basis of charge, molecular size or shape, pI, or hydrophobicity. The same CE instrument can be used for zone electrophoresis, IEF, sieving separations, isotachophoresis, and chromatographic techniques such as MEKC and capillary electrokinetic chromatography. This section provides a brief description of each separation mode. Zone electrophoresis, IEF, and sieving are the primary modes used for protein separations, and these will be discussed in detail in the following sections.

Capillary Zone Electrophoresis (CZE)

In CZE, the capillary, inlet reservoir, and outlet reservoir are filled with the same electrolyte solution. This solution is variously termed *background electrolyte*, *analysis buffer*, or *run buffer*. In CZE, the sample is injected at the inlet end of the capillary, and components migrate toward the detection point according to their mass-to-charge ratio by the electrophoretic mobility and separations principles outlined in the preceding text. It is the simplest form of CE and the most widely used, particularly for protein separations. CZE is described in "Capillary Zone Electrophoresis."

Capillary Isoelectric Focusing (CIEF)

IEF is similar in concept to conventional gel IEF; a stable pH gradient is formed in the capillary using carrier ampholytes, and proteins are focused in the gradient at their pIs. The major difference in performing IEF in the capillary format rather than slab gel is the requirement for mobilizing focused protein zones past the detection point. IEF is described in "Capillary Isoelectronic Focusing."

Capillary Sieving Techniques

Sieving techniques are required for separation of species that have no difference in mass-to-charge ratio. This includes native proteins composed of varying numbers of identical subunits, protein aggregates, and sodium dodecylsulfate (SDS)–protein complexes. Sieving systems include cross-linked or linear polymeric gels cast in the capillary or replaceable polymer solutions. Sieving techniques are described in "Sieving Separations."

Capillary Electrochromatography (CEC)

CEC, or capillary electrokinetic chromatography, is a chromatographic technique performed with CE instrumentation. It employs fused silica capillaries packed with 1.5 to 5 μm microparticulate porous silica beads, usually derivatized with a hydrophobic ligand such as C18. Mobile phases are similar to those used for conventional reversed-phase HPLC, e.g., mixtures of aqueous buffers and an organic modifier such as acetonitrile. The silica surface of the derivatized beads has a sufficient density of ionized silanol groups to generate a high electroendosmotic flow when a voltage is applied to the system, thus pumping mobile phase through the column. In contrast to the pressure-driven flow of HPLC, the EOF-driven flow of CEC generates negligible pressure and allows the use of small-diameter beads to improve efficiencies and resolution. In addition, EOF is pluglike rather than laminar in nature, so efficiencies in CEC can be much higher than in HPLC. Like MEKC, CEC is used primarily for small-molecule analysis. Some of the obstacles to using CEC for reversed-phase protein applications include forming reproducible solvent gradients and improving chromatographic supports to maintain stable EOF and increase protein recovery.

Micellar Electrokinetic Chromatography (MEKC)

MEKC is a mode of CE that applies mostly to the analysis of small molecules. The technique is usually performed in uncoated capillaries under alkaline conditions to generate a high electroendosmotic flow. Like CEC, MEKC is a chromatographic technique in which sample components are separated by differential partitioning between two phases: the analysis buffer and a pseudostationary phase usually consisting of detergent micelles. The main differences between MEKC and true chromatography are the presence of the pseudostationary phase and the fact that molecules are transported by an electric field rather than the pressure used in chromatography. The analysis buffer contains a surfactant at a concentration above its critical micelle concentration (CMC). The most widely used MEKC system employs SDS as the surfactant. The sulfate groups of SDS are anionic, so both surfactant monomers and micelles have electrophoretic mobility counter to the direction of EOF. Sample molecules will be distributed between the bulk aqueous phase and the micellar phase depending upon their hydrophobicity (Figure 9.3). Hydrophilic neutral species with no affinity for the micelle will remain in the aqueous phase and reach the detector in the time required for EOF to travel the effective length of the column. Hydrophobic neutral species will spend varying amounts of time in the micellar phase depending on their hydrophobicity, and their migration will therefore be retarded by the anodically moving micelles. Charged species will display more complex interactions because they have the potential for electrophoretic migration and electrostatic interaction with the micelles in addition to hydrophobic partitioning. The selectivity of MEKC can be expanded with the introduction of chiral selectors or chiral surfactants to the system. MEKC is used almost exclusively for small molecules such as drugs and metabolites; it has been used occasionally for peptides.

The use of MEKC for the analysis of proteins has been reported in the literature,[17] but the separation mechanism described in the preceding text is not applicable. SDS micelles range from 3 to 6 nm in diameter, which is too small to accommodate molecules larger than about 5 kDa. However, proteins can bind tenaciously to SDS monomers and micelles via hydrophobic, hydrophilic, and electrostatic interactions. Evidence indicates that SDS–protein complexes consist of protein-enclosed micelles distributed along the protein chain.[18] Perhaps differential binding of SDS at low concentration by different proteins can provide or improve resolution for some protein samples. Under this concept, those proteins that bind more SDS will acquire a higher rate of mobility. One concern would be the reproducibility of the method because variations in detergent concentration (run-to-run, day-to-day) might change the mobility of the high SDS-binding proteins accordingly. Too much SDS (or a long incubation time) can mask the natural charges of the polypeptides, thus giving them a near-equal mass-to-charge ratio. Other detergents can improve performance by increasing sample solubility and by minimizing sample–capillary wall interaction.

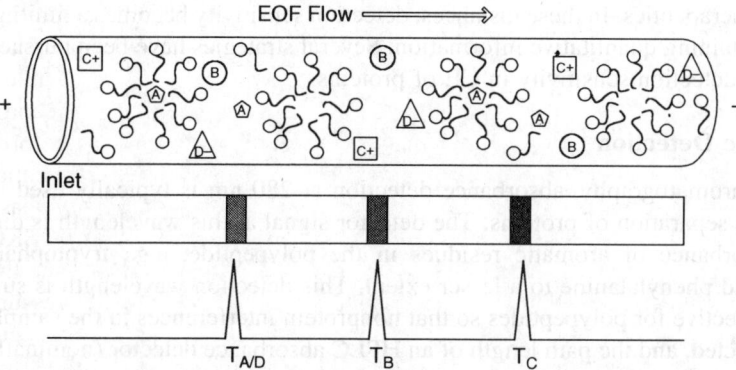

Figure 9.3 An MEKC system (top) includes detergent micelles that possess a hydrophobic core and an aqueous phase represented by the buffer solution. The hydrophobicity of the micelle depends on the chain length of the detergent. The migration profile during analysis of samples depends on charge and degree of interaction with the micelles (bottom). Because EOF is faster than the migration of most analytes, the detector is placed near the negative electrode (cathode) for capillaries with negative wall charge. Micelles of SDS exhibit a strong negative charge and migrate toward the positve electrode (anode). However, the force of EOF carries the negatively charged detergent against its electrophoretic mobility and toward the cathode. In this example, analyte C, with a migration time of T_C, has a positive charge, and therefore its mobility is a combination of electrophoretic mobility toward the cathode and EOF. Analyte B is a neutral molecule interacting loosely with the micelle and thus exhibits a longer migration time (T_B) as it is carried by EOF alone. Analyte A, a neutral molecule that interracts with the hydrophobic core of the SDS micelles, migrates toward the anode but is eventually carried to the cathode by EOF. Thus, T_A is later than T_B. Like the negatively charged SDS, analyte D also migrates toward the anode before being overcome by EOF. Migration of samples that do not interact with the detergent is based solely on electrophoretic mobility and EOF. A high degree of selectivity is achieved because of the combination of electrophoretic mobility and partitioning within the micelles.

Although protein behavior in SDS-containing buffers is qualitatively different from small molecules, applications of MEKC-type conditions have been applied to many protein separations. In applications using uncoated capillaries, protein–wall interactions are eliminated because of the anionic character of SDS–protein complexes. In applications using coated capillaries with no EOF, the high electrophoretic mobility of SDS–protein complexes can decrease analysis time.

DETECTION OF PROTEINS

CE is increasingly applied to analysis of proteins present at trace levels in biological materials or to determination of impurities and degradation products in formulations of protein therapeutics. In these instances, detection sensitivity becomes a limiting factor in obtaining quantitative information. Several strategies have been pursued to increase detection sensitivity in CE of proteins.

Absorbance Detection

In liquid chromatography, absorbance detection at 280 nm is typically used to monitor the separation of proteins. The detector signal at this wavelength is due to the absorbance of aromatic residues in the polypeptide, e.g., tryptophan, tyrosine, and phenylalanine to a lesser extent. This detection wavelength is sufficiently selective for polypeptides so that nonprotein interferences in the sample are not detected, and the path length of an HPLC absorbance detector (nominally 1 cm) provides sufficient signal for satisfactory sensitivity. However, reduction of the detector path length to 25 to 75 μm in CE cuts the detector signal by a factor of 100 to 400. Therefore, detection at 280 nm rarely provides sufficient signal for satisfactory sensitivity. Instead, detection at 200 nm is typically employed where proteins exhibit 50- to 100-fold greater absorbance.

As in HPLC, absorbance detection is used in the vast majority of CE applications, and all commercial CE systems employ UV or UV-Vis absorbance as the primary mode of detection. All commercial CE absorbance detectors employ on-tube detection in which a section of the capillary itself is used as the detection cell. In accordance with Beer's law, the sensitivity of a concentration-sensitive detector is a direct function of the length of the light path. Therefore, in comparison to an HPLC detector with a 1-cm path length, detector signal strength is reduced 200-fold in a CE system equipped with a 50-μm I.D. capillary. Concentration sensitivity can be improved by employing focusing lenses to collect light at the capillary lumen, by detecting at low wavelengths (where most analytes have greater absorbance), and by using sample-focusing techniques during the injection process. However, even under ideal conditions, the concentration limit of detection (CLOD) is about 10^{-7} M.

Several commercial CE systems incorporate absorbance spectra detectors. Absorbance spectra detection enables on-the-fly acquisition of spectra as analytes migrate through the detection point; this information can assist in the identification of peaks based on spectral patterns, in the detection of peak impurities by variation in spectral profiles across a peak, or in the determination of the absorbance maximum of an unknown compound. Two different designs are used to accomplish this type of detection in CE instruments. In photodiode array (PDA) detectors, the capillary is illuminated with full-spectrum source light; the light passing through the capillary is dispersed by grating onto an array of photodiodes that individually sample a narrow spectral range. In fast-scanning detectors, monochromatic light is collected from the source using a movable grating and slit assembly and directed to the capillary; light transmitted by the capillary is detected by a single photodiode. Scanning is accomplished by rapidly rotating the grating through an angle to "slew" across the desired spectral range.

Fluorescence Detection

The high sensitivity and selectivity of fluorescence detection make this the obvious choice for improving detection of proteins. Three approaches have been used: direct detection of intrinsic protein fluorescence, indirect fluorescence detection, and protein derivatization for fluorescence detection.

Fluorescence detection offers the possibility of high sensitivity and, in the case of complex samples, improved selectivity. However, this mode of detection requires that the analyte exhibit native fluorescence or contain a group to which a fluorophore can be attached by chemical derivatization. Because only tryptophan and tyrosine exhibit significant native fluorescence, fluorescence detection of proteins usually requires derivatization.

When compared to fluorescence detectors for HPLC, the design of a fluorescence detector for CE presents some technical problems. In order to obtain acceptable sensitivity, it is necessary to focus sufficient excitation light on the capillary lumen. This is difficult to achieve with a conventional light source but is easily accomplished using a laser. The most popular source for laser-induced fluorescence (LIF) detection is the argon ion laser, which is stable and relatively inexpensive. The 488-nm argon ion laser line is close to the desired excitation wavelength for several common fluorophores. The CLOD for a laser-based fluorescence detector can be as low as 10^{-12} M.

For SDS–protein complexes, the proteins can be derivatized with fluorescent dyes prior to the analysis. Derivatized proteins can be detected in the attomole (amol) level, and because the complex is resolved by sieving, multiple reaction products are detected as peak broadening instead of multiple peaks.[19]

Intrinsic Protein Fluorescence

Compared to absorbance detection, direct detection of proteins rich in aromatic amino acids by the intrinsic fluorescence of tryptophan and tyrosine residues provides enhanced sensitivity without the complexity of pre- or postcolumn derivatization. The optimal excitation wavelengths for these amino acids are in the 270- to 280-nm range.

Indirect Protein Fluorescence

A simple alternative to direct detection of intrinsic protein fluorescence detection is the technique of indirect fluorescence detection proposed by Kuhr and Yeung.[20] In this approach, the analysis buffer contains a fluorescent anion that produces a high background fluorescence signal. Nonfluorescent analyte anions displace the fluorescent species, producing a zone of reduced signal. Sensitivity in indirect fluorescence detection is determined by the dynamic reserve (ratio of signal intensity to signal fluctuation, S/N) and the displacement ratio. Using a coated capillary and salicylate as the background fluorophore, a detection limit of 100 amol was demonstrated for lysozyme with a stabilized HeCd laser providing excitation at 325 nm and collection of emission at 405.1 nm.

Protein Derivatization for Fluorescence Detection

Some precolumn derivatization procedures described for absorbance detection can also be applied to fluorescence detection. In fact, many of the reagents used for absorbance detection (OPA, NDA, fluorescamine) are also highly fluorescent. Although many analytes contain reactive groups (e.g., amino, carboxyl, hydroxyl), most derivatization chemistries are limited by such disadvantages as slow reaction kinetics, complicated reaction or cleanup conditions, poor yields, interference by matrix components, derivative instability, and interference by reaction side products or unreacted fluorescence agent. In addition, most proteins possess multiple reactive sites, and incomplete derivatization yields a family of products varying in the number of fluorophores. The reactive sites are usually side-chain amino groups, and the derivatized products (which vary in mass and charge) may be resolved into multiple peaks or migrate as a single broad peak. Loss of efficiency and resolution of multiple species has been observed in CZE separation of proteins following precolumn derivatization with fluorescein isothiocyanate[21] and OPA.[22] Bardelmeijer et al.[23] provide an extensive review of fluorescent labeling of proteins.

Mass Spectrometry

With the increasing need to obtain absolute identification of separated components and the gradual price reduction of mass spectrometers, there is a growing demand for direct coupling of CE with MS instruments. On-line coupling of a CE system to a mass spectrometer enables molecular weight and structural information to be obtained for separated components.[24] The most frequent configuration is introduction of the capillary outlet into an electrospray interface (ESI) coupled to the mass spectrometer. In this configuration, the outlet electrode of the CE is eliminated and the MS becomes the ground. Because the volumetric flow out of the capillary is negligible or nil, separated components are usually transported from the capillary to the electrospray using a liquid sheath flow.

Compatibility with MS requires the use of volatile buffer systems such as acetic and formic acids or their ammonium salts for low-pH separations, or ammonium carbonate for high-pH applications. This limits the choice of CE separation modes and selectivity of the CE separation system. The background electrolyte causes discrimination against analyte ions by charge competition in the electrospray, thus reducing overall sensitivity. Using low concentrations of electrolytes in the analysis buffer can minimize this problem. Another factor in reducing sensitivity is the dilution of ions by the sheath flow liquid at the CE–MS interface. This can be minimized by reducing sheath flow rates to 2 to 5 μl/min and by using sheath liquids composed of low concentrations of electrolyte in organic solvents (e.g., acetic acid in methanol).

Samples collected during CE analysis can be used for off-line mass analysis by matrix-assisted laser desorption ionization (MALDI) and time-of-flight (TOF) MS.

MINIMIZATION OF NONSPECIFIC PROTEIN–WALL INTERACTIONS

Interactions of proteins with the surface of the capillary, especially silica columns, are not different from those encountered when using unmodified chromatographic silica supports. These interactions are often strong or irreversible and have been major obstacles to successfully applying CE to protein separations. Bare silica capillaries contain weakly acidic silanol groups on the surface that ionize rapidly above pH 3. The charge density on the wall increases until the silanol groups are fully dissociated at pH 10. Under these conditions, proteins with basic amino acid residues positioned on the protein surface can participate in electrostatic interactions with ionized silanols. This results in band broadening, tailing, and, in the case of strong interaction, reduced detector response or complete absence of peaks. Such interactions change the state of the capillary wall during an analysis and can alter the magnitude of EOF from run to run, resulting in poor reproducibility. Three strategies have been employed to minimize protein–wall interactions: operation at pH extremes, use of buffer additives, and use of wall-coated capillaries. The simple solution of using "HPLC-like" conditions (e.g., high salt-eluting conditions for ion exchangers) is often not suitable for CE because of negative effects to other aspects of the system (e.g., excessive electric current).

Operation at pH Extremes

The simplest approach to minimizing protein–wall interaction is to use a buffer pH at which interactions do not occur. At acidic pH the silanols on the surface of the capillary are protonated, and the net charge of the proteins is positive. At high pH, the wall is negatively charged, and so are the sample components. Both conditions result in electrostatic repulsion. Problems associated with operation at pH extremes include the potential instability of proteins (denaturation, degradation, and precipitation) and the limited pH range in which to achieve resolution. Additionally, operation at extreme pH does not eliminate all nonspecific interactions.

Use of Buffer Additives

Buffer additives can overcome some of the protein interactions with the capillary wall. Some of these additives are widely used in HPLC to elute proteins off the chromatographic supports. Precautions should be taken when selecting buffer additives, especially in the areas of detector interference and their impact in the conductivity of the buffer. A list of additives used for various CZE applications can be found in Table 9.1.

Table 9.1 Additives Used in CZE

Additive	Application	Reference
Betaine	Basic proteins	154
Cadaverine	Acidic and basic proteins	159
Ethylene glycol (20%)	Acidic and basic proteins, serum proteins	160
Cationic and zwitterionic fluorosurfactants	Acid and basic proteins	155–158, 161
2-(N-cyclohexylamino)ethanesulphonic acid (CHES)	Insulins	162
N,N-bis(2-hydroxyethyl)-2-aminoethane-sulphonic acid (BES)	Insulins	162
3-[(1,1-dimethyl-2-hydroxyethyl)amino]-2-hydroxypropanesulphonic acid (AMPSO)	Insulins	.162
3-(cyclohexylamino)-2-hydroxy-1-propanesulphonic acid (CAPSO)	Insulins	162
N-alkyl-N,N-dimethylammonio-1-propane sulfonic acid	Basic proteins	163
1,4-diaminobutane	Protein glycoforms	148, 152, 164
Spermine,spermidine	Protein glycoforms	165
Triethylamine	Basic proteins	149, 151, 166
Triethanolamine	Basic proteins	149, 151
Galactosamine	Basic proteins	150
Glucosamine	Basic proteins	150
Trimethylammonium propylsulfonate (TMAPS), trimethylammonium chloride (TMAC)	Monoclonal antibody	167
Trimethylammonium propylsulfonate (TMAPS)	Acidic, neutral, and basic proteins	168
(Trimethyl)ammonium butylsulfonate (TMABS)	Acidic, neutral, and basic proteins	168
2-hydroxyl-3-trimethylammonium propylsulfonate (HTMAPS)	Acidic, neutral, and basic proteins	168
3-(dimethyldodecylammonio)-propanesulfonate	Basic proteins	169
Cetyltrimethylammonium bromide	Acidic and basic proteins	170, 171
Chitosan	Basic proteins	172
Amino acids	Basic proteins	149
Hexamethonium bromide, hexamethonium chloride	Protein glycoforms	153
Decamethonium bromide	Protein glycoforms	153
Polydimethyldiallylammonium chloride	Basic proteins	173, 174
Phytic acid	Acidic and basic proteins	175–177
Ethylenediamine	Basic proteins	148
1,3-diaminopropane	Basic proteins	148
N,N-diethylethanolamine	Basic proteins	149
N-ethyldiethylamine	Basic proteins	149
Morpholine, tetraazomacrocycles	Basic proteins	178

Capillary Coatings and Other Surface Modifications

Capillary coatings restrict or eliminate access of sample molecules to the capillary wall. Surface modifications to the capillary wall may be grouped into two categories: covalent and dynamic coatings. Covalent coatings are attached to the wall through a chemical bond, whereas dynamic coatings physically interact with the silica or a first layer deposited over the silica. Depending on the strength of the interaction of the dynamic coatings, it might be necessary to add the wall-interacting compound in the run or conditioning buffer.

The presence of a coating modifies or eliminates EOF. Thus, degradation of the coating through physical loss or chemical changes leads to loss of efficiency and poor reproducibility. There are many coating chemistries described in the literature, which reflect the continuous search for more stable coatings and the inadequacy of any single approach to provide satisfactory results for all applications. Coating chemistries have been extensively reviewed previously.[25–27]

When developing a CE method that requires coatings, the end use of the method itself must first be defined. If the information to be gained is for research, the use of any chemistry is considered to achieve results. On the other hand, if the technique will be used in production or quality control environments, the capillary coating chemistries are limited to those commercially available. An in-house-produced coated capillary or buffer with additives must have a production protocol and quality control release testing. The benefits of having an optimal method must be weighed against the cost, and it is hard to justify all the extra work required if there is a simpler option available.

SAMPLE INJECTION AND PREPARATION

Sample Injection

Sample injection is an event common to all modes of CE except CIEF, and the most relevant aspects of this step are described here. Because the total volume of the capillary is very low, sample injection in CE requires the precise introduction of very small amounts of analyte at the capillary inlet. There are two basic procedures for sample injection: electromigration and hydraulic displacement. All commercial instruments offer electromigration, or electrophoretic, injection and at least one type of displacement injection.

In electrophoretic injection, the capillary inlet is immersed in the sample solution and a voltage is applied for a determined period of time. The amount of sample introduced into the capillary depends on the voltage and the time it was applied. Sample injection is a compromise between detection and resolution, and its parameters are often best determined experimentally. If detection is not a problem, resolution can be greatly improved by maintaining the sample "plug" as narrow as possible. If EOF is present, sample ions will be introduced by a combination of electrophoretic mobility and EOF; under these conditions, this injection mode is generally termed electrokinetic injection.

Electrophoretic injection can be used as a means for zone sharpening or sample concentration if the amount of ions, particularly salt or buffer ions, is lower in the sample than the running buffer. Because sample ions enter the capillary based on mobility, low-mobility ions will be loaded to a lesser extent than high-mobility ions. For this reason, the presence of nonsample ions will reduce injection efficiency, so electrophoretic injection is very sensitive to the presence of salts or buffers in the sample matrix. The disadvantages of electrophoretic injection argue against its use in routine analysis except in cases where displacement injection is not possible, e.g., in capillary gel electrophoresis (CGE) or when sample concentration by stacking is necessary.

Displacement injection is usually the preferred method because analyte ions are present in the sample zone in proportion to their concentration in the bulk sample. In addition, injection efficiency is less sensitive to variations in sample ionic strength. However, it should be noted that the presence of high salt can affect detector response and variations in the sample viscosity due to temperature, or the presence of viscosity-modifying components can affect displacement injection efficiency.

For quantitative purposes, the peak area must be corrected for mobility when using displacement injection. This correction is necessary because peak area is a result of sample response and time. Analytes with lower mobility spend more time in front of the detector, thus generating larger relative area counts. Such correction is not necessary for electrophoretic injections.

Sample Preparation

Sample preparation is often a parameter whose impact on the analysis is overlooked or underestimated. In reality, the composition of the sample matrix is often key to the quality of the data obtained at the end of the analysis. An important consideration for biopharmaceutical molecules is to minimize sample preparation because the impact of sample manipulations must be evaluated during methods validation.

In some cases, sample preparation for CZE requires only the dilution of the sample, mostly to accommodate detection (for signal and linearity of response). However, as was previously mentioned, sample characteristics such as viscosity, buffer composition (pH and excipients), and salt content can especially affect electrophoretic injection and performance.

Sample preconcentration techniques are used with two purposes: (1) to increase concentration in order to achieve detection and (2) to eliminate disturbances of the electrophoretic system during hydraulic or electrokinetic sample introduction when the conductivity of the sample is significantly different from that of the analysis buffer. It is important to keep sample manipulations and modifications to a minimum, and a rule of thumb is to prepare the sample so that its composition is at the same pH as the analysis buffer. It is also advantageous to reduce the sample buffer composition to a conductivity approximately 10% of the background electrolyte. The two common reasons to modify sample composition, and the common methods used are given in the following subsections.

Preconcentration to Improve Detection

Several strategies have been described for the preconcentration of sample components present at low concentrations. These techniques include zone sharpening,[28,29] on-line packed columns,[30] and transient capillary isotachophoresis (cITP).[31,32] Other standard laboratory techniques are often used, including solid-phase extraction, protein precipitation, ultrafiltration, etc. Two important points to keep in mind when selecting a concentration protocol are the sample requirements of the method and the potential selectivity on relative concentrations of sample components. The latter point applies to purity and concentration analysis.

Preconcentration to Regulate Sample Conductivity

The introduction of a zone of different conductivity into a capillary produces an uneven distribution of the electric field. If the sample zone represents a small portion of the capillary, the electrical current usually "recuperates," presumably by diffusion of charge to the level normally seen in the analysis buffer. However, when the sample zone is of significant length due to overloading, the migration time can be severely affected[33] and, in extreme cases, the current drops to zero. Caution should be taken when using large bore and short columns because they can be easily overloaded. For this reason it is necessary to preconcentrate diluted samples to increase their conductivity, thus improving reproducibility and often resolution. The length of the sample zone that can be tolerated without major adverse effects depends on the difference in conductivity between the sample zone and the analysis buffer. In addition, due to uneven distribution of the electric field along the capillary, fluctuations in current are magnified when using higher voltage.

CAPILLARY ZONE ELECTROPHORESIS

In slab gel electrophoresis, it is necessary to use a medium that prevents convective disturbances during the analysis. CZE is analogous to native electrophoresis performed in low-percentage acrylamide or agarose gels, but in CE, convection is minimized by the low electric current generated and the high heat-dissipation characteristics of the capillary. Thus, separations can be performed in solution. CZE is often referred to as *free zone electrophoresis*, where "free" is used to specify that the buffer is free of stabilizing media.

In native gel electrophoresis and CZE, the sample components are resolved by their differences in electrophoretic mobility or mass-to-charge ratios. Electrophoretic analysis under native conditions in gel electrophoresis is not as widely used as SDS-PAGE. In gels, disadvantages of native analysis are the low field strength that is used and the usually low charge density of native proteins. Both factors contribute to long analysis time. Native gels often require several hours for completion of the analysis. In the capillary format, high electric fields allow fast analysis time, and the coupling of on-line detection provides fast quantitative results. At the same time, CZE increases the range of application to smaller (e.g., inorganic ions, peptides) and larger sample species such as nano- or microparticles (Figure 9.4).

Manipulation of buffer pH or use of additives can easily vary separation selectivity. In contrast to gel electrophoresis, a single capillary can be used to evaluate the effect of buffer composition or pH on resolution. In CZE the capillary is simply filled with a fresh electrolyte of the chosen composition between analyses.

CZE offers several advantages in comparison to other CE separation modes, including its inherent simplicity: a single buffer is used throughout the capillary and electrode vessels, and sample is introduced as a zone or plug at one end. Capillary preparation often involves just filling the capillary with the separation buffer, although uncoated capillaries generally require prior washing or conditioning steps.

Developing a CZE Method

The development of a CZE separation requires the definition of instrumental and chemical parameters. There have been several attempts to model the behavior of proteins when placed under the influence of an electric field.[34,35] However, difficulties in estimating the charge of the polypeptide as opposed to free amino acids require that instrumental parameters including temperature, injection mode, detection, and power supply settings be determined experimentally. The main chemical parameter that affects CZE is buffer composition (including the type of buffer, additives, pH, etc.). The selection of the appropriate capillary type is intimately linked to the selection of buffer composition. Some parameters (e.g., temperature) need to be defined experimentally, whereas others (e.g., buffer pH) can be defined by the known characteristics of the sample such as solubility, pI, etc. The following subsections provide some general guidelines to consider during the development of a CZE method.

Buffer Selection

Buffer pH defines the net charges of the sample components and therefore the magnitude and direction of their mobilities. Ideally, the run buffer selected for CE should have good buffering capacity in the pH range that provides optimal resolution, be "transparent" to the detector system, and possess low conductivity at a concentration that still allows good pH buffering. Because the sample characteristics dictate the optimal pH range to be used (i.e., the range of maximum mobility difference for the components of interest), no one buffer is suitable for all applications. The pH of the buffer should be at least one pH unit above or

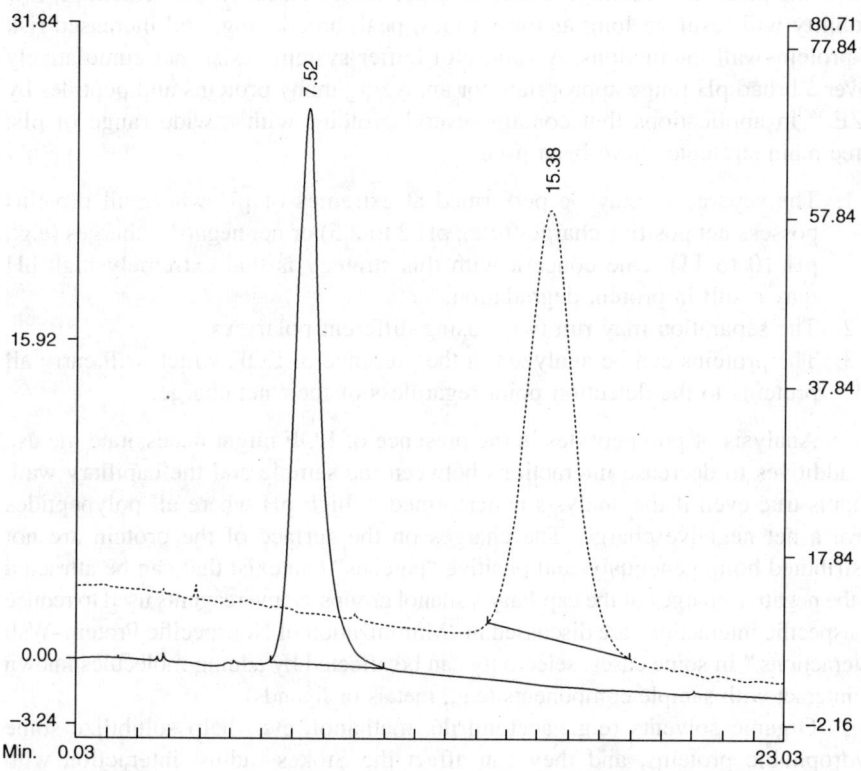

Figure 9.4 Overlay of electropherograms for Hepatitis B surface antigen (HBsAg) and HBsAg chemically linked to immunostimulatory oligonucleotides (ISS) as analyzed by CZE. HBsAg self-assembles into viruslike particles with diameters ranging from 20 to 30 nm. Because of its particle nature, HBsAg is difficult to characterize by analytical techniques such as chromatography (in which the sample particles get stuck in filters and small spaces between packing particles) and gel electrophoresis (in which the sample does not penetrate the pores of the separation matrix). Because CZE can be performed in capillaries of various internal diameters filled with buffer, HBsAg is easily analyzed by this technique. The broken line represents the unmodified particle, whereas the solid trace represents the particle conjugated to oligonucleotide. Notice that the sample conjugated to ISS has an increased mobility (shorter migration time) than the same particle before the attachment of ISS. The increased mobility is due to the high negative charge density of the oligonucleotide compared to its mass, which imparts a higher charge-to-mass ratio to the conjugate. The antigen and conjugate were analyzed using 50 mM phosphate buffer, pH 6.5 in a 40-cm × 50-μm coated capillary. The sample was introduced by applying pressure (5 psi) for 1 sec, and the electric field was set to 15 kV with the negative electrode in the injection port. The capillary temperature was maintained at 20°C, and the detection was performed by UV absorption at 200 nm.

below the pI of the protein of interest; at pH values closer to the protein pI, low mobility will result in long analysis times, peak broadening, and increased risk of protein–wall interactions. A variety of buffer systems exist that cumulatively cover a broad pH range appropriate for analyzing many proteins and peptides by CZE.[36] In applications that contain several proteins with a wide range of pIs, three main strategies have been used:

1. The separation may be performed at extremes of pH where all proteins possess net positive charges (e.g., pH 2 to 2.5) or net negative charges (e.g., pH 10 to 11). One concern with this strategy is that extremely high pH may result in protein degradation.
2. The separation may run twice using different polarities.
3. The proteins can be analyzed in the presence of EOF, which will carry all proteins to the detection point regardless of their net charge.

Analysis of polypeptides in the presence of EOF might necessitate the use of additives to decrease interactions between the sample and the capillary wall. This is true even if the analysis is performed at high pH where all polypeptides have a net negative charge. The charges on the surface of the protein are not distributed homogeneously, and positive "patches" can exist that can be attracted to the negative charges of the capillary's silanol groups. Some reagents used to reduce nonspecific interactions are discussed in "Minimization of Nonspecific Protein–Wall Interactions." In some cases, selectivity can be affected by adding molecules known to interact with sample components (e.g., metals or ligands).

Organic solvents (e.g., acetonitrile, methanol, etc.) help solubilize some hydrophobic proteins, and they can affect the Stokes radius, interaction with counterions, and the ionization state of the polypeptide. For more hydrophobic samples, the use of detergents and/or other organic solvents may prove beneficial.

Capillary Selection

Fused silica capillaries are almost universally used in capillary electrophoresis. The inner diameter of fused silica capillaries varies from 20 to 200 μm, and the outer diameter varies from 150 to 360 μm. Selection of the capillary inner diameter is a compromise between resolution, sensitivity, and capacity. Best resolution is achieved by reducing the capillary diameter to maximize heat dissipation. Best sensitivity and sample load capacity are achieved with large internal diameters. A capillary internal diameter of 50 μm is optimal for most applications, but diameters of 75 to 100 μm may be needed for high sensitivity or for micropreparative applications. However, capillary diameters above 75 μm exhibit poor heat dissipation and may require use of low-conductivity buffers and low field strengths to avoid excessive Joule heating.

Selection of capillary length is dictated by the type of capillary used and the required resolution. When using coated capillaries with insignificant EOF, separations can be achieved with relatively short capillaries of 20 to 30 cm effective length from inlet to detection point. When using uncoated capillaries under conditions in which there is appreciable EOF, longer lengths of 50 cm or greater may be needed to achieve a separation, particularly for basic proteins that are migrating toward the detector under the combined forces of EOF and electrophoresis.

Capillary length is one of the parameters commonly used to improve resolution, but in our experience, more dramatic effects can be accomplished by changing the mobility of the sample components through manipulation of pH or the use of buffer additives. In fact, we usually develop a method in the shortest capillary possible, and only when the separation is adequate do we increase the capillary length for final optimization. This practice saves time in methods development because more data can be collected if the analysis time is short.

High-Voltage Parameters

High-voltage parameters include mode of operation, field strength, and polarity. Most CE systems can be operated in constant voltage or constant current, and some instruments allow the use of constant power. The majority of protein separations reported in the literature have been performed in constant voltage mode. Selection of field strength impacts the analysis time and resolution. Operation at high field strength reduces analysis time but increases band broadening due to thermal effects (which can be minimized by using low-conductivity buffers); operation at low field strength reduces heating but increases analysis time and band broadening due to diffusion. The latter effect is probably small for proteins because of their low diffusion coefficients. In our experience, operation at 400 to 600 V/cm provides optimal separations in terms of speed and resolution. When high voltages >1000 V/cm are used, the areas surrounding the electrodes must be kept clean and well sealed to avoid electric arcing. Selection of high-voltage polarity depends on the sample composition and capillary type. For example, when using uncoated capillaries under neutral to alkaline conditions, positive (inlet side) to negative (detector side) polarity is employed so that EOF will carry all samples toward the detection point.

Temperature

The use of temperature in most CZE applications is to maintain a constant and reproducible environment from one analysis to the next. Several applications also use temperature to improve resolution or provide additional characterization such as elucidation of protein structure.

Applications

A number of CZE applications exist for the separation of proteins and other molecules in purity analysis, structural studies, binding and equilibrium determinations, in-process product analysis, and mobility measurements. The following applications illustrate the use of CZE for both research and routine QC analysis.

Development and Validation of a CE Method
for Characterization of Protegrin IB-367

An example of a simple CZE method for peptide analysis and characterization is the one developed for protegrin IB-367.[37] IB-367 is a peptide containing 17 amino acid residues that possess antimicrobial properties, and it is being developed for treatment of oral mucositis associated with aggressive cancer chemotherapy as well as other topical applications. This polycationic product was chemically synthesized using solid-phase and purified by preparative reversed-phase HPLC. IB-367 is rich in cysteine and arginine residues.

Optimized electrophoresis parameters included resolution, reproducibility, and minimal analysis time. The CZE method employs a 100-mM phosphate buffer, pH 2.6, and a 50-μm I.D. × 45-cm effective length uncoated capillary. Because the pH of the background electrolyte is 2.6, most silanols should be fully protonated on the capillary wall, thus minimizing undesired sample–capillary wall interactions. The capillary was conditioned with 0.1 M NaOH, water, and 0.1 M HCl between runs prior to a final water rinse and fill with the run buffer. The capillary was thermostated to 25°C, and detection was performed by UV-absorption at 200 nm. The sample was injected using pressure at 0.5 psi for 5 sec, and the run voltage was set to 20 kV. The voltage was selected by creating an Ohm's plot (voltage vs. electric current) and defining the highest voltage in the linear range of the plot. The linear range of an Ohm's plot defines the range of voltage at which there is still efficient Joule heat dissipation.

Both CE and HPLC methods were capable of resolving the IB-367 peptide from impurities and degradation products. However, the CE method provided better separation and resolution between this polycationic peptide and truncated analogs than HPLC methods. In addition, the CE methods resolved the potential impurities and degradation products from each other, whereas the HPLC methods failed to separate some truncated species.

The CE method was validated in terms of accuracy, precision, linearity, range, limit of detection, limit of quantitation, specificity, system suitability, and robustness. Improved reproducibility of the CZE method was obtained using area normalization to determine the purity and levels of potential impurities and degradation products of IB-367 drug substance. The internal standard compensated mainly for injection variability. Through the use of the internal standard, selected for its close mobility to IB-367, the method achieved reproducibility in relative migration time of 0.13% relative standard deviation (RSD), and relative peak area of 2.75% RSD.

System suitability tests serve to define the level of electrophoretic performance necessary to ensure valid CE assay results. System suitability of the method was evaluated by analyzing the symmetry of the IB-367 peak, theoretical plates of the capillary, and resolution between IB-367 and IB-300, the closest peak to IB-367. The sample concentration of the method was selected at approximately 0.5 mg/ml to assure symmetry below 3.5 and to assume sufficient sensitivity for detecting low concentrations of impurities. At 0.5 mg/ml, the limit of detection for impurities was 0.1%. The following parameters were summarized from runs with different lots of capillary on a 0.5-mg/ml IB-367 solution:

- Asymmetry <3.5 (T = $W_{5\%}/2f$), where T is the tailing factor, $W_{5\%}$ is peak width at 5% peak height, and f is the width at 5% peak height measured from the leading edge to a vertical line extrapolated from the apex of the peak.
- Theoretical plates >50,000 (N = $16(t_M/W)^2$, where t_M is the migration time and W the peak width, both in minutes.
- Resolution >2 (Rs = $2(t_{M2} - t_{M1})/(W_2 + W_1)$), where t_{M1} and t_{M2} are the migration times of the two peaks being measured, and W_1 and W_2 are their peak widths, respectively. Units are all in minutes. Note that the values of the later migrating peak are entered first.

The accuracy of the method was evaluated by assaying six independently prepared solutions of IB-367 against two standard solutions of the same lot as external standards. The mean of 102.3% met the criteria set in validation protocol (97 to 103%).

The intermediate precision (day 2) of the method was evaluated by assaying six independently prepared solutions of IB-367 against two reference standard solutions used as external standards on 2 d. The RSD of 2.1% for the accuracy test (RSD) met the criteria set for repeatability (<4%). The RSDs of 1.9% for the precision test on day 2 and 2.5% for total 12 samples on 2 d met the criteria set for intermediate precision by the validation protocol (<4%).

The linearity and range were determined using solutions of IB-367 at 0.1%, 1%, 10%, 50%, 80%, 100%, and 150% of the specified IB-367 concentration (0.5 mg/ml in water) assayed in duplicate. The method was linear in the 10 to 150% range.

The truncated peptide analogs were used to demonstrate the specificity of the method and to evaluate the limit of quantitation of potential impurities. Potential impurities were spiked into a solution of IB-367 at 0.05%, 0.1%, 0.2%, 0.5%, and 1% to assay the linearity of potential impurities at low concentrations. The method exhibited acceptable linearity for impurities from 0.05 to 1%. The relative response factors of these analogs were assessed to determine area normalization feasibility.

The limit of detection for IB-367 was 0.1% (100% being 0.5 mg/ml) for a peak height at least twice the noise level. The limit of quantitation was determined to be 0.5% by the criteria of signal-to-noise ratio of at least ten. Similar values were obtained for the limits of detection and quantitation of potential impurities and degradation products. The potential degradation products and impurities were synthesized and used to show that they were all resolved from the product peak. The impurities were defined as IB-468 ([des-Arg^1]), IB-469 ([des-Arg^{1}Gly^2]), and IB-300 ([des-Arg^{1}Gly^{2}Gly^2]).

The robustness of the method was evaluated by four electrophoretic parameters. The relative migration time of IB-367 and its impurities and the resolution between the four peptides were monitored by changing the buffer strength, the pH of the analysis buffer, the separation voltage, and the capillary temperature. Because of the relative nature of the results, changes in any of the parameters listed in the preceding text were compensated by the presence of the internal standard, thus rendering the method robust.

Enzyme Assays

Application of CE to determine enzyme activity typically involves off-line incubation of enzyme and substrate with timed injections of the reaction mixture into the capillary to separate and quantitate the low-molecular weight substrate, intermediates, and product. A CE method termed electrophoretically mediated microassay (EMMA) is based on electrophoretic mixing of enzyme and substrate[38–41] under conditions where the mobility of enzyme and product are different. The capillary was prefilled with all of the required components for the assay (buffer, substrate), and the enzyme was introduced at the capillary inlet. Both product and enzyme were transported to the UV-Vis detector, and product was detected at a selective wavelength where the enzyme did not interfere. The relative effective mobility of enzyme and product were adjusted by manipulating the rate of EOF using covalent (e.g., epoxy polymer) or dynamic (e.g., nonionic surfactant adsorbed onto an octadecyl layer) coatings. Highest sensitivity was achieved by operation in zero potential mode; in this mode, enzyme was first injected and mixed with substrate, and then voltage was turned off for a fixed incubation period to accumulate product. In the final step, potential was reapplied to transport product to the detection point. A limitation of the zero potential mode was band broadening caused by diffusion of the product.

Affinity Capillary Electrophoresis

Affinity capillary electrophoresis (ACE), reviewed by Shimura and Kasai,[42] is a method for studying receptor–ligand binding in free solution using CE. The technique depends upon a shift in the electrophoretic mobility of the receptor upon complexation with a charged ligand. Pure receptor preparations or accurate concentration values are not required because only migration times are measured.

In a typical ACE experiment, the receptor is injected into a capillary containing free ligand at a variety of concentrations. Depending upon the kinetics of the on and off processes, incremental shifts in migration times are observed. If association and dissociation have slow kinetics, the sample will be resolved into bound and free components, and binding constants can be calculated from the peak areas at different ligand concentrations in the electrophoresis buffer. In the case of fast kinetics, only one peak will be observed, and affinity is determined by the change in migration time due to variations in time spent in bound and free states during migration through the capillary. The maximum mobility shift will occur as the free receptor becomes fully saturated, and the relationship between mobility shift and ligand concentration can be used to determine the receptor–

ligand association constant. Scatchard analysis of migration time shifts (usually normalized to a neutral EOF marker or a nonbinding reference protein) in response to ligand concentration is used to estimate the ligand–receptor binding or dissociation constant. Mobility corrections using a neutral marker (mesityl oxide) could compensate for variable EOF to yield accurate estimates of binding constants.[43]

Affinity complexation can be detected by an indirect mode termed vacancy affinity capillary electrophoresis, or VACE.[44] The capillary is filled with buffer containing a mixture of receptor and ligand, a plug of buffer is injected at the capillary inlet, and voltage is applied. Differential migration of ligand and receptor through the buffer zone produces zones deficient in ligand and receptor, resulting in the appearance of two negative peaks at steady state. In the VACE experiment, the concentration of ligand or receptor is varied, and the mobility of the two peaks relative to a neutral marker is monitored. The magnitude of the peaks provides information about the degree of complexation, e.g., the receptor-deficient peak reflects the concentration of free receptor and the ligand-deficient peak indicates the level of free ligand. Similarly, the shift in mobility of each peak provides information on the receptor–ligand association constant and the number of receptor binding sites. Using vancomycin and the dipeptide *N*-acetyl-D-alanyl-D-alanine as the receptor–ligand pair, these authors compared the ACE and VACE approaches and found satisfactory agreement of the measured association constants.

ACE has been used to characterize protein–sugar interactions,[45] DNA binding to an anti-DNA monoclonal antibody,[46] antibody–antigen,[47] antisteroidal inflammatory drugs,[48] and prion protein in sheep brain preparations.[49]

In capillary immunoelectrophoresis, antigen–antibody interactions are used to affect resolution.[50] This method can be easily modified to quantitate antigen or antibody, or to study the interaction of the two molecules. To increase the detection sensitivity, one of the polypeptides can be labeled with a fluorescent dye, or the antigen and antibody can be labeled with two different dyes for dual fluorescence. Under these conditions, the free and bound forms may be analyzed simultaneously. If the amount of antibody is not high enough to derivatize, the antibody can be labeled indirectly by using a labeled ligand such as protein A. Capillary immunoelectrophoresis has been used to study interactions between antibody and corresponding antigen or hapten with results comparable to those obtained by ELISA. Concentration of the antigen or antibody must be selected carefully to prevent precipitation, especially when using polyclonal antibodies.

Analysis of Protein Folding

Determination of protein folding is important in various areas of biotechnology and thus is an area of active research for biopharmaceuticals. Crucial parameters such as activity and stability are related to protein folding. Because improper protein folding occurs with high frequency in proteins produced by recombinant technology and proteins tend to denature upon storage, any technique capable of elucidating tertiary structure or able to monitor its changes will find immediate use in areas such as drug screening, formulation, and QC/QA.

CE has been shown to be a valuable tool for analysis of protein folding. Unlike techniques such as chromatography and gel electrophoresis, CZE is performed in free solution, and migration is a function of the intrinsic properties of the molecule. The ability of CZE to distinguish different folding states of a protein depends upon changes in solvent-accessible charge, and the migration rate thus reflects a cross section of the conformational states. Moreover, peak shape can provide information on the distribution of the protein among folding states.

Rush et al.[51] first described the effect of thermally induced conformational changes on migration behavior of α-lactalbumin. A sigmoidal dependency of the viscosity-corrected mobility on temperature was observed. Transition temperature also agreed closely with that determined by intrinsic fluorescence measurements.

Strege and Lagu[52] used CZE to monitor reformation and interchange of disulfide bonds during reoxidation of reduced trypsinogen. In this study, CE was performed under low-pH conditions to minimize protein–wall interactions for this basic protein. A population of refolding intermediates distributed between native and unfolded trypsinogen was resolved. Resolution was further improved by addition of ethylene glycol and sieving polymers.

These authors also monitored the transition of bovine serum albumin from the native folded state to the unfolded state using CZE in the presence of increasing amounts of urea.[53] The resulting plot of EOF-corrected migration time vs. urea concentration was similar to urea denaturation profiles obtained with other techniques. Using CE in the presence of urea, Kilár and Hjertén[54] detected intermediate unfolding states of transferrin as distinct peaks and were able to resolve unfolding intermediates for each of the five transferrin glycoforms (2, 3, 4, 5, and 6-sialotransferrin). Hilser et al.[55] monitored the migration behavior of lysozyme as a function of capillary temperature and observed a sigmoidal behavior characteristic of a protein-unfolding transition. Calculation by van't Hoff analysis of the transition temperature, entropy, and enthalpy of protein unfolding yielded values in close agreement with those determined by differential scanning calorimetry and confirmed that the temperature-dependent decrease in electrophoretic mobility represented a two-state thermal denaturation. CE has also been used as a confirmatory technique to assess the conformational states of proteins eluted from reversed-phase HPLC columns.[56]

Using CZE, Rochu et al.[57,58] analyzed the thermal denaturation of β-lactoglobulin, which exhibits various oligomeric states depending on the protein concentration, pH, and temperature. A commercial CE instrument was modified by connecting the liquid temperature control to an external water bath providing accurate temperature control up to 95°C. Under various pH conditions, transition temperature (T_m), enthalpy change (ΔH), and entropy change (ΔS) associated with thermal denaturation were determined. The technique is unique in its ability to estimate the heat capacity change (ΔC_p). CZE performed in the presence of EOF was used to determine the stability curves of proteins.

In this application, the authors used an uncoated capillary and a neutral marker to monitor changes in EOF. Because viscosity and electrophoretic mobility change with temperature, the simplest way of determining mobility changes due to unfolding is to estimate the mobility of the unfolding protein to a molecule, in this case the neutral marker, which does not change conformation. Because the data are greatly affected by temperature, the thermostating of the capillary needs to be tightly controlled to within less than $0.1°C$.

The van't Hoff plots for thermal denaturation of proteins are linear in the transition region, thus allowing the enthalpy change (ΔH_m) of unfolding at the transition temperature (T_m) to be estimated. Because of the change in free energy in (ΔG) = 0 at T_m (reversible process), the entropy of unfolding (ΔS_m) at the transition midpoint can be calculated from:

$$\Delta S_m = \Delta H_m / T_m$$

For the ΔG of denaturation transition, the temperature (T) data were fitted to the Gibbs–Helmholtz equation:

$$\Delta G_{(T)} = \Delta H_m (1 - T/T_m) - \Delta C_p [(T - T_m) + T\ln(T/T_m)]$$

where ΔH_m is the enthalpy change at T_m and ΔC_p is the change in heat capacity between the native and denatured state. ΔC_p for the unfolding reaction was calculated using the Kirchoff equation:

$$\Delta C_p = d(\Delta H)/d(T)$$

as the slope of the plot of ΔH_m vs. T_m, measured at different pH values of phosphate buffer.

Glycoproteins

Glycoproteins often exist as multiple glycoforms sharing a common amino acid primary sequence but differing in the number, location, and structure of carbohydrate groups attached to the polypeptide chain. The importance of glycosylation patterns in the biological activity of glycoproteins has generated strong interest in methods for separation of glycoforms, particularly in the case of therapeutic glycoproteins. Variation in the number of sialic acid residues confers charge microheterogeneity to glycoproteins; gel electrophoresis and IEF have been used successfully for their characterization. Therefore, CZE and CIEF are obvious candidates for automated analysis of protein glycoforms. A typical strategy for optimizing CZE separations of glycoforms starts with determination of conditions that yield the best resolution of glycoforms. Enzymatic cleavage (e.g., with neuraminidase) of carbohydrate moieties and the subsequent disappearance of peaks in the electropherogram following enzyme treatment confirm their identity as glycoforms.

Separation of protein glycoforms was first described by Kilár and Hjertén,[59] who used CZE with a Tris-borate + EDTA (pH 8.4) buffer and a coated capillary to separate the di-, tri-, tetra-, penta-, and hexasialo isoforms of iron-free human serum transferrin. Yim[60] used CZE to resolve glycoforms of recombinant tissue plasminogen activator (rtPA), a 60-kDa glycoprotein containing complex N-linked oligosaccharides attached to the polypeptide chain at two (type II) or three (type I) sites. Using an ammonium phosphate buffer at pH 4.6 containing 0.01% reduced Triton X-100 + 0.2 M ε-aminocaproic acid (EACA, added to stabilize solubility of the protein) in a linear polyacrylamide-coated capillary, approximately 15 glycoforms were partially resolved. Recently, Thorne et al.[61] at the same institution have expanded this study and found that other ω-amino carboxylic acids were less effective than EACA in achieving glycoform resolution, and the addition of the Tween 80 surfactant was necessary to obtain good recovery.

Purification and Process Monitoring

Because of the speed and high resolution of CZE separations as well as the small sample volumes required to yield information about complex protein samples, CE is increasingly being used to assess protein purity in multistep purification protocols in laboratory, pilot plant, and process scales. Similarly, it is being considered as a candidate for monitoring fermentation.

McNerney et al.[62] described separation of recombinant human growth hormone (rhGH) and its variants from very crude mixtures of *E. coli* cell extracts using CZE in a phosphate-deactivated capillary. The 18-h deactivation procedure included washing the capillary with 0.1 M nitric acid and 0.1 M sodium hydroxide to remove contaminants prior to conditioning with the run buffer (250 mM sodium phosphate + 1% propylene glycol). This method allowed separation of a variety of rhGH variants including deamidated and dideamidated rhGH, desPhe- and desPhePro rhGH, and 2-chain rhGH. In addition, the method could detect changes in fermentation conditions that affected rhGH production. Washing with 3 M guanidine HCl + 0.2 M sodium phosphate between runs was required to remove adsorbed contaminants. The phosphate-deactivated column provided superior resolution and reproducibility compared to bare fused-silica or PVA-coated capillaries.

Purification of murine antiheparin monoclonal antibody produced in cell culture was monitored by Malsch et al.[63] using a CZE method with a borate or boric acid buffer (pH 9) in an uncoated capillary.

Kundu et al.[64] used MEKC conditions to assess the purity of two recombinant proteins: a cytomegalovirus-CMP-KDO synthetase fusion protein expressed in *E. coli* and a hepatitis C viral protein expressed in CHO cells. Proteins were prepared in a 10-mM Tris–1% SDS buffer (pH 8.5) and analyzed in a 10-mM borate–100-mM SDS buffer (pH 9.5) in uncoated capillaries. The level of impurities, which varied with the method of protein production, agreed within ±5% with results obtained by densitometric scanning of SDS-PAGE gels of the same materials.

Lipoproteins

Serum lipoproteins have been analyzed by isotachophoresis, typically after staining with a lipophilic dye such as Sudan Black.[65,66] Apolipoproteins can be analyzed by CZE as reported by Tadey and Purdy.[67,68] They were able to resolve apoA-I, apoA-II, apoB-100, and apo-B48 from HDL and LDL preparations using uncoated capillaries with a 30-mM borate buffer (pH 9) containing SDS. Other detergents were less effective, although either SDS or cetyl trimethylammonium bromide provided good resolution of VLDL apolipoproteins with polyacrylamide-coated capillaries. Lehmann et al.[69] developed a method for direct analysis of apoA-I in serum using an uncoated 50-cm × 50-μm capillary and a proprietary buffer. Dilution of the serum sample in the buffer allowed direct injection and resolution of apoA-I from all other serum proteins. Correlation of apoA-I levels in patient sera correlated well with values from nephelometric determinations. Using a 50-mM borate buffer containing 3.5 mM SDS and 20% (v/v) acetonitrile, Cruzado et al.[70,71] compared CZE with reversed-phase HPLC for separation of apolipoproteins A-I and A-II. The HPLC method resolved apoA-I and apoA-II into three and two isoforms, respectively, whereas CZE could not. However, the apoA-I and apoA-II isoforms overlapped and could not be resolved by HPLC, thus preventing analysis of the two apolipoproteins in mixtures. Therefore, CZE was the preferred technique for quantitation of apoA-I and apoA-II in HDL. CE values for these two proteins determined in nondelipidated HDL fractions obtained from serum controls by density gradient centrifugation agreed well with immuno-based assay values.

CAPILLARY ISOELECTRIC FOCUSING

Background

Capillary isoelectric focusing (CIEF) combines the high resolving power of conventional gel IEF with the advantages of CE instrumentation. Just as in gel IEF, proteins are separated according to their pI in a pH gradient generally formed by carrier *ampholytes* (amphoteric electrolytes) when an electric potential is applied (Figure 9.5). The use of small-diameter capillaries allows the efficient dissipation of Joule heat and permits the application of high voltage for a rapid focusing of the protein zones. The resolving power of CIEF is usually higher than most protein analysis techniques, including other modes of CE. The introduction of CIEF expanded the use of IEF to include the analysis of peptides, amino acids, and other small organic zwitterions.

As in conventional IEF, the high resolving power of CIEF depends upon the focusing effect of the technique. At steady state, the ampholytes form a stable pH gradient within which proteins become focused at the positions where their net charges are zero, i.e., where pH = pI. Diffusion of a protein toward the anode (positive electrode) will result in the acquisition of positive charge, and the

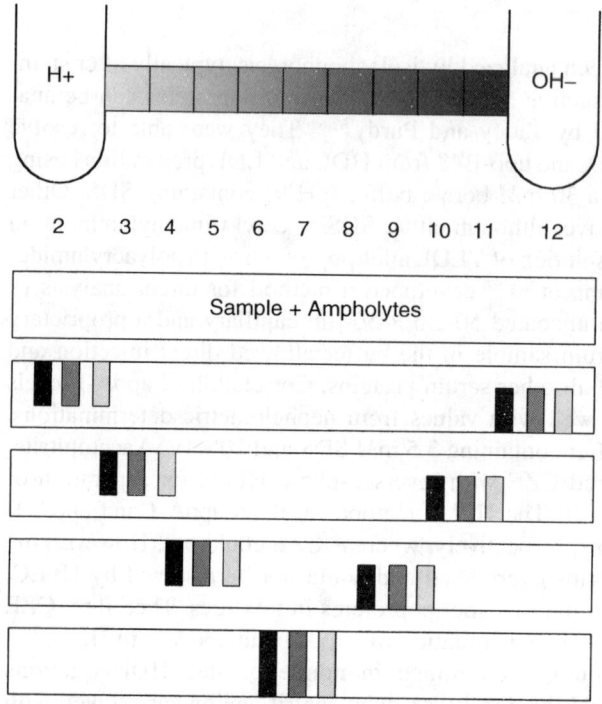

H+   OH−

2   3   4   5   6   7   8   9 10 11   12

Sample + Ampholytes

Figure 9.5 Generation of a pH gradient by ampholytes within a capillary flanked by an acid as anodic solution and base as cathodic solution. Ampholyte solutions are composed of high numbers of low-molecular weight amphoteric electrolytes (from which the name is derived) with slightly different pI values. Because ampholytes possess buffering capacity, they maintain a pH value in the specific area occupied by the different molecular species. The sample, which is also amphoteric, focuses in between ampholytes with higher and lower pI. To achieve resolution, there must be at least one ampholyte with a pI intermediate to the two sample components of interest.

molecule will return to the focused zone (attracted by the cathode). Similarly, diffusion toward the cathode (negative electrode) will result in acquisition of negative charge, causing back-migration to the pI zone. As long as the electric field is applied, electrophoretic migration counters the effects of diffusion. Because detection in CIEF is performed on-line, the electric field is maintained throughout the analysis, and resolution is usually very high. CIEF usually produces more complex patterns than conventional IEF because smaller peptides (and even aromatic amino acids) that do not stain well or diffuse out of the gel are also detected. CIEF offers resolution comparable to slab gel IEF but not as high as immobilized pH gradients.[72]

For all CE instruments that use on-line detection at a fixed point along the capillary, CIEF must include a means of transporting the focused zones past the detection point. This process, commonly referred to as *mobilization*, can occur as an independent stage or be combined with the focusing of the sample–ampholyte matrix components. The forces used to achieve mobilization are various, and in some instances they are applied in combinations of two or more simultaneously.

There are several protocols for performing CIEF that can generally be classified in two groups: two-step CIEF and single-step CIEF. Two-step CIEF is characterized by the performance of focusing and mobilization as two distinct stages. Mobilization can be achieved by ion addition or by applying a hydraulic force such as pressure, vacuum, or gravity. This method requires electroosmosis to be eliminated or reduced to a very low level. In single-step CIEF, focusing occurs while the nascent protein zones are being transported toward the detection point. The forces used to transport the focused zones are the same as those for multiple-step CIEF, but in this case EOF can also be used alone or in combination with any of the other forces.

Sample Preparation and Injection

Sample preparation for CIEF includes selection of the appropriate ampholyte composition, adjustment of sample salt levels, and dilution or concentration of the sample to the proper protein levels required for detection.

Excessive sample ionic strength due to the presence of salts (including buffer) or ionic detergents will interfere with the IEF process, greatly increase focusing times, and cause peak broadening during mobilization. Elevated current due to the presence of salt can increase the risk of precipitation as proteins become concentrated in focused zones. Dilution, dialysis, gel filtration, or ultrafiltration should be used to desalt samples with salt concentration of 50 mM or greater. The practical upper limit for ionic strength is 30 to 50 mM.

The ampholyte composition should be selected based upon the desired separation range. For separating complex samples containing proteins with widely different pI, or to estimate the pI of an unknown protein, a wide-range ampholyte blend such as pH 3 to 10 is appropriate. The final ampholyte concentration should be between 1 and 2% (w/v). In situations in which enhanced resolution of proteins with similar pI values is desired, the use of narrow-range ampholyte mixtures may be considered. Narrow-range ampholyte mixtures generating gradients spanning 1 to 3 pH units are available from several commercial sources. Improved results have been obtained by adding various ratios (10 to 80%) of narrow-range ampholytes to a "base" of broad-range ampholytes.

In order to detect proteins at the basic end of the gradient during cathodic mobilization, it is necessary that the pH gradient span only the effective length of the capillary, e.g., the distance from the capillary inlet to the detection point. In cases where the total capillary length is much greater than the effective length, many sample components may focus in the "blind" segment distal to the monitor point and be undetected during mobilization. A basic compound such as

N,N,N′,N′-tetramethylethylenediamine (TEMED) can be used to block the distal section of the capillary. As a rule of thumb, the ratio of TEMED concentration (% v/v) to ampholyte concentration should be approximately equal to the ratio of the "noneffective" (or blind) capillary length to total length.

The final protein concentration in the sample and ampholyte mixture will depend upon sensitivity requirements and the solubility of protein components under focusing conditions. As an approximation, a final concentration of 0.5 mg/ml per protein should provide adequate sensitivity and satisfactory focusing and mobilization performance. However, many proteins may still precipitate during focusing at this starting concentration because the final protein concentration in the focused zone may be as high as 200 mg/ml. Immunoglobulins, membrane proteins, and high-molecular weight or hydrophobic proteins generally have a higher risk of precipitation during CIEF. In such cases, the use of very dilute protein solutions may be required.

Prior to injection, the prepared sample should be centrifuged for 2 to 3 min in a microcentrifuge to remove any particulate material and to degas the solution. This practice is particularly important if the sample contains protein aggregates or other large particles, and when polymers are used to increase the viscosity of sample and ampholyte solution.

Focusing

Although there are several approaches to generating pH gradients, the most widely used to date is with the use of carrier ampholytes. Ampholytes are mixtures of a high number of synthetic chemical species that posses slightly different pIs. Carrier ampholytes are oligoamino acids and oligocarboxylic acids with different pI values.[73] The number of ampholytes per pH unit has been calculated to range between 50 and 1000. Besides having different pI values, carrier ampholytes must be good buffers and conductors at their pI so that they can carry the electric current while also maintaining a steady pH gradient.

During the performance of a CIEF analysis, the capillary is first filled with the sample and ampholyte mixture. The focusing step begins with the immersion of the capillary in the anolyte (dilute phosphoric acid) and catholyte (dilute sodium hydroxide) solutions followed by application of high voltage. Typically, the catholyte solution is 20 to 40 mM NaOH, and the anolyte is half the catholyte molarity, e.g., 10 to 20 mM phosphoric acid. It is important that the catholyte be prepared fresh because sodium hydroxide solutions will gradually take up carbon dioxide from the atmosphere.

For narrow-bore capillaries (e.g., 50 μm I.D.), field strengths of 300 to 900 V/cm or greater can be used. Our experience indicates that about 600 V/cm is optimal. Upon application of high voltage, the charged ampholytes migrate in the electric field to generate a pH gradient that is defined by the composition of the ampholyte mixture. A pH gradient develops with low-pH components toward the anode (+), and high-pH components toward the cathode (−). At the same time, protein components in the sample migrate until a steady state is reached, at which point each protein becomes focused in a narrow zone at its pI (Figure 9.5). Focusing is achieved rapidly, typically within a few minutes in short capillaries, and is accompanied by an exponential drop in current. Focusing is usually considered to be complete when the current has dropped to a level approximately 10% of its initial value for samples containing low salt and the rate of change approaches zero. It is generally not advisable to prolong focusing beyond this point because the risk of protein precipitation increases with time.

Two-Step CIEF

Because most commercial CE instruments use on-line detection at a fixed point along the capillary, CIEF must include a means of transporting the focused zones past the detection point. Mobilization has been regarded for the most part as a stage of little importance in the overall performance of the CIEF process, but now it has been shown that mobilization conditions can be manipulated to improve resolution and reproducibility. Three approaches have been used to mobilize focused zones. In *chemical mobilization* (ion addition), changing the chemical composition of the anolyte or catholyte causes a shift in the pH gradient, resulting in electrophoretic migration of focused zones past the detection point.[74,75] In *hydraulic mobilization*, focused zones are transported past the detection point by applying pressure[76,77] or vacuum[78] at one end of the capillary, or by volume height differential of the anolyte and catholyte levels (siphon or gravity).[79] In *electroosmotic mobilization*, focused zones are transported past the detection point by electroosmotic pumping.[80–82] Mobilization by EOF is used only in single-step CIEF.

Chemical Mobilization

At the completion of the focusing step, high voltage is turned off and the anolyte or catholyte is replaced by the mobilization reagent. High voltage is again applied to begin mobilization. As in focusing, field strengths of 300 to 900 V/cm can be used for mobilization, with optimum separations achieved in capillaries with small I.D. using a field strength of about 600 V/cm. The choice of anodic vs. cathodic mobilization and the composition of the mobilizing reagent depend upon the pIs of the protein analytes and the goals of separation. Because the majority of proteins have pIs between 5 and 9, cathodic mobilization (mobilization toward the cathode) is most often used. The most common chemical mobilization method is the addition of a neutral salt such as sodium chloride to the anolyte or catholyte; sodium serves as the nonproton cation in anodic mobilization and chloride functions as the nonhydroxyl anion in cathodic mobilization. A suggested, cathodic mobilization reagent is 80 mM NaCl in 40 mM NaOH. At the beginning of mobilization, current initially remains at the low value observed at the termination of focusing, but gradually begins to rise as the chloride ions enter the capillary. Later in mobilization, when chloride is present throughout the tube, a rapid rise in current signals the completion of mobilization. The electrical current at the end of mobilization using NaCl is much higher than the current observed at the beginning of focusing. When using NaCl for mobilization, set the current limit to a value (100 to 150 μA) that will ensure that excessive current heat does not damage the capillary coating.

Use of zwitterions is an alternative approach that provides more effective mobilization of protein zones across a wide pH gradient.[83] For example, cathodic mobilization with a low-pI zwitterion enables efficient mobilization of proteins with pIs ranging from 4.65 to 9.60. The proposed mechanism for zwitterion mobilization couples a pH shift at the proximal end of the tube with a displacement effect at the distal end as the zwitterion forms an expanding zone within the gradient at its pI. Effective zwitterion mobilization depends on the selection of the appropriate mobilization reagent.

Hydraulic Mobilization

Hydraulic mobilization utilizes positive pressure or negative pressure (vacuum) as the force that transports the focused protein zones toward the detection point. During hydraulic mobilization, it is necessary to apply an electric field across the capillary in order to maintain focused protein zones.[78] The main disadvantage of this type of mobilization is the parabolic shape of the hydrodynamic flow profile, which can decrease resolution. For this reason, only weak forces are used.

From an instrument perspective, the simplest hydraulic approach to transport focused zones to the detector is by gravity mobilization.[79] In this technique, focused proteins are transported toward the detection point using a difference in the levels of anolyte and catholyte contained in the reservoirs. The force generated by the liquid-height difference can be manipulated to be extremely small compared with pressure or vacuum. Flow velocity can also be modulated by changing the capillary dimensions or, in the case of large-bore capillaries, with internal diameters greater than 50 μm, by the addition of viscous polymers.

Single-Step CIEF

Single-step or dynamic CIEF is a variation of CIEF in which focusing occurs while sample proteins are simultaneously transported toward the detection point by EOF. Single-step CIEF was first developed as a means to perform CIEF in uncoated capillaries. Uncoated capillaries are inexpensive and have long lifetimes limited only by column plugging. However, sample recovery can be a problem, and capillaries with wall modifications (such as C8) or additives are more commonly used to reduce, but not eliminate, EOF. An advantage of single-step CIEF is simplification of the protein pattern because focusing peaks, generated by proteins migrating past the detection point during the focusing step, are eliminated. However, because the capillary may be only partially filled with sample and ampholyte solution, some resolution and sensitivity are sacrificed.

Approaches to single-step CIEF in the presence of EOF include partial and full capillary injection. In the partial injection method, the sample and ampholyte mixture is introduced as a plug at the inlet of the capillary prefilled with catholyte. Successful application of partial capillary injection depends upon optimization of the polymer concentration (which modulates EOF), ampholyte concentration, and sample load to minimize protein adsorption and modulate EOF level so that focusing approaches completion before the detection point is reached.

The second single-step CIEF method consisting of full capillary injection was described by Mazzeo and Krull.[80,82] In initial studies using uncoated capillaries, methylcellulose was added to modulate EOF and TEMED was used to block the detector-distal capillary segment. This approach was successful only for neutral and basic proteins due to variations in the rate of EOF during the separation. As the separation progressed, the drop in average pH due to mobilization of the basic segment of the pH gradient into the catholyte resulted in diminished EOF. This, in turn, caused peak broadening and poor resolution for acidic proteins. Improved mobilization of acidic proteins was achieved using commercial C8-coated capillaries in which EOF varied less with pH.[84] However, pH-dependent variation of EOF was still significant enough that plots of pI vs. migration time were not linear over broad pH ranges.[85] Use of multiple internal standards was recommended for accurate pI determination with this method.

Optimal separation of proteins spanning the whole pH range is difficult with EOF-driven CIEF. As for all variations of IEF, salt concentration (higher than 10 mM, in this case) greatly diminishes resolution. The concentration of NaOH (catholyte) has a significant effect on migration times of the protein zones, mainly by affecting the rate of electroosmosis. Longer analysis times were observed at higher concentrations of NaOH with an improvement in resolution. On the other hand, higher concentrations of anolyte (phosphoric acid) shortened the analysis time and diminished resolution. Migration was also affected by the concentration of the ampholytes used (1, 2.5, and 5%), with slower mobilization at lower concentrations (1%). The initial length of capillary occupied by the sample affects migration times and resolution, with longer sample zones providing better resolution at the expense of analysis time.

Capillary Selection

To obtain good resolution and reproducibility when performing CIEF with chemical mobilization, it is essential to reduce EOF to a very low level. In the presence of significant levels of EOF, stable focused zones are not maintained. This results in band broadening and, in some instances, multiple peaks caused by incomplete fusion of the nascent protein zones focusing from both capillary ends. Therefore, the use of coated capillaries is necessary for optimal use of chemical mobilization. A viscous polymeric coating is recommended for greatest reduction in EOF. In addition, the use of neutral, hydrophilic coating materials reduces protein–wall interactions.

The low level of EOF in coated capillaries permits separations to be carried out with very short effective capillary lengths. Earlier work using chemical mobilization was performed using capillaries as short as 11 cm with internal diameters up to 200 μm.[76] More recently, 12- to 25-cm capillaries with internal diameters of 25 or 50 μm have been used.[83] Theoretically, resolution in CIEF should be independent of capillary length because the number of ampholyte species remains constant with only the amount changed. In practice, however, resolution is diminished with very short capillaries, small sample injections, and very dilute sample and ampholytes, particularly in single-step CIEF. Because analysis time increases with capillary length, problems associated with protein precipitation are more severe in longer capillaries.

The length of capillaries used for single-step CIEF is very important, especially when EOF is the driving force. The capillary length must be optimized according to the size of the injection and the velocity of EOF or flow due to hydraulic forces so that the sample will not reach the detection point before it has finished focusing.

Detection

UV-Vis Absorption

Most applications published to date employ on-line detection of mobilized proteins by absorption in the ultraviolet or visible spectrum at a fixed point along the capillary. The strong absorbance of the ampholytes at wavelengths below 240 nm makes detection of proteins in the low-UV region impractical. Therefore, 280 nm is generally used for absorbance detection in CIEF. This results in as much as a 50-fold loss in detection signal relative to the detection at 200 nm, but the high protein concentrations in focused peaks more than compensate for the loss of sensitivity imposed by detection at 280 nm. Because ampholytes may still be detected at 280 nm, care should be exercised when analyzing dilute or low-absorbance proteins.[86] In some instances, proteins possess chromophores that can be detected in the visible range of the spectrum, e.g., hemoglobin and cytochromes.

Concentration Gradient Detection

The use of concentration gradient detectors has been extensively reported.[87–96] The system incorporates a capillary mounted in a holder that aligns the column to a HeNe laser beam. A positioning sensor located at the exit side of the laser beam detects deflections generated by the passage of substances with a refractive index different than that of the background buffer. The main advantage of this detector is its universality in detecting sample components. Although ampholytes may produce signals during the mobilization step, the derivative nature of the detector enables recognition of the sharp bands generated by the protein zones against the background of the broader zones produced by the ampholytes. The detection system can also be built to scan along the capillary, performing detection of focused protein zones without mobilization. Optimization produces fast analysis times (2 min) and detection limits in the 1- to 5-mg/ml range. Use of capillary arrays greatly improves throughput. An important application for concentration gradient detection involves peptides, many of which do not contain aromatic amino acids required for UV detection at 280 nm. Using this system, peptides produced by the tryptic digestion of bovine and chicken cytochrome C were analyzed.

Laser-Induced Fluorescence (LIF)

LIF is a highly sensitive mode of detection, but lasers that emit at a visible wavelength often require derivatization of the sample prior to analysis. Because chemical modification of the sample molecules can change the pI, often producing multiple peaks, derivatization is not widely used. Instead, this problem is solved by using a laser that emits in the UV range. LIF detection of tagged antibodies directed against the protein of interest can also be used.[97]

CIEF and Mass Spectrometry

The separation power of CIEF often generates a high number of peaks even when relatively pure samples are analyzed. As already discussed, one of the advantages of CIEF is its potential micropreparative capabilities. Capillary IEF allows the collection of fractions that can be further analyzed by other methods. Some of the most widely used characterization tools include MS, peptide mapping, and amino acid analysis.

Foret et al.[98] collected fractions of model proteins and variants of human hemoglobins after fractionation by CIEF, and then analyzed them by matrix-assisted laser desorption–time-of-flight–mass spectrometry (MALDI-TOF-MS). As the authors point out, MS is an orthogonal method to CIEF because it separates according to molecular mass.

Optimizing CIEF Analysis

Resolution in CIEF

As previously mentioned, resolution in CIEF strongly depends on the ampholyte composition. The estimated maximum resolving power of IEF is 0.02 pH units when carrier ampholytes are used to create the pH gradient.[99] The Law of Monotony[99] formulated by Svensson in 1967 states that a natural pH gradient increases continually and monotonically from the anode to the cathode; that the steady state does not allow for reversal of pH at any position along the gradient; and that two ampholytes (in stationary electrolysis) cannot be completely separated from each other unless the system contains a third ampholyte of intermediate pH (or pI). The latter explains why better resolution is obtained when mixing ampholytes from different vendors and production batches; as the number of ampholytes species increases, the chance that one or more ampholytes have intermediate pI relative to those of the sample components also increases.

Additives and Protein Precipitation

A major problem in CIEF is the precipitation of proteins as the focusing step concentrates the sample components. Precipitation may be due possibly to protein denaturation, hydrophobic interactions, and electrostatic interactions resulting from salt removal. Precipitation in CIEF is manifested by current fluctuation and loss, by variations in peak heights or migration patterns, and by spikes in the electropherogram generated as protein aggregate particulates transit to the detection point. This results in poor pattern reproducibility, variable migration times, variable peak areas affecting quantitation, capillary clogging, slow mobilization, and other undesirable effects. Solubilizing agents are commonly used in CIEF to prevent precipitation. Other additives include hydrophilic polymers, usually used for fluid stabilization, to reduce EOF, and to increase viscosity during hydraulic mobilization.

Denaturing CIEF

Poor protein solubility under CIEF conditions has limited the number of applications developed for this technique. Solubilization in some cases may be a problem even before the CIEF analysis is initiated.[100] Some proteins, e.g., membrane proteins, are difficult to analyze in typical electrophoresis buffers. These polypeptides, however, can be rendered soluble in the presence of additives such as SDS. Unfortunately, SDS or similar ionic detergents cannot be used in CIEF because they eliminate the amphoteric properties of proteins. An additive that can be used to increase protein solubility is urea. Unfortunately, hydrophobic interactions and hydrogen bonds play a major role in protein tertiary structure, and because urea disrupts these forces, protein denaturation occurs. When the protein is denatured, all hydrophobic residues are exposed, increasing the possibility of hydrophobic interactions and aggregation. For this reason, detergents are used in combination with urea. Not all detergents are suitable for denaturing CIEF. Some detergents used with success include Tritons (reduced form), Nonidet P-40, CHAPS, octyl glucoside, and lauryl maltoside. Typical concentrations range from 0.1 to 5%.

The use of urea must be approached with caution, because urea solutions often contain ammonium cyanate, the concentration of which increases with temperature and pH. This contaminant can react with the amino group of lysines and the amino terminus of the polypeptide chain, thus leading to artifact peaks. This effect is minimized by the presence of ampholytes, whose primary amines are cyanate scavengers, and by deionizing the urea solution with a mixed-bed resin prior to adding the ampholytes and detergent.

Other CIEF Parameters

Applied Voltage: Theoretically, the best resolution is obtained at high voltages. In practice, variation of the electric field intensity under typical analysis conditions for CIEF (300 to 1000 V/cm) is a parameter of relatively minor importance in optimizing resolution. Field strength should be kept low enough to avoid excessive Joule heating, particularly at the beginning of focusing. If reduction in analysis time is desired, high field strengths will shorten focusing time but will effect mobilization time only when chemical mobilization or single-step CIEF in the presence of EOF is used.

Capillary Temperature: Protein conformation and solubility are affected by the temperature of the solution. As described above, focused proteins may tend to precipitate during the CIEF process as they become highly concentrated at their pI with low concentration of salt ions. The temperature of the capillary can be manipulated to increase their solubility, but it should not be too high as protein denaturation may occur. Higher temperature is used mainly in combination with other solubilizing agents, and temperature may be used to increase the solubility of additives such as urea rather than the solubility of the proteins.

Because pI, mobility, and viscosity are all affected by temperature, the use of internal standards is recommended. However, the use of synthetic pI standards to estimate pIs should be approached with caution because the pIs of the protein sample and standards may not be affected equally by temperature. Temperature has a direct effect on viscosity and therefore all effects of viscosity on the CIEF process apply as temperature changes.

The Use of Internal Standards: Important characteristics of internal standards include high purity, stability, high absorption at the detection wavelength, nonreactivity with sample components and ampholytes, and availability of species with known pI values spanning the pH range of interest. Protein standards are widely used in slab gel IEF, and they are available from multiple commercial sources. However, they are only available as premixed solutions of a fairly high number of proteins, and they are intended to be applied in a single lane of a slab gel. For CIEF, it is preferable to combine the sample and standard, but the complexity of the resulting pattern often makes identification of the compounds of interest very difficult, if not impossible. Single-protein standards usually lack the necessary purity and appear as several bands that complicate the separation pattern.

A family of substituted aromatic aminophenol compounds has been synthesized,[101] and they possess all of the desirable characteristics of internal standards, including very high UV absorption at 280 nm. They can be introduced into the capillary as a secondary injection that occupies only a small portion of the capillary. By not premixing the sample with the standards, the unused portion of the uncontaminated sample can be recovered. Furthermore, the standards are small molecules that can be removed easily by dialysis (an important factor to consider when collecting fractions). Figure 9.6 displays a CIEF analysis of recombinant human monoclonal antibody rhuMAB against HER2 flanked by synthetic 7.9, 8.4, and 10.1 pI markers. Table 9.2 shows the reproducibility (% RSD) obtained for two hemoglobin variants by analyzing the migration times and using the pI standards to calibrate the protein zones.[102,103] The percent RSD values reported for pI were extremely good (0.061%) as compared with migration times percent

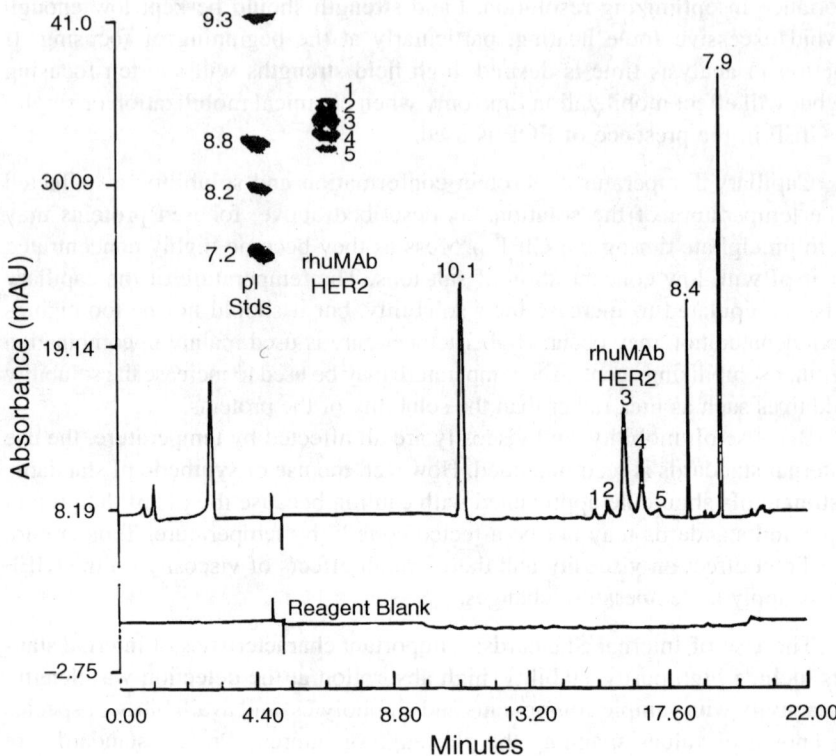

Figure 9.6 Capillary IEF and slab gel IEF of antibodies of rhuMAbHER2. There was good correlation between the number of bands obtained by the two techniques. Notice the excellent resolution of the sample components achieved by CIEF. In this application, the sample was bracketed by synthetic pI markers of known pI that are not only extremely useful in the determination of pI, but can also be used for correction of migration time variations. During methods development, the pI standards can be prepared in ampholytes at a higher concentration (e.g., 10X) and introduced as a second injection after the capillary is first filled with sample prepared in ampholytes. Using this approach, the combination of pI markers can be optimized with a minimal usage of sample (see Reference 112).

RSD (1.1%). Software packages that identify the internal standards, plot their migration time as a standard curve and automatically report the pI of "unknowns," make the CIEF process simpler, more powerful, and reliable.

CIEF as a Micropreparative Technique

Ever since CE was introduced, the desire to collect fractions of pure sample components has existed. The extremely small quantities injected into the capillary in CZE has limited the number of applications in which sufficient amount can

Table 9.2 Migration Time and Isoelectric Point Reproducibility for Human Hemoglobins A and S Analyzed by CIEF

Migration Time Reproducibility			Isoelectric Point Reproducibility		
Run	Hb A	Hb S	Run	Hb A	Hb S
1	19.16	18.77	1	7.39	7.22
2	19.01	18.63	2	7.38	7.23
3	18.88	18.50	3	7.38	7.22
4	18.74	18.36	4	7.38	7.22
5	18.63	18.26	5	7.38	7.22
Average	18.88	18.50	Average	7.38	7.22
Std. Dev.	0.21	0.20	Std. Dev.	0.004	0.004
% RSD	1.11	1.10	% RSD	0.06	0.06

Source: From T. Wehr, R. Rodriguez-Diaz, and M. Zhu, Chromatographic Science Series, Vol. 80 (1999). With permission.

be recovered. CIEF not only utilizes the whole length of the capillary for injection, but protein zones are highly concentrated during the analysis. Thus, CIEF has potential as a micropreparative technique, especially for enzymes, because only very small quantities of these polypeptides are required for enzymatic reactions. It is noteworthy that denaturation may occur when collecting fractions into NaOH (or any other extreme pH solution). Buffered catholyte should be used to avoid protein denaturation.

Applications

A variety of CIEF applications have been published, including the analysis of human transferrin isoforms,[104] recombinant proteins (e.g., human recombinant tissue plasminogen activator,[105] human growth hormone,[106] γ-globulins,[106] and hemoglobin variants.[107,108] As indicated by these reports, CIEF analyses are used to characterize proteins, as well as to determine their purity. It has been suggested that conventional IEF in gels can distinguish conformational states of proteins,[109] although the same has not yet been reported for CIEF. CIEF is a powerful tool for the detection of protein modifications, such as deamidation, deletions, insertions, proteolytic clips, N- or C-terminal modifications, and glycosylation.[73]

Immunoglobulins

Immunoglobulins in the form of monoclonal antibodies are manufactured commercially for therapeutic and diagnostic uses. Major areas of consideration in these applications are quality control and bioactivity. Separation of immunoglobulins has proven to be a challenge even for well-established techniques such as HPLC. Difficulties arise due to the large size of antibodies and their surface properties, which increase their tendency to interact with proteins and matrix.

Monoclonal antibodies have been shown to possess microheterogeneity due to posttranslation modifications such as glycosylation. Gel IEF is used routinely to analyze different batches of antibodies, but this type of analysis presents several drawbacks already discussed. CIEF can be applied to the analysis of antibodies with special consideration to maintaining antibody solubility. The use of additives, short focusing time, low sample concentration, and other precautions mentioned in the subsection titled "Optimizing CIEF Analysis" are a major part of method development to achieve reproducible high-resolution separations. CIEF of monoclonal antibodies with pIs near neutral pH was carried out in single-step mode in the presence of EOF and using protein markers as internal standards.[110]

CIEF was also used to follow the production of recombinant antithrombin III (r-AT III) in cultures of hamster kidney cells.[111] r-AT III inhibits serine proteases such as blood factors (IXa, Xa, and XIa) and thrombin. Interference by the media from which the samples were collected posed some difficulties because some of the media components have similar characteristics to those of the compounds of interest. CIEF was used to determine the pIs of the separated components after sample purification by HPLC. Three major peaks showed pIs of 4.7, 4.75, and 4.85, and three minor peaks had pIs of 5.0, 5.1, and 5.3. These data closely resembled the data already published for serum AT III based on conventional IEF.

The feasibility of using CIEF for analysis of monoclonal antibodies in a quality control environment was demonstrated for recombinant humanized monoclonal antibody HER2 (rhuMAbHER2) by Hunt et al.[112] This protein is present at increased levels in certain breast cancers. Besides primary structure heterogeneity present in the 214-residue light chains and 449- or 450-residue heavy chains, rhuMAbHER2 can exhibit charge differences due to deamidation or C-terminal clipping. Resolution of the five observed components was optimized by mixing Pharmalyte 8–10.5, Bio-Lyte 3–10, and Bio-Lyte 7–9 ampholytes in an 8:1:1 ratio. Figure 9.6 shows a very good correlation of the IEF and the CIEF pattern obtained for rhuMAbHER2. The pIs of the major peaks were determined through the use of internal standards, and the values obtained correlated well with the values obtained from gel IEF. The method was capable of revealing differences due to storage at 5 and 37°C, and the increased acidic peaks observed were consistent with protein deamidation. Intra-assay reproducibility ranged from 0.7 to 0.9% RSD for migration time, 0.8 to 3% RSD for peak area, and from 1 to 3.7% for area percent. Interassay reproducibility for migration time varied from 0.4 to 0.6% RSD, 1.2 to 3.2% for peak area, and 1.1 to 4.2% for area percent. All analyses were performed in the same coated capillary.

Due to the generally low concentration of contaminants, an important parameter for QC labs is the limit of detection. For rhuMAbHER2 this limit was estimated to be 2 ppm. Care should be exercised when analyzing low-concentration samples because ampholytes may show residual absorption even at 280 nm.[86] This problem can be reduced by decreasing the ampholyte concentration to 0.5% w/v.

Glycoform Analysis

A glycoprotein may vary in the location, length, and composition of sugar moieties attached to the polypeptide chain. The saccharide component of these glycoforms may play important roles in cell recognition, protein function, stability, solubility, and immunogenicity. In the development and manufacture of recombinant protein therapeutics, the distribution of glycoforms can therefore determine the efficacy and stability of the product. An understanding of the sugar content of recombinant proteins is particularly important because the glycosylation pattern is defined by an organism other than the end user. Introduction of carbohydrate groups can produce subtle changes in the protein pI, so IEF is a standard method for characterization of glycoforms; CIEF provides an automated quantitative method for glycoform analysis. Applications of glycoform analysis include determination of hemoglobin A1c to monitor diabetes mellitus and determination of elevated transferrin glycosylation as an indicator of alcoholism and pregnancy.

An example of two-step CIEF applied to the analysis of glycoforms is the fractionation of human recombinant tissue plasminogen activator (rtPA).[105] This activator is a protein that degrades blood clots, and its recombinant form is produced for the treatment of myocardial infarction. This 59-kDa glycoprotein possesses three N-glycosylation sites. Type I rtPA is glycosylated at all three sites (residues 117, 184, and 448), whereas type II rtPA is glycosylated at two sites (residues 117 and 448). Although rtPA was purified extensively to yield high purity of the polypeptide, in some instances up to 20 peaks were observed during CIEF. Treatment of rtPA with neuraminidase, an enzyme that removes sialic acid residues, has greatly simplified the pattern and suggests that heterogeneity is due to the variation of sialylation. CIEF performance was suitable for validation of the technique as a routine test.[113]

CIEF analysis of rtPA in the presence of urea was also carried out in an uncoated capillary using pressure mobilization.[114] The final urea concentration used was 4 M, and EOF was reduced by adding polymers to the reagents and sample (0.4% hydroxypropylmethyl cellulose produced better results than polyethylene glycol). A one-step CIEF method described by Moorhouse et al.[115] for the analysis of rtPA produced a constant residual EOF in a neutral capillary. The sample was prepared by dilution to 125 to 250 µg of protein per milliliter in 3% ampholytes 3 to 10 and 5 to 8 (1:1) containing 7.5% TEMED and 4 M urea. Results obtained by CIEF correlated well with those generated by IEF, and the analysis was completed in less than 10 min.

Protein Concentration and Dynamics of Interaction

It is well known that a very important feature of many biological systems is specific recognition at the molecular level. Antibodies as a group are widely used for molecular recognition, e.g., affinity assays. This feature can be used by labeling an anti-human growth hormone antibody fraction with a fluorescent tag (tetramethylrhodamine-iodoacetamide) to detect the presence of growth hormone to a level of 0.1 ng/ml. An application of CIEF that exploits the concentration effect of the technique with the advantages of affinity interaction and the detection power of laser-induced fluorescence has been developed.[116]

SIEVING SEPARATIONS

The performance of electrophoresis in narrow-bore capillaries obviated most of the functions of gels in electrophoresis, e.g., elimination of convection through rapid dissipation of Joule heat and reduced diffusion through short analysis time. However, another important feature of gels is their capability to actively partic-ipate in the separation process by providing a sieving media that differentially affects the migration velocity of sample components according to molecular size. Macromolecules such as nucleic acids and SDS–protein complexes exhibit no significant mobility differences during free-zone electrophoresis and require the presence of an interactive sieving separation matrix to achieve resolution.

Size-based analysis of SDS–protein complexes in polyacrylamide gels (SDS-PAGE) is the most common type of slab gel electrophoresis for the char-acterization of polypeptides, and SDS-PAGE is one of the most commonly used methods for the determination of protein molecular masses.[117] The uses for size-based techniques include purity determination, molecular size estimation, and identification of posttranslational modifications.[118,119] Some native protein studies also benefit from size-based separation, e.g., detection of physically interacting oligomers.

Due to the importance and broad spectrum of sieving applications, many groups[120–122] have attempted to adapt gels to the capillary format. Unfortunately, there are several technical difficulties that have limited the use of gel-filled capillaries, including short lifetimes due to bubble formation or contamination after repeated runs. For polypeptides, gel-filled capillaries have poor UV absor-bance detection sensitivity because only higher wavelengths can be used. The main advantage of gel-filled capillaries is higher resolution.

An alternative to gel-filled capillaries is the use of polymer solutions.[123] In this chapter, gel-filled capillaries are those containing matrices that are not replaceable. Replaceable matrices are referred to as "polymer solutions." Gels are typically polymerized *in situ*, whereas polymer solutions are pumped into the capillary and are usually replaced between each injection. The use of replaceable gels has been variously termed *entangled polymer CE*, *nongel sieving*, and *dynamic sieving*. The polymer solutions have been referred to as *replaceable gels* and *physical gels*. The advantage of polymer solution is that the whole content of the capillary is replaced in between runs, thus eliminating aggregates that move extremely slowly, do not penetrate the gel, accumulate in the gel matrix, and cause deteriorating results. This practice allows for the analysis of "dirty" samples during process development.

Many polymers have been tested for the analysis of SDS–protein complexes, and the main difference observed is in the resolution (or resolution range) from polymer to polymer and among molecular weight distributions of the same polymer that can be used to improve the resolution. Interestingly, studies on resolution show that it can increase or decrease with increasing temperatures depending on the type of sieving polymer used. Although we occasionally prepare buffers with dissolved polymers for research purposes, commercially available solutions are preferred for routine applications and testing.

Low-melting agarose has been proposed as a sieving medium for CE.[124] This material can be introduced into the capillary by pressure at a temperature above its melting point of 25.6°C and then induced to form a gel by dropping the capillary temperature below the melting point. Following separation, the capillary temperature can be raised and the gel extruded from the capillary by pressure, and then replaced with fresh uncontaminated agarose prior to the next injection. Unfortunately, the pore size of agarose gels is too large to provide sufficient sieving for most proteins.

Analysis of Native Proteins

Native proteins consisting of varying numbers of identical subunits or protein conjugates made up of monomers joined by cross-linking agents may be difficult to resolve by free-zone electrophoresis or CIEF, but they can be easily separated by sizing methods. In some instances, it is desirable to maintain the tertiary and quaternary structures of proteins, which are lost when polypeptides are denatured. In these cases, sieving of native proteins can be performed using either gel-filled capillaries[122,124] or polymer solutions.[125] Examples of size-based analysis of native proteins by CE include bovine serum albumin (BSA), rat liver proteins,[125] and human serum albumin.[124]

Analysis of SDS–Protein Complexes

SDS binds in an approximately stoichiometric fashion to polypeptide chains, with roughly one SDS molecule bound per two amino acid residues[126] or an average of 1.4 g of SDS per gram of protein. Assuming that the contributions of the charged amino acid side chains are low relative to that of the surfactant phosphate groups, the SDS–protein complexes possess the same charge-to-mass ratio independent of polypeptide chain length. SDS complexes of proteins with molecular masses greater than 10 kDa exhibit identical mobility in free solution,[127] although proteins that are not fully complexed with SDS may exhibit variable mobility and may be resolved into multiple species.[128] For smaller proteins, the intrinsic charge of the polypeptide has a more pronounced effect, introducing larger errors when estimating the molecular weight. Most polymer solutions described to date do not provide enough resolution when separating proteins smaller than 10 kDa. In theory, these molecules can be easily resolved by gel-filled capillaries with high-percentage polyacrylamide.

An important advantage of using SDS to denature polypeptides is the solubilizing power of the detergent. This property allows for the study of proteins (e.g., membrane proteins) that easily precipitate under most other conditions.

Size-based analysis by CE provides similar information and comparable limits of detection to analysis by SDS-PAGE with Coomassie blue staining.[120,129] The performance of both electrophoretic techniques for the analysis of polypeptides is far superior to size exclusion chromatography. Figure 9.7 shows the separation of SDS-complexed recombinant protein standards by CE.

Protein separations by CE are often negatively affected by sample–capillary wall interactions and require additives or surface modifications to eliminate undesirable interactions (see "Minimization of Nonspecific Protein–Wall Interactions"). This problem is minimal or nonexistent for SDS–protein complexes. Untreated silica possesses negative charge at pH above 2 to 3, and the SDS–protein complex is also anionic at almost any pH. Electrostatic repulsion between the silica wall and SDS–protein complexes eliminates protein adsorption. Nevertheless, in cases where EOF limits resolution or introduces migration time variations, internally coated capillaries can be used.[130] Capillaries coated with linear polyacrylamide through C-Si bonds were found to be more stable than capillaries coated through siloxane groups.[131–133] Uncoated capillaries have been used with a linear polyacrylamide-sieving matrix that provides sufficient viscosity (>100 cP) to prevent extrusion of the sieving medium from the column by EOF.[117] In some instances, the sieving matrix acts as a surface coating.[134]

Theory

The analysis of SDS–protein complexes is based upon all complexes having the same mobility in free solution and a sieving media providing resolution based on size. Nevertheless, the theory of resolution is important to explore anomalous behavior of sample components, or when the user desires to develop his or her own separation system. Because resolution of sample components is based on molecular size alone, the separation mechanisms include the behavior of a polyionic molecule and the restrictions that the gel imposes on larger analytes. Size-dependent separations can be used to estimate the molecular weight of analytes. Molecular mass can be calculated once the mobility differences of the sample components are cancelled. These mobility variations can be eliminated by two approaches. One is mathematical, and it is performed after mobility measurements in gels of various concentrations (see "Molecular Weight Determination and Ferguson Analysis"). The other involves the binding of a charged ligand that masks the native charge of the protein. The most common practice for the latter approach is the binding of SDS by proteins at a constant ratio, rendering the mass-to-charge relationship close to a constant for most proteins 10 kDa to above 200 kDa. Although the presence of a sieving media suggests size-based separations only, electrophoretic mobilities are affected by differences in shape, size, or net charge.

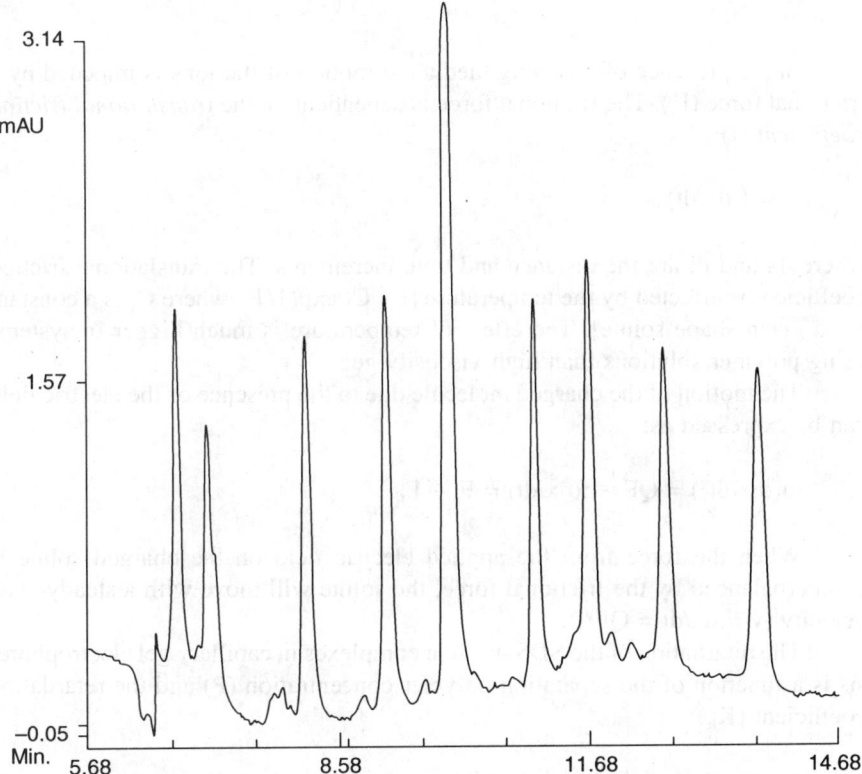

Figure 9.7 Capillary SDS of prestained recombinant protein standards (Bio-Rad Laboratories). The front dye and buffer ions were removed by buffer exchange into CE-SDS sample buffer using a BioSpin 6 (Bio-Rad Laboratories) as suggested by the manufacturer. Due to the high resolving power of CE-SDS as well as detection by UV-absorption, blends of naturally occurring proteins generate a number of peaks that can make peak identity difficult. Recombinant standards, on the other hand, produce sharp peaks and clean electropherograms. Molecular weight of sample components: (1) 10 kDa, (2) 15 kDa, (3) 25 kDa, (4) 37 kDa, (5) 50 kDa (this peak is present at a higher concentration for easy visual reference), (6) 75 kDa, (7) 100 kDa, (8) 150 kDa, and (9) 250 kDa. The proteins were resolved using the Bio-Rad CE-SDS analysis kit and a 50-μm × 30-cm uncoated capillary thermostated at 20°C. The sample was injected by applying 5 psi of pressure for 6 sec, and detection was performed at 220 nm. The polarity was set with the negative electrode at the detector end of the capillary and the electric field as 15 kV constant voltage.

The following explanation of the behavior of analytes during CE-SDS in the presence of sieving media is found in a monograph by Guttman:[135] the electric force (F_e) that a particle experiences when placed in an electric field depends on the net charge (Q) of the particle and on the intensity of the electric field (E):

$$F_e = QE$$

In the presence of a sieving media the motion of the ions is impeded by a frictional force (F_f). The frictional force is dependent on the *translational friction coefficient* (f):

$$F_f = f(dx/dt)$$

where dx and dt are the distance and time increments. The translational friction coefficient is affected by the temperature [$f = C_1 \exp(1/T)$, where C_1 is a constant for a given shape solute]. The effect of temperature is much higher in systems using polymer solutions than high-viscosity gels.

The motion of the charged molecule due to the presence of the electric field can be expressed as:

$$m(d^2x/dt^2) = QE - f(dx/dt) = F_e - F_f$$

When the force from the applied electric field on the charged solute is counterbalanced by the frictional force, the solute will move with a steady-state velocity (v = dx/dt = QE/f).

The retardation of the SDS–protein complexes in capillary gel electrophoresis is a function of the separation polymer concentration (P) and the retardation coefficient (K_R):

$$\mu = \mu_0 \exp(-K_R P)$$

where μ is the apparent electrophoretic mobility and μ_0 is the free solution mobility of the analyte.

According to these equations, for a given separation system, the main parameters involved in the separation of SDS–protein complexes are the electric force, the frictional force, and the retardation coefficient. These parameters are in turn affected by the strength of the electric field, molecular charge, analyte shape and size, polymer concentration, and temperature.

Sieving Mechanisms

When the mobility of a molecule is plotted against the gel concentration (Ferguson plots), three regimes are distinguishable. In the Ogston regime the average pore size of the matrix is similar to that of the hydrodynamic radius of the migrating analyte. It is in this region that true sieving occurs, and therefore the retardation coefficient (K_R) is proportional to the molecular mass of the analyte [$\mu \sim \exp(M_r)$]. Under these conditions the logarithm of mobility for a given analyte is a linear function of the molecular mass, and the Ferguson plots are linear. Typically, the plots intersect each other at zero gel concentration.

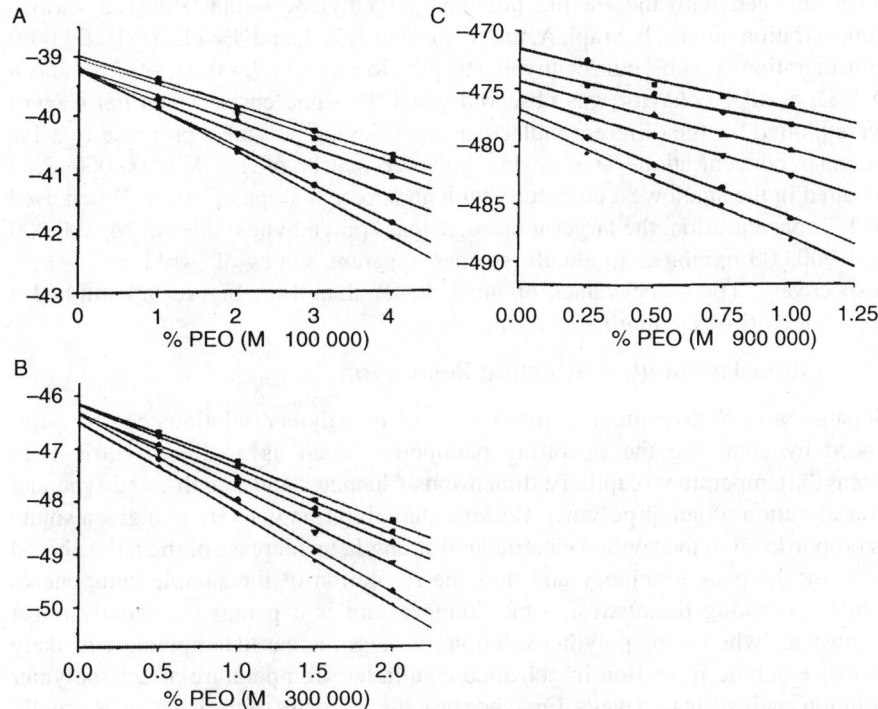

Figure 9.8 Ferguson plots for polyethylene oxide used as sieving polymer. The graphs represent plots of the polymer concentration and the log of electrophoretic mobility for six different samples. Graphs A, B, and C represent polymers with MW of 100, 300, and 900 kDa, respectively (From A. Guttman, *Electrophoresis, 16*: 611 (1995). With permission.)

The Ogston theory assumes that the migrating particles behave as unperturbed spherical objects and that the gel or sieving matrix has similar pore sizes as the analyte. Molecules with large Stokes radii such as the flexible chain of biopolymer molecules (DNA and SDS–protein complexes) can still migrate through pores much smaller than their size would permit. This can be explained by the reptation model (reptation-without-stretching regime), which describes the migration of the polyelectrolyte as a "head first, snakelike" motion through the pores of the sieving media. This model suggests an inverse relationship between the mobility and the molecular mass of the analyte.

When high electric fields are applied, molecular migration is explained by the reptation-with-stretching model. A common way to recognize the Ogston theory, the reptation, or the reptation-with-stretching regimes, is to plot the log solute mobility vs. the log solute's molecular mass curves. Pure reptation is suggested when the values of these slopes are close to −1. Figure 9.8 shows the plots obtained with the sieving polymer polyethylene oxide (PEO) at various concentration ranges. In graph A, the six plots at 1, 2, 3, and 4% PEO (M_r 100,000) concentrations exhibit consecutively steeper slopes (−0.07, −0.01, −0.014, and − 0.018). Similar behavior was observed when the same curves (data not shown) were plotted for the different molecular mass sieving polymers prepared to a 1% polymer concentration. The sieving polymer matrix of 1% M_r 100,000 PEO resulted in the shallowest curvature, with an apparent slope of −0.07. When used in 1% concentration, the larger molecular-mass polyethyne oxides of M_r 300,000 and 900,000 exhibited gradually steeper apparent slopes of −0.11 and −0.15, respectively. The slope values of much lower than 1 in Figure 9.8 implied a reptation-with-stretching mechanism.

Other Parameters Affecting Resolution

Separation of SDS–protein complexes in gel or polymer solutions can be influenced by changing the operating parameters, such as applied electric field strength, temperature, capillary dimensions (diameter and length), and type and concentration of gel or polymer. Because the migration velocity of a given solute is proportional to the applied electric field strength, an increase of the latter should increase the peak efficiency and thus the resolution of the sample components while shortening the analysis time. Temperature is a parameter varied almost exclusively when using polymer solutions because increased temperature is likely to cause bubble formation in gel-filled capillaries. Temperature affects polymer solution analysis in two ways. First, because the viscosity of the solution is usually lower at higher temperatures, an increase in temperature decreases the friction of the migrating SDS–protein complexes. Second, the temperature changes can influence the structure of the dissolved polymer, resulting in differences in pore structure. Experimental data suggest that oriented arrangements (channel-like structures) in concentrated polymer solutions play an important role in the sieving of SDS–proteins complexes. This means that any changes that result in the generation of these organized structures is likely to enhance separation efficiency.

As for CZE, sieving separation of SDS–protein complexes improves at the expense of analysis time when using longer capillaries. According to theoretical formulations, a higher applied electric field results in improved separation efficiency, and the capillary should be short enough to allow a high number of volts per centimeter field to be used. In practice, however, the improvements in separation efficiency are modest or nonexistent. Manipulating the length of the capillary usually produces more tangible effects on resolution than variations on the electric field. Once again, the I.D. of the capillary is a compromise between detection and Joule heat generation and dissipation. A larger sample volume can be loaded into a capillary with a larger I.D. without compromising resolution, and the increased path length of the detection window generates a better signal. For polymer solutions, selection of the capillary I.D. needs to take into consideration the viscosity of the solution. Narrow-bore capillaries may require high pressure and prolonged purging time in between analyses to replace the capillary contents, unnecessarily increasing total analysis time.

Gel-Filled Capillaries: The first described use of gel-filled capillaries for analysis of SDS-denatured proteins was in 1983 by Hjerten.[122] Since then, most reports employed either of two types of gels: polyacrylamide cross-linked with bis-acrylamide,[121] and linear polyacrylamide. Both gels are polymerized *in situ* because their high viscosities preclude pumping them into the narrow-bore column. The chemical or cross-linked gels have a well-defined pore structure largely determined by the concentrations of the monomer and cross-linker (most commonly acrylamide and *N,N'*-methylenebisacrylamide, respectively). Because one of the drawbacks of gel-filled capillaries is poor lifetime, several studies have aimed to increase the useful life of the columns. It was found that a lower degree of cross-linking correlated with longer column lifetime.[136] Thus, linear polyacrylamide gels with no cross-linking the capillaries were introduced. These gels were also more compatible with high electric fields than cross-linked gels.[120] Unfortunately, even gels with zero cross-linking could not be used for more than 20 to 40 runs.

Based on the high resolving power of gels compared to polymer solutions,[120] efforts were made to adapt gels to the capillary format. In CGE one of the main disadvantages of using cross-linked polyacrylamide is the lack of flexibility during the separation and injection process. Sample plugs in the gel-filled capillary may result in bubble formation and poor separation efficiencies. Manufacture and shelf life of gel-filled capillaries are also challenges. Chemical gels are heat sensitive; at slightly higher than room temperature, bubbles can form. Other disadvantages of gel-filled columns include a short lifetime, low reproducibility, and poor detection sensitivity due to high UV absorption of the gel matrix. Protein detection in gels is usually accomplished at 280 nm, but the extinction coefficients of proteins are 20 to 50 times higher at 214 nm. At 214 nm there is also less variability in the intensity of absorbance of proteins.[136] Another drawback of gel-filled capillaries is that the composition of the gel-filled capillary cannot be changed. Coupled with short capillary lifetimes and shelf life, this adds to the cost of the method.

Polymer Solutions: The advantages of using polymer solutions to achieve size-based separations include increased reproducibility (because the capillary's content is replaced at each run), increased capillary lifetime, the possibility of using polymers with low absorption in the 200- to 220-nm range, and ease of storage and handling. Another important advantage of polymer solutions is the possibility of using the less problematic, often more reproducible, pressure injection method of sample injection. Separation parameters are simple to optimize by changing polymer type and concentration, buffer pH, viscosity, and conductivity. As stated above, the main drawback of polymer solutions is that resolution is not as high as that obtained with gel-filled capillaries or SDS-PAGE. Polymer solutions can be applied when the difference in molecular weight between the product and contaminants or degradation products is at least 5%. SDS-PAGE can be used during methods development for comparative purposes.

Several types of polymers have been shown to be suitable for separation of a broad molecular weight range of polypeptides. Most polymers used to date are noncross-linked linear or branched polymers. Because the polymers are in solution, the structure of their pores is flexible and dynamic. Care should be exercised when selecting polymers and optimizing analysis conditions,[137] because separation parameters such as temperature do not affect all polymers equally.[138–140] Resolution also depends on the type, size, and concentration of the polymer used. Under optimized conditions, polypeptides differing by as little as 4% in molecular mass can be resolved.[134] Resolution achieved using polymer-sieving CE is comparable with that obtained using a 12%T polyacrylamide slab gel. ·

Early reports on the use of polymer solutions for the analysis of SDS complexes included dextran and polyethylene glycol (PEG).[136] Both of these polymers are practically transparent at 214 nm and thus greatly improve detection over polyacrylamide gels. Migration time (MT) reproducibility is of prime importance in this technique because migration times are used to estimate the molecular size of proteins. The use of a replaceable matrix increases MT reproducibility, and RSD values as low as 0.3% were obtained using dextrans. Similar values were obtained for PEG matrices. Using polymer solutions, the life of the column was also extended to over 300 analyses.

One important consideration when selecting a polymer is the viscosity of the final solution. In most cases, low viscosity is desired for easy replacement of the capillary content in between analyses. However, uncoated capillaries will exhibit higher EOF with lower-viscosity polymers.[139]

Because of the properties of both systems, polymer solutions can be used at any stage of the production process, whereas gel-filled capillaries can be used in late stages when clean samples are available and higher resolution might be required.

Molecular Weight Determination and Ferguson Analysis

Molecular weight (MW) determinations are easily performed when analyzing polypeptides in the presence of SDS and sieving media. The MW is obtained by comparing the mobility of protein standards of known MW and the sample of interest. Plotting the log of the MW against the migration time yields a near-linear relationship. This linearity is observed within a range that depends on the type of polymer used for sieving and on the analysis conditions. The protein standard curve may introduce errors in the estimation of MW if the binding of detergent by the protein is anomalous (e.g., membrane proteins, glycoproteins, or highly basic proteins). Because detergent binding directly affects protein mobility by changing the mass-to-charge ratio, the MW discrepancy originates from differences in free-solution mobilities of the different polypeptides.[141]

A method used to correct such errors is to electrophorese the polypeptide in a series of gels of varying concentration. The size effect is then canceled mathematically by constructing Ferguson plots. In electrophoresis, Ferguson analyses are performed for two main reasons: to optimize resolution by determining the appropriate gel concentration and to estimate the molecular weight of proteins. In CE, Ferguson plots are made by measuring the migration times at different polymer concentrations and constructing a universal calibration curve[142] by plotting the logarithms of the relative migrations as a function of polymer concentration. According to Ferguson, the logarithm of the protein's mobility varies linearly as a function of the gel concentration employed. The slope of this mobility line yields a parameter called the retardation coefficient (K_r), which is proportional to the square of the radius of the protein. Universal standard curves are constructed by plotting the logarithm of known protein MW as a function of the square roots of the retardation coefficients. The slope of the curve represents the retardation coefficient, whereas the intercept at zero polymer concentration corresponds to the free-solution mobility of a protein. Different intercepts at zero polymer concentration is an indication of differences in free-solution mobilities of the SDS–protein complexes.

One important advantage of CE is that the mobility of the protein in free solution can be determined experimentally. If the SDS–protein complexes do not have the same mobility in free solution, a Ferguson plot should be constructed. Proteins with similar molecular radii show the same slope, independently of where they intersect on the concentration axis. Ferguson analysis for traditional SDS-PAGE is time consuming, especially because the analysis is best performed using at least six different gel concentrations. Consequently, this method of analysis was practically abandoned until the use of CE with replaceable polymer networks made the Ferguson analysis more feasible. In CE, Ferguson plots can be generated automatically by using different dilutions of the sieving buffer.[142,143]

Practical Considerations in the Analysis of SDS–Protein Complexes Using Polymer Solutions

Commercial kits for analysis of SDS–protein complexes using entangled-polymer sieving systems are currently available from Bio-Rad Labortories, Beckman Instruments, Sigma, and others. This discussion is based on the authors' experience with the replaceable polymer sieving system from Bio-Rad, which employs a proprietary hydrophilic sieving polymer in 0.4 M Tris borate buffer (pH 8.5) containing 0.1% SDS. The chain length and concentration of the sieving polymer were formulated to provide resolution of SDS–protein complexes over a MW range of 14 to 200 kDa. The buffer also contains a low concentration of the polymer modified by charged functional groups. The combination of the high-viscosity sieving polymer and the cationic modified-polymer additive serves to reduce the electroendoosmotic flow in an uncoated capillary to less than 5×10^{-5} cm^2 V^{-1}sec^{-1}. Because proteins that have been complexed with SDS are strongly anionic, they do not adsorb to the capillary wall under the alkaline run conditions, allowing analyses to be performed in uncoated capillaries in the absence of significant EOF.

Sample preparation for CE-SDS analysis is essentially the same as for SDS-PAGE. Protein samples are diluted 1:1 in a Tris-HCl + SDS (pH 9.2) sample preparation buffer. If the proteins are to be analyzed under reduced conditions, an appropriate reducing agent such as β-mercaptoethanol (final concentration 2.5%) or dithiothreitol (15 mM) is added. The Bio-Rad CE-SDS kit employs benzoic acid at a final concentration of 50 μg/ml as the internal standard. Because the marker is a small molecule not subject to sieving, it cannot be used to correct for variations in migration time due to changes such as new polymer or solution batch, temperature, or polymer concentration. For this purpose, it is necessary to find a protein whose migration time differs from that of the main protein and its contaminants or degradation products. In addition, always analyze the sample and internal standard separately to determine peak identities.

After mixing the sample, buffer, and internal standard, the mixture should be heated at 95 to 100°C for 10 to 12 min to complex proteins with SDS. For new proteins, it is advisable to prepare several vials of the same solution and heat them for various lengths of time in order to check for heat degradation artifacts. It is our experience that heating in a water bath is necessary; use of a contact heating block is not always effective and may result in reduced separation efficiency.

The presence of salt in the sample will interfere with the injection process, and the highest sensitivity will be obtained if the sample salt concentration is less than 50 mM. Samples containing higher salt or buffer concentrations should be desalted.

Entangled-polymer sieving buffers are quite viscous, and bubbles are frequently trapped in the bottom of the buffer reservoir vials when the analysis buffer is pipetted into them. In this situation, the capillary orifice and high-voltage electrode will not contact the buffer, resulting in erratic current and failed analysis. To prevent this, the buffer vials should be centrifuged for at least 2 min at full speed in a microcentrifuge immediately prior to installing them in the CE instrument.

The Bio-Rad CE-SDS analysis buffer is designed for use with uncoated capillaries, and no prior capillary conditioning is required. However, the capillary should be purged with acid and base wash solutions for the appropriate length of time before replenishing the run buffer prior to each injection. These purge cycles serve to sweep residual buffer and any remaining sample components from the capillary. Fresh run buffer is then introduced. The viscosity of the CE-SDS analysis buffer is approximately 43 cP, and the purge times given by the manufacturer are calculated for the capillary length and instrument purge pressure of 100 psi. If the CE instrument employs lower purge pressures, or different capillary dimensions are used, purge times should be modified appropriately.

Because of the high viscosity of the entangled-polymer solutions, the buffer can be retained on the outer surfaces of the capillary and electrodes after the replenishment step, resulting in carryover of the buffer into the sample solution during injection. This will reduce injection efficiency and compromise sensitivity. To prevent this, the capillary and electrode surfaces should be immersed in one or two vials of wash solution (water or diluted sample preparation buffer) without application of pressure.

In CZE, electrophoretic or electrokinetic injection is usually not the preferred injection mode because of electrophoretic bias: sample ions of low mobility will migrate more slowly in the injection process and therefore will be at lower relative concentrations in the starting zone. In the case of SDS–protein complexes, all sample components will have approximately the same mass-to-charge ratio because of the constant charge density of SDS on the protein. Therefore, all SDS–protein complexes will be loaded with the same efficiency using electrokinetic injection. The high ionic strength of the analysis buffer (0.4 M Tris-borate) provides a stacking effect, thus decreasing the starting zone width and increasing zone concentration. Consequently, electrokinetic injection is the preferred mode for this technique. However, if the sample contains appreciable salt concentration and it is not practical to desalt the sample, pressure injection may be used. The high viscosity of the run buffer requires sufficient injection times to introduce enough material. For example, in the case of a 24-cm $\times$ 50-μm capillary, an injection of 12 sec at 5 psi is necessary to inject a 0.3-mm sample zone. If sample salt concentration is greater than 50 mM, even pressure injection will not provide satisfactory sensitivity.

The great advantage of entangled-polymer systems is their transparency in the low UV range. However, the absorbance of the buffer and sample components such as Tris and SDS contribute appreciable background signal below 210 nm. Detection at 220 nm reduces background interference without significant loss in protein response. A protocol for detection of proteins using precolumn derivatization and LIF[144] is described in the following subsection, "Analysis of rMAbs."

Operation at a field strength of 625 V/cm provides satisfactory resolution with short run times; typical current is approximately 20 μA using a 24-cm $\times$ 50-μm capillary. The capillary should be thermostatted close to ambient temperature (e.g., 20°C) for good reproducibility. It is extremely important not to expose the capillary tips to drying conditions when using entangled-polymer sieving buffers. The sieving polymer will precipitate and plug the capillary.

Repeatability of migration times and peak areas for eight protein standards using electrokinetic injection are presented in Table 9.3. Migration time precision was approximately 0.5% RSD and peak area precision varies by 1 to 3%. Peak area precision using pressure injection was comparable (data not shown). The ability to acquire quantitative information on protein concentration is considered a major advantage of CE compared to SDS-PAGE; in the latter the staining response has only narrow linear ranges, depends on operator technique, and is subject to batch-to-batch variability of the stain. In contrast, protein response using polymer sieving CE with UV detection at 220 nm is linear over three orders

Table 9.3   Migration Time Reproducibility (n = 10) for CE-SDS Using
Electrokinetic Injection

Protein	Migration Time, %RSD	Peak Area, %RSD
Lysozyme	0.35	1.51
Trypsin inhibitor	0.40	0.85
Carbonic anhydrase	0.45	1.24
Ovalbumin	0.55	1.57
Serum albumin	0.56	3.30
Phosphorylase B	0.55	3.32
β-Galactosidase	0.59	3.05
Myosin	0.66	1.76

Source: From T. Wehr, R. Rodriguez-Diaz, and M. Zhu, Chromatographic Science
Series, Vol. 80 (1999). With permission.

of magnitude and quite reproducible (Table 9.3). In this system, the detection
limit for carbonic anhydrase (S/N = 3) is 0.5 µg/ml.

Using the polymer sieving system described in the preceding text, the log
of protein MW is correlated with migration time. MW can be determined directly
by comparing to migration times of standard proteins. However, small variations
in migration times can introduce significant error in the calculated MW value.
More reliable estimates may be obtained by normalizing the migration times of
protein standards and samples to that of an internal standard.

Applications

Analysis of rMAbs

Recombinant monoclonal antibodies (rMAbs) are therapeutic biomolecules with
potential applications covering a broad spectrum of indications. Classical
approaches to the analysis of antibodies include extensive use of SDS-PAGE to
monitor consistency, purity, and stability of these molecules. Often, detection
problems arise if the antibodies are not concentrated enough in final formulations.
Coomassie blue staining provides relative quantitative results, but lacks the sensitivity to detect contaminants present in small amounts. Alternatively, silver
staining is more labor intensive, requires more skill, and it is nearly impossible
to obtain consistent quantitation. With the interest generated by the therapeutic
use of rMAbs, a replacement analytical approach for the analysis of size-based
rMAb variants is needed. Hunt and Nashabeh[144] built upon the early research of
Gump and Monning[145] and Wise et al.[146] to develop a CE method using laser-
induced fluorescence to increase the detection limit to values comparable to those
obtained with silver staining. The assay was developed and validated according
to the guidelines of the International Committee on Harmonization (ICH) for use
in routine lot release testing of an rMAb pharmaceutical.

In this assay, rhuMAbs produced in transfected Chinese hamster ovary
(CHO) cells were analyzed using a commercially available SDS–protein analysis

kit. The rMAb in solution is derivatized with a neutral fluorophore, e.g., 5-carboxytetramethylrhodamine succinimidyl ester. Perhaps the key to the success of this method is the use of a neutral fluorophore. One of the dangers of using precolumn derivatization of proteins is the high potential for the generation of multiple species, which occurs because the reaction produces a distribution of products. Molecules with a different number of neutral fluorophores will vary slightly in mass, but the resolving power of CE with polymer solutions is not sufficient to separate the components with little variation in Stokes radii into multiple peaks. Thus, only a small increase in peak width is observed. The hydrophobic character of the fluorophore may also bind SDS to maintain a more constant mass-to-charge ratio and limit peak broadening. The increase in mass can also be compensated by the elimination of positive charge, such as an amino group that reacts with the fluorophore.

Recombinant MAb samples (2.5 mg) were buffer exchanged into 800 µl of 0.1 M sodium bicarbonate, pH 8.3, using an NAP-5 column. A measure of 10 µl of 5-TAMRA.SE (1.4 mg/ml) dissolved in DMSO was then added to 190 µl of rMAb solution, and the resultant mixture was incubated at 30°C for 2 h. After incubation, 190 µl of the antibody–dye conjugate was loaded into a second NAP-5 column and collected into 700 µl of 0.1 M sodium bicarbonate, pH 8.3.

The labeled sample was mixed with SDS-containing sample buffer and incubated at 90°C for 3 min. The samples were incubated with and without reducing agent. The capillary was conditioned by rinsing sequentially with 0.1 M NaOH, 0.1 M HCl, and running buffer containing a hydrophilic polymer as sieving matrix. The sample was injected electrophoretically, applying 10 kV for 15 sec. The applied voltage during the analysis was 15 kV, and the temperatures of the capillary and sample compartment were maintained at 20°C. For comparative purposes, the same unlabeled samples were analyzed by SDS-PAGE using a 5 to 20% gel. Voltage was applied for 3 h (20 mA), and the proteins were visualized by an Oakley silver stain. Figure 9.9 shows a comparison of CE using UV and LIF, both of which are comparable to SDS-PAGE (not shown).

The authors noted the importance of removing the unreacted fluorophore with a second NAP-5 column because prolonged incubation with the dye resulted in protein aggregation and peak broadening during the analysis. The ratio of dye to protein also must be optimized because high relative amounts of dye increased detectable aggregates. CE-LIF of derivatized rMAb using the labeling conditions described in this study allowed the detection of rMAb at a low nanomolar concentration (9 ng/ml) as determined by a signal-to-noise ratio of 2.5. This level of detection compares well with silver staining and is a 140-fold improvement over detection by UV absorption.

To estimate the recovery of the protein from the capillary, a sample was labeled with ^{125}I to track mass balance. A nonradiolabeled sample was used to show that both samples produced the same CE profile. A baseline value or control was obtained by performing the assay through the injection step, followed by a high-pressure rinse to expel the entire capillary contents into a collection vial. A

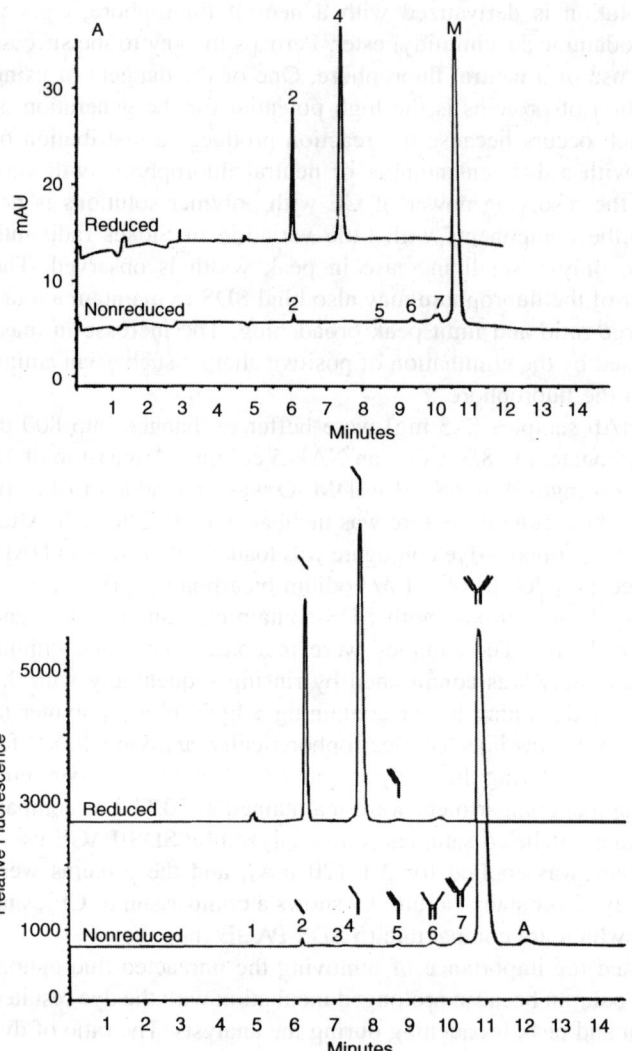

Figure 9.9 Comparison of CE-SDS using UV and LIF detection. A therapeutic recombinant monoclonal antibody (rMAb) was analyzed by CE-SDS with or without a reduction step. The top electropherogram depicts an overlay of reduced and nonreduced samples using detection by UV absorbance at 220 nm. The bottom electropherogram shows the same samples as detected by LIF using a 3.5-mW argon laser (set at 488 nm excitation wavelength, and 520 nm emission wavelength). The samples for CE-SDS with LIF detection were derivatized prior to analysis with 5-TAMRA-SE. The slight increase in the peak width of the monomer (M) for the LIF-detected sample could be the result of heterogeneous derivatization or sample overloading (which can occur when using more sensitive detectors even if the injection is constant). The numbers identifying the peaks correspond to observed bands in SDS-PAGE analysis with silver-stain visualization (see Reference 144).

second analysis was performed under standard protocol conditions. After the last peak had migrated past the detector, the voltage was disconnected, and the capillary content was purged into a collection vial. The counts per minute of both samples were determined using a gamma counter. Comparison of the control and experimental samples resulted in a mean recovery of 95% (n = 3). The authors noted that such recovery studies are often not performed for SDS-PAGE, in which selective loss of material may occur in the stacking gel.

The precision of the assay for nonreduced samples was demonstrated by the evaluation of six independent sample preparations on a single day (repeatability) and the analysis of independent sample preparations on three separate days by two different analysts (intermediate precision). The RSD values for the migration time were 0.9%. The RSD values for peak area percent of the main peak and the minor peaks in the profile were 0.6 and 12.6%, respectively. The higher variability observed with the minor peaks was determined to be primarily related to the sample heating during preparation for the analysis. These results demonstrate that the use of uncoated fused-silica capillaries in combination with a sieving matrix can provide adequate precision and analyte recovery.

During methods development, the authors noticed changes in the peak distribution of the samples. A time–temperature study of sample heating demonstrated that the sample was being degraded at high temperature. Because heating is used to speed up the unfolding of proteins for SDS binding, the best conditions for this rMAb were determined to be 37°C for 8 to 15 min. It is a common practice when performing SDS-PAGE to evaluate the effect of heating on sample profile, especially when working with a new protein. As shown in this application, it is also a useful practice for CE. The main drawback of too little heating temperature or time is the incomplete unfolding of the polypeptide, which results in peak broadening and anomalous migration in CE and in SDS-PAGE.

Recombinant Proteins

SDS-PAGE is widely used in the biopharmaceutical industry to monitor the purification and to estimate the purity of recombinant protein therapeutics. However, SDS-PAGE is a labor-intensive and (at best) semiquantitative technique that is not ideally suited for high-throughput analysis in commercial laboratories. Kundu et al.[147] have evaluated a commercially available entangled-polymer sieving kit for CE of SDS–proteins and compared the results with SDS-PAGE and Coomassie Brilliant Blue (CBB) using a 72-kDa viral-CKS fusion protein. Protein purity was assessed from densitometric scanning of CBB-stained gels and from integration of electropherograms monitored at 220 nm. The authors demonstrated that purity levels determined by the two methods were in good agreement, and that SDS-PAGE quantitation was limited by the nonlinearity of CBB staining at low concentration. Molecular mass estimates obtained using MALDI-TOF MS, CE-SDS, and SDS-PAGE were found to differ from the theoretical molecular mass by 100, 600, and 1400 mass units, respectively.

References

1. P.D. Grossman and J.C. Colburn (Eds.), *Capillary Electrophoresis, Theory and Practice*, Academic Press, San Diego CA, 1992.
2. S.F.Y. Li (Ed.), *Capillary Electrophoresis: Principles, Practice, and Applications*, J. Chromatogr. Library, Vol. 52, Elsevier Science, Amsterdam, 1992.
3. R. Weinberger, *Practical Capillary Electrophoresis*, Academic Press, Boston, 1992.
4. James P. Landers (Ed.), *Handbook of Capillary Electrophoresis*, 2nd ed., CRC Press, Boca Raton, FL, 1996.
5. P. Camilleri (Ed.), *Capillary Electrophoresis: Theory and Practice*, CRC Press, Boca Raton, FL, 1993.
6. N.A. Guzman, *Capillary Electrophoresis Technology*, Marcel Dekker, New York, 1993.
7. J. Vindevogel, *Introduction to Micellar Electrokinetic Chromatography*, Hüthig, Heidelberg, Germany, 1992.
8. S.M. Lunte and D.M. Radzik (Eds.), *Pharmaceutical and Biomedical Applications of Capillary Electrophoresis, Progress in Pharmaceuical and Biomedical Analysis*, Vol. 2, Elsevier Science, Oxford, 1996.
9. W.G. Kuhr, *Anal. Chem., 62:* 403R (1990).
10. W.G. Kuhr and C.A. Monnig, *Anal. Chem., 64:* 389R (1992).
11. C.A. Monnig and R.T. Kennedy, *Anal. Chem., 66:* 280R (1994).
12. M.V. Novotny, K.A. Cobb, and J. Liu, *Electrophoresis, 11:* 735 (1990).
13. Z. Deyl and R. Struzinsky, *J. Chromatogr., 569:* 63 (1991).
14. C. Schöneich, S.K. Kwok, G.S. Wilson, S.R. Rabel, J.F. Stobaugh, T.D. Williams, and D.G. Vander Velde, *Anal. Chem., 65:* 67R (1993).
15. C. Schöneich, A.F.R. Hühmer, S.R. Rabel, J.F. Stobaugh, S.D.S. Jois, C.K. Larive, T.J. Siahaan, T.C. Squier, D.J. Bigelow, and T.D. Williams, *Anal. Chem., 67:* 155R (1995).
16. É. Szökö, *Electrophoresis, 18:* 74 (1997).
17. M.A. Strege and A.L. Lagu, *J. Chromatogr. A, 780:* 285 (1997).
18. K. Ibel, R.P. May, K. Kirschner, H. Szadkowski, E. Mascher, and P. Lundahl, *Eur. J. Biochem., 190:* 311 (1990).
19. E.L. Gump and C.A. Monning, *J. Chromatogr. A, 715:* 167 (1995).
20. W.G. Kuhr and E.S. Yeung, *Anal. Chem., 60:* 2642 (1988).
21. D.F. Swaile and M.J. Sepaniak, *J. Liq. Chromatogr., 14:* 869 (1991).
22. B. Nickerson and J.W. Jorgenson, *J. Chromatogr., 480:* 157 (1989).
23. H.A. Bardelmeijer, H. Lingeman, C. de Ruiter, and W.J.M. Underberg, *J. Chromatogr. A, 807:* 3 (1998).
24. J. Cai and J. Henion, *J. Chromatogr., 703:* 667 (1995).
25. T. Wehr, *LC GC, 11:* 14 (1993).
26. K.A. Turner, *LC GC, 9:* 350 (1991).
27. J.R. Mazzeo and I.S. Krull, *Biochromatography, 10:* 638 (1991).
28. R.L. Chien and D.S. Burgi, *J. Chromatogr., 559:* 141 (1991).
29. P. Jandik and W.R. Jones, *J. Chromatogr., 546:* 431 (1991).
30. N.A. Guzman, M.A. Trebilock, and J.P. Advis, *J. Liq. Chromatogr., 14:* 997 (1991).
31. S. Hjertén, K. Elenbring, F. Kilár, J.L. Liao, A.J.C. Chen, C.J. Siebert, and M. Zhu, *J. Chromatogr., 403:* 47 (1987).
32. F. Foret, E. Szoko, and B.L. Karger, *Electrophoresis, 14:* 417 (1993).

33. T. Hirokawa, A. Ohmori, and Y. Kiso, *J. Chromatogr. A, 634:* 101 (1993).
34. V. Dolnik, *Electrophoresis, 18:* 2353 (1997).
35. V. Dolnik, *Electrophoresis, 20:* 3106 (1999).
36. G.M. Janini and H.J. Issaq, *Chormatographia Suppl., 53:* S-18 (2001).
37. J. Chen, J. Fausnaugh-Pollitt, and J. Gu, *J. Chromatogr. A, 853:* 197 (1999).
38. J. Bao and F.E. Regnier, *J. Chromatogr., 608:* 217 (1992).
39. D. Wu and F.E. Regnier, *Anal. Chem., 65:* 2029 (1993).
40. B.J. Harmon, D.H. Patterson, and F.E. Regnier, *Anal. Chem., 65:* 2655 (1993).
41. D.H. Patterson, B.H. Harmon, and F.E. Regnier, *J. Chromatogr. A, 732:* 119 (1996).
42. K. Shimura and K. Kasai, *Anal. Biochem., 251:* 1 (1997).
43. F.A. Gomez, L.Z. Avila, Y.H. Chu, and G.M. Whitesides, *Anal. Chem., 66:* 1785 (1994).
44. M.H.A. Busch, H.F.M. Boelens, J.C. Kraak, and H. Poppe, *J. Chromatogr. A, 775:* 313 (1997).
45. S. Honda, A. Taga, K. Suzuki, S. Suzuki, and K. Kakehi, *J. Chromatogr., 597:* 377 (1992).
46. N.H.H. Heegaard, D.T. Olsen, and K.-L.P. Larsen, *J. Chromatogr. A, 744:* 285 (1996).
47. R.G. Nielsen, E.C. Rickard, P.F. Santa, D.A. Sharknas, and G.S. Sittampalam, *J. Chromatogr., 539:* 177 (1991).
48. P. Sun, A. Hoops, and R.A. Hartwick, *J. Chromatogr. B, 661:* 335 (1994).
49. M.J. Schmerr, K.R. Goodwin, R.C. Cutlip, and A.L. Jenny, *J. Chromatogr. B, 681:* 29 (1996).
50. F.T.A. Chen, *J. Chromatogr. A., 680:* 419 (1994).
51. R.S. Rush, A. Cohen, and B.L. Karger, *Anal. Chem., 63:* 1346 (1991).
52. M.A. Strege and A.L. Lagu, *J. Chromatogr. A., 652:* 179 (1993).
53. M.A. Strege and A.L. Lagu, *Am. Lab., 26:* 48C (1994).
54. F. Kilár and S. Hjertén, *J. Chromatogr., 638:* 269 (1993).
55. V.J. Hilser, G.D. Worosila, and E. Freire, *Anal. Biochem., 208:* 125 (1993).
56. R.T. Bishop, V.E. Turula, and J.A. de Haseth, *Anal. Chem., 68:* 4006 (1996).
57. D. Rochu, G. Ducret, and P. Masson, *J. Chromatogr. A, 838:* 157 (1999).
58. D. Rochu, G. Ducret, F. Ribes, S. Vanin, and P. Masson, *Electrophoresis, 20:* 1586 (1999).
59. F. Kilár and S. Hjertén, *J. Chromatogr., 480:* 351 (1989).
60. K. Yim, *J. Chromatogr., 559:* 401(1991).
61. J.M. Thorne, W.K. Goetzinger, A.B. Chen, K.G. Moorhouse, and B.L. Karger, *J. Chromatogr. A, 744:* 155 (1996).
62. T.M. McNerney, S.K. Watson, J.-H. Sim, and R.L. Bridenbaugh, *J.Chromatogr. A, 744:* 223 (1996).
63. R. Malsch, T. Mrotzek, G. Huhle, and J. Harenberg, *J. Chromatogr. A, 744:* 215 (1996).
64. S. Kundu, C. Fenters, M. Lopez, B. Calfin, M. Winkler, and W.G. Robey, *J. Capillary Electrop., 6:* 301 (1996).
65. D. Josíc, A. Böttcher, and G. Schmitz, *Chromatographia, 30:* 703 (1990).
66. G. Schmitz and C. Möllers, *Electrophoresis,15:* 31 (1994).
67. T. Tadey and W.C. Purdy, *J. Chromatogr., 583:* 111 (1992).
68. T. Tadey and W.C. Purdy, *J. Chromatogr. A, 652:* 131 (1993).
69. R. Lehmann, H. Liebich, G. Grübler, and W. Voelter, *Electrophoresis, 16:* 998 (1995).

70. I.D. Cruzado, A.Z. Hu, and R.D. Macfarlane, *J. Capillary Electrop., 3:* 25 (1996).
71. I.D. Cruzado, S. Song, S.F. Crouse, B.C. O'Brien, and R.D. Macfarlane, *Anal. Biochem., 243:* 100 (1996).
72. T.J. Pritchett, *Electrophoresis, 17:* 1195 (1996).
73. X. Liu, Z. Sosic, and I.S. Krull, *J. Chromatogr. A, 735:* 165 (1996).
74. S. Hjertén and M. Zhu, *J. Chromatogr., 347:* 265 (1985).
75. S. Hjertén, J.-L. Liao, and K. Yao, *J. Chromatogr., 387:* 127 (1987).
76. S. Hjertén and M. Zhu, *J. Chromatogr., 346:* 265 (1985).
77. T.L. Huang, P.C.H. Shieh, and N. Cooke, *Chromatographia, 39:* 543 (1994).
78. S.-M. Chen and J.E. Wiktorowicz, *Anal. Biochem., 206:* 84 (1992).
79. R. Rodriguez and C. Siebert, poster presentation, 6th International Symposium on Capillary Electrophoresis, San Diego, CA, 1994.
80. J.R. Mazzeo and I.S. Krull, *J. Microcol. Sep., 4:* 29 (1992).
81.· W. Thormann, J. Caslavska, S. Molteni, and J. Chmelik, *J. Chromatogr., 589:* 321 (1992).
82. J.R. Mazzeo and I.S. Krull, *Anal. Chem., 63:* 2852 (1991).
83. M. Zhu, R. Rodriguez, and T. Wehr, *J. Chromatogr., 559:* 479 (1991).
84. J.R. Mazzeo and I.S. Krull, *J. Chromatogr., 606:* 291 (1992).
85. C. Schwer, *Electrophoresis, 16:* 2121 (1995).
86. A.B. Chen, C.A. Rickel, A. Flanigan, G. Hunt, and K.G. Moorhouse, *J. Chromatogr. A, 744:* 279 (1996).
87. N. Wu, P. Sun, J.H. Aiken, T. Wang, C.W. Huie, and R. Hartwick, *J. Liq. Chromatogr., 16:* 2293 (1993).
88. J. Wu and J. Pawliszyn, *J. Chromatogr., 608:* 121 (1992).
89. J. Wu and J. Pawliszyn, *Electrophoresis, 14:* 469 (1993).
90. J. Wu and J. Pawliszyn, *Electrophoresis, 16:* 670 (1995).
91. J. Wu and J. Pawliszyn, *J. Chromatogr. B, 657:* 327 (1994).
92. L. Vonguyen, J. Wu, and J. Pawliszyn, *J. Chromatogr. B, 657:* 333 (1994).
93. J. Wu and J. Pawliszyn, *Anal. Chem., 66:* 867 (1994).
94. J. Wu and J. Pawliszyn, *Anal. Chem., 64:* 224 (1994).
95. J. Wu and J. Pawliszyn, *Anal. Chem., 64:* 219 (1992).
96. J. Wu and J. Pawliszyn, *J. Liq. Chromatogr., 16:* 3675 (1993).
97. L. Fang, R. Zhang, E.R. Williams, and R.N. Zare, *Anal. Chem., 66:* 33696 (1994).
98. F. Foret, O. Muller, J. Thorne, W. Gotzinger, and B.L. Karger, *J. Chromatogr. A, 716:* 157 (1995).
99. P.G. Righetti, *Isoelectric Focusing: Theory, Methodology, and Applications*, Elsevier Biomedical Press, Amsterdam, 1989.
100. T. Rabilloud, *Electrophoresis, 17:* 813 (1997).
101. K. Slais and Z. Fiedl, *J. Chromatogr. A, 661:* 249 (1994).
102. R. Rodríguez-Díaz, M. Zhu, and T. Wehr, *J. Chromatogr. A, 772:* 145 (1997).
103. R. Rodríguez-Díaz, M. Zhu, V. Levi, and T. Wehr, presented at the 7th Symposium on Capillary Electrophoresis, Würzburg, Germany (1995).
104. F. Kilar and S. Hjerten, *Electrophoresis, 10:* 23 (1989).
105. K.W. Yim, *J. Chromatogr., 559:* 401 (1991).
106. T. Wehr, M. Zhu, R. Rodriguez, D. Burke, and K. Duncan, *Am. Biotech. Lab.,* September 1990.
107. G.L. Klein and C.R. Joliff, in J.P. Landers (Ed.), *Handbook of Capillary Electrophoresis*, CRC Press, Boca Raton, FL, 1993, p. 452.

108. P. Ferranti, A. Malorni, P. Pucci, S. Fanali, A. Nardi, and L. Ossicini, *Anal. Biochem.*, *194:* 1 (1991).

109. J.W. Drysdale, P.G. Righetti, and H.F. Bunn, *Biochem. Biophys. Acta*, *229:* 42 (1971).

110. S. Kundu and C. Fenters, *J. Cap. Elec.*, *2:* 6 (1995).

111. O-S. Reif and R. Freitag, *J. Chromatogr. A*, *680:* 383 (1994).

112. G. Hunt, K.G. Moorhouse, and A.B. Chen, *J. Chromatogr. A*, *744:* 295 (1996).

113. J.M. Thorne, W.K. Goetzinger, A.B. Chen, K.G. Moorhouse, and B.L. Karger, *J. Chromatogr. A, 744:* 155 (1996).

114. J. Kubach and R. Grimm, *J. Chromatogr. A*, *737:* 281 (1996).

115. K.G. Moorhouse, C.A. Eusebio, G. Hunt, and A.B. Chen, *J. Chromatogr. A, 717:* 61 (1995).

116. K. Shimura and B.L. Karger, *Anal. Chem.*, *66:* 9 (1994).

117. D. Wu and F. Regnier, *J. Chromatogr.*, *608:* 349 (1992).

118. K. Tsuji, *J. Chromatogr. A, 652:* 139 (1993).

119. A. Guttman, *Electrophoresis, 16:* 611 (1995).

120. K. Tsuji, *J. Chromatogr. B, 662:* 291 (1994).

121. A.S. Cohen and B.L. Karger, *J. Chromatogr.*, *397:* 409 (1987).

122. S. Hjerten, *Electrophoresis '83*, H. Hirai (Ed.), Walter de Gruyter, New York, 1994, pp. 71–79.

123. M. Zhu, D.L. Hansen, S. Burd, and F. Gannon, *J. Chromatogr.*, *480:* 311 (1989).

124. S. Hjerten, T. Srichaiyo, and A. Palm, *Biomed. Chromatogr.*, *8:* 73 (1994).

125. M. Zhu, V. Levi, and T. Wehr, *Am. Biotech. Lab.*, *11:* 26 (1993).

126. C. Tanford, *The Hydrophobic Effect: Formation of Micelles and Biological Membranes*, 2nd ed., John Wiley & Sons, New York, 1980, pp. 159–164.

127. M.R. Karim, S. Shinagawa, and T. Takagi, *Electrophoresis, 14:* 1141 (1994).

128. K. Sasa and K. Taked, *J. Colloid Interface Sci.*, *147:* 516 (1993).

129. K. Benedek and S. Thiede, *J. Chromatogr. A, 676:* 209 (1994).

130. P.C.H. Shieh, D. Hoang, A. Guttman, and N. Cooke, *J. Chromatogr. A, 676:* 219 (1994).

131. M. Nakatani, A. Shibukawa, and T. Nakagawa, *Biol. Pharm. Bull.*, *16:* 1185 (1993).

132. M. Nakatani, A. Shibukawa, and T. Nakagawa, *Anal. Sci.*, *10:* 1 (1994).

133. M. Nakatani, A. Shibukawa, and T. Nakagawa, *J. Chromatogr. A, 672:* 213 (1994).

134. W.E. Werner, D.M. Demorest, J. Stevens, and J.E. Wiktorowicz, *Anal. Biochem.*, *212:* 253 (1993).

135. A. Guttman, *Electrophoresis, 17:* 1333 (1996).

136. K. Ganzler, K.S. Greve, A.S. Cohen, B.L. Karger, A. Guttman, and N.C. Cooke, *Anal. Chem.*, *64:* 2665 (1992).

137. A. Guttman, P. Shieh, D. Hoang, J. Horvath, and N. Cooke, *Electrophoresis, 15:* 221 (1994).

138. K. Tsuji, *J. Chromatogr. A, 661:* 257 (1994).

139. E. Simo-Alfonso, M. Conti, C. Gelfi, and P.G. Righetti, *J. Chromatogr. A, 689:* 85 (1995).

140. A. Guttman, J. Horvath, and N. Cooke, *Anal. Chem.*, *65:* 199 (1993).

141. K. Benedek and S. Thiede, *J. Chromatogr. A, 676:* 209 (1994).

142. W.E. Werner, D.M. Demorest, and J.E. Wiktorowicz, *Electrophoresis, 14:* 759 (1993).

143. A. Guttman, P. Shieh, J. Lindahl, and N. Cooke, *J. Chromatogr. A, 676:* 227 (1994).

144. G. Hunt and W. Nashabeh, *Anal. Chem.*, *71:* 2390 (1999).

145. E.L. Gump and C.A. Monning, *J. Chromatogr. A, 715:* 167, (1995).
146. E.T. Wise, S. Navjot, and B.L. Hogan, *J. Chromatogr. A, 746:* 109 (1996).
147. S. Kundu, C. Fenters, M. Lopez, A. Varma, J. Brackett, S. Kuemmerle, and J.C. Hunt, *J. Cap. Elec., 4:* 7 (1997).
148. L. Song, Q. Ou, and W. Yu, *J. Liq. Chromatogr., 17:* 1953 (1994).
149. N. Cohen and E. Grushka, *J. Chromatogr. A, 678:* 167 (1994).
150. D. Corradini, A. Rhomberg, and C. Corradini, *J. Chromatogr. A, 661:* 305 (1994).
151. D. Corradini, G. Cannarsa, E. Fabbri, and C. Corradini, *J. Chromatogr. A, 709:* 127 (1995).
152. J.P. Landers, R.P. Oda, B.J. Madden, and T.C. Spelsberg, *Anal. Biochem., 205:* 115 (1992).
153. R.P. Oda, B.J. Madden, T.C. Spelsberg, and J.P. Landers, *J. Chromatogr. A, 680:* 85 (1994).
154. M.M. Bushey and J.W. Jorgenson, *J. Chromatogr., 480:* 301 (1989).
155. A. Emmer, M. Jansson, and J. Roeraade, *J. Chromatogr., 547:* 544 (1991).
156. A. Emmer and J. Roeraade, *J. Liq. Chromatogr., 17:* 3831 (1994).
157. A. Emmer, M. Jansson, and J. Roerrade, *J. High Res. Chromatogr. 14:* 778 (1991).
158. E.L. Hult, A. Emmer, and J. Roeraade, *J. Chromatogr. A, 757:* 255 (1997).
159. V. Rohlicek and Z. Deyl, *J. Chromatogr., 494:* 87 (1989).
160. M.J. Gordon, K.-J. Lee, A.A. Arias, and R.N. Zare, *Anal. Chem., 63:* 69 (1991).
161. W.G.H.M. Muijselaar, C.H.M.M. de Bruijn, and F.M. Everaerts, *J. Chromatogr., 605:* 115 (1992).
162. G. Mandrup, *J. Chromatogr., 604:* 267 (1992).
163. B.Y. Gong and J.W. Ho, *Electrophoresis 18:* 732 (1997).
164. M. Taverna, A. Baillet, D. Biou, M. Schlüter, R. Werner, and D. Ferrier, *Electrophoresis, 13:* 359 (1994).
165. M.E. Legaz and M.M. Pedrosa, *J. Chromatogr. A, 719:* 159 (1996).
166. H.G. Lee and D.M. Desiderio, *J. Chromatogr. B, 691:* 67 (1997).
167. N. Guzman, J. Moschera, K. Iqbal, and A.W. Malick, *J. Chromatogr., 608:* 197 (1992).
168. X.-H. Fang, T. Zhu, and V.-H. Sun, *J. High Res. Chromatogr., 17:* 749 (1994).
169. M.A. Strege and A.L. Lagu, *J. Liquid Chromatogr., 16:* 51 (1993).
170. M.A. Strege and A.L. Lagu, *J. Chromatogr., 630:* 337 (1993).
171. A. Cifuentes, M.A. Rodriguez, and F.G. García-Montelongo, *J. Chromatogr. A, 742:* 257 (1996).
172. Y.J. Yao and S.F.Y. Li., *J. Chromatogr. A, 663:* 97 (1994).
173. N. Cohen and E. Grushka, *J. Cap. Elec., 1:* 112 (1994).
174. M. Morand, D. Blaas, and E. Kenndler, *J. Chromatogr. B, 691:* 192 (1997).
175. G.N. Okafo, H.C. Birrell, M. Greenaway, M. Haran, and P. Camilleri, *Anal. Biochem., 219:* 201 (1994).
176. G.N. Okafo, A. Vinther, T. Kornfelt, and P. Camilleri, *Electrophoresis, 16:* 1917 (1995).
177. J.R. Veraart, Y. Schouten, C. Gooijer, and H. Lingeman, *J. Chromatogr. A, 768:* 307 (1997).
178. A. Cifuentes, J.M. Santos, M. de Frutos, and J.C. Diez-Mesa, *J. Chromatogr. A, 652:* 161 (1993).
179. T. Wehr, R. Rodriguez-Diaz, and M. Zhu, Chromatographic Science Series Vol. 80, Marcel Dekker, 1999.

10

Mass Spectrometry for Biopharmaceutical Development

Alain Balland and Claudia Jochheim

INTRODUCTION

Recent advances in mass spectrometry (MS) techniques have radically changed the analysis of biomolecules. MS has become the analytical method of choice for discovery and characterization of molecules with therapeutic value. Technological breakthroughs in the discovery area are now increasingly applied in the process development field and have recently entered the production process in manufacturing and quality control (QC) areas. In this presentation, after a review of the current state of the art, we would like to demonstrate how MS methods are influencing the development and manufacturing of therapeutic molecules.

MS METHODS

Ionization Methods

MS involves the separation of ions based on their mass-to-charge ratio (m/z). The concept was invented a century ago[1] with a dramatic impact on analytical chemistry.[2,3] The fundamental principle of MS requires vaporization of the molecules in the gas phase and in ionization. Early ionization methods such as electron impact (EI) and chemical ionization (CI)[4,5] were limited to small organic molecules that were volatile and stable to heat and amenable to transfer into high vacuum. Introduction of the fast-atom-bombardment (FAB) method of ionization[6] allowed the expansion of the MS field to polar, thermally labile molecules with masses around a few thousand Daltons. Using this ionization technique, Barber et al. proved that peptides and small proteins could be analyzed by MS.[7]

Electrospray Ionization

The revolution in the analysis of biomolecules resulted from the introduction of soft-ionization techniques such as electrospray ionization (ESI) and matrix-assisted laser desorption ionization (MALDI). The concept of ESI developed in the 1960s[8] was applied by Fenn and coworkers for the transfer of large biomolecules into a mass spectrometer.[9] The importance of this contribution is underscored by the recent award of the Nobel Prize in chemistry to Fenn for his pioneer work. In the electrospray technique, ions are formed in solution, then transferred to the gas phase at atmospheric pressure. The ions are analyzed as positively or negatively charged species, according to the pH of the solution. The liquid is passed through a needle maintained at high voltage (3 to 4 kV) and the ions accumulate in small charged droplets until the repulsive columbic forces exceed the surface tension force, generating smaller droplets. The potential applied between the tip of the electrode and the instrument inlet drives the flow of ions into the mass spectrometer while a counter flow of gas facilitates solvent evaporation. The formation of the gas phase ions is still under debate between two theories. The original charged-residue model of Dole et al.[8] proposes continuous droplet breakup and evaporation until a single ion is formed. An alternative mechanism, the ion evaporation model, suggests field desorption of gaseous ions from the small droplets.[10] The theory of electrospray MS has been reviewed in several publications.[11-15]

Matrix-Assisted Laser Desorption/Ionization

The other soft-ionization technique, MALDI, was introduced by Tanaka et al. and further developed by Hillenkamp and Karas in the late 1980s.[16,17] Koichi Tanaka was awarded the 2002 Nobel Prize in chemistry for this innovation. The process involves introduction of the molecule in the mass spectrometer from the solid phase. An aromatic compound is used to assist the energy transfer from the ultraviolet laser beam to the molecule. Conditions for soft ionization, which result in desorption of intact molecules, are obtained by mixing the analyte with a large excess of UV-absorbing matrix and allowing the mixture to crystallize on a target plate. Pulses of energy from the UV laser, of a duration of 100 nsec or less to limit thermal decomposition, are absorbed by the matrix and transferred to the molecule, resulting in its desorption into the gas phase. Nitrogen lasers are normally used and can produce the energy of a few hundred microjoules at 337 nm at a repetition rate of 20 Hz. Nd-YAG lasers are used to speed up the process with their repetition rate of a few KHz. Although still under debate, the ionization process is believed to occur in the gas phase in the plume of ions created by the irradiation. The fact that MALDI mainly detects singly charged species supports this hypothesis. The low mass range of the spectrum is not available for measurement due to intense signals of matrix ions. The MALDI process is sensitive to size, and a decrease of ionization efficiency associated with peak broadening is observed with larger biomolecules. Principles and theory have been reviewed in several publications.[18-20]

Mass Spectrometers

ESI-Quadrupole Instruments

ESI has been traditionally used with quadrupole mass analyzers,[21] composed of four parallel rods that separate ions by a combination of direct current (DC) and radio frequency (RF) voltages. For particular fields, ions with a specific range of m/z values have stable paths through the filter, and all other ions are not transmitted. The resolving power of the quadrupole mass filter increases with mass as higher mass ions have decreased velocities and spend a longer time in the analyzer. As a result, commercial quadrupoles scan mass ranges up to 4000 at unit mass resolution. Another important feature of quadrupoles is their high ion transmission, especially efficient in the selected ion monitoring mode where detection limits can reach the femtogram range. ESI of a protein generates a series of multiply-charged ions proportional to the number of charged residues. In the case of an acidic solution, the basic residues, Lys, Arg, and N-terminal primary amine, contribute to an envelope of charges along the molecule. As the number of charged residues statistically varies with the size of the molecule, the envelope of ions remains in a limited m/z range, allowing the analysis of molecules of vastly different molecular masses, notably large biomolecules. Although the mass range, resolving power, and accuracy of quadrupoles are limited compared to magnetic sector mass analyzers, their compactness, high transmission, tandem mass spectrometry (MS/MS) potential, and low cost make them the most popular analyzer among commercial instruments.[22] A crucial application of quadrupoles is their use for MS/MS. Application of this technique to biomolecules in recent years has transformed the biological sciences. In a triple-quadrupole instrument, the first quadupole is used to select a single precursor ion, the second is operated in RF-only mode as a collision cell to fragment the precursor ion, and the third measures the m/z values of the product ions. Collision-induced dissociation (CID) with an inert gas is efficient in these instruments at low collision energies (<100 eV) and for ions of limited size (<2000 Da).

ESI–Ion Trap Instruments

In the ion trap technology, ions are captured in three-dimensional electric fields. The continuous beam of ions fills the trap up to the limit of their space charge. When additional electric fields are applied, ions are ejected sequentially and detected. Accumulation of ions in the trap results in high sensitivity for these instruments. The trap can be operated in MS and MS/MS modes. In the latter, the ions of interest are maintained in the trap, whereas the other ions are excluded. Sequential fragmentation steps can be performed to generate MS^n spectra, highly valuable for structural characterization studies.

The sensitivity, compactness, automation, and low prices of ion trap instruments made them very popular in biological MS.[23] Limitations of ion traps include low resolution and mass accuracy at high m/z. In addition, in MS/MS mode, the lower end of the fragment mass range cannot be visualized. Recent developments in the linear geometry of ion traps are aimed at improving on those limitations.

MALDI–Time-of-Flight Instruments

MALDI sources are usually associated with time-of-flight (TOF) mass analyzers that measure the flight time of ions in a field-free drift tube. Ions of different masses extracted from the source simultaneously will take different lengths of times to hit the detector. The time to reach the detector is related to m/z, the acceleration potential, and the length of the tube. In principle, ions of the same m/z with the same kinetic energy should be detected at the same time. Accuracy and resolution are actually limited by practical lengths of flight tubes and spatial dispersion of the ions in the plume. The importance of MALDI in the structural characterization of biomolecules has significantly increased due to two technical advances: reflectron and delayed extraction. The energy spread of the ions of the same m/z can be compensated by a reflectron, an electrostatic lens that reflects the ions back into the flight tube. Ions of higher energy travel further in the reflectron, whereas ions of lower energy travel less. Ions of the same m/z are therefore focused at the exit of the reflectron mirrors and reach the detector at the same time. Delayed-extraction technology[24–26] applies a time lag between creation of ions and their extraction. Delayed extraction compensates for the decreased resolution linked to continuous extraction by focusing the ions at the start. Peptides are analyzed by MALDI-TOF with delayed extraction and reflectron, with resolution of 10,000 to 20,000 and accuracy of a few parts per million (ppm). Other advantages of TOF instruments are their very fast cycle time and high sensitivity. A disadvantage of MALDI-TOF instruments can be their limited potential for tandem MS. MS/MS experiments are performed using a technique called postsource decay (PSD), which has low sensitivity and mass accuracy. The method is based on metastable decay in the flight tube, a fragmentation difficult to control, which does not offer a good alternative to CID.

Orthogonal and Hybrid Instruments

Recent developments in instrumentation take advantage of specific features of each technique to improve the overall performance of mass spectrometers. Orthogonal injection was used to connect the electrospray source and TOF instrument.[27,28] The continuous beam of ions generated by the electrospray source had to be transformed in a series of pulses injected in a perpendicular direction into the TOF region. This technique gave ESI access to the unlimited mass range allowed by TOF analyzers and therefore overcame traditional limitations in mass range observed in ESI–quadrupole and ESI–ion trap instruments. As a result, ESI-TOF instruments are especially suited for the analysis of large biomolecules such as noncovalent protein complexes that can reach several hundred thousand Daltons. In addition, the fast-duty cycle of TOF analyzers makes ESI-TOF instruments the preferred choice for coupling MS to ultrafast separation techniques such as capillary HPLC and capillary electrophoresis (CE, reviewed in Reference 29). Further arrangement of ionization techniques and mass analyzers led to the so-called hybrid instruments. The next logical step was to link quadrupole and TOF technologies in order to offer improved MS/MS capabilities to TOF instruments. The modern quadrupole–TOF instruments combine a first quadrupole mass analyzer where the parent ion is selected, a second quadrupole used as collision cell, and a TOF analyzer where resulting fragment ions are separated.[30,31] These instruments offer high sensitivity, high resolving power, and accuracy; therefore, they have supplanted triple-quadupole instruments for MS/MS measurements of biomolecules. Further developments of this concept recently are leading to the development of instruments that combine MALDI and ESI sources, quadrupoles, and TOF mass analyzers. MALDI-Q-TOF instruments can exploit the advantages of MALDI sources, sensitivity, high throughput, and automation in MS mode with quadrupole-selected MS/MS fragmentation for sequence identification.[32] MALDI-TOF-TOF instruments combine two TOF tubes with a fragmentation chamber and use high-energy fragmentation to get structural identification.[33] Ion source-interchangeable MALDI-ESI-Q-TOF instruments combining flexibility and high performance have recently been commercialized.

Fourier Transform Mass Spectrometers

Recent instrumentation development in biological MS involves Fourier transform ion cyclotron resonance (FT-ICR), mass spectrometry, a technique that traps ions by a combination of electric and magnetic fields. This technique, developed in the 1960s for small organic molecules, is progressively moving into the macromolecule realm. FT-ICR offers unsurpassed resolution and mass accuracy for the analysis of biomolecules.[34-36] Although the instruments' complexity and prices, and the skills required for their operation currently limit their use to specialized laboratories, they represent the future of bio–MS and their use is bound to explode in the coming years.

MS in Combination with Separation Techniques

Gas chromatography coupled with mass spectrometry (GC/MS) remains a workhorse for identifying unknown species in organic chemistry and in the clinical laboratory.[37] However, in the protein and peptide world, liquid chromatography (LC) is the most applied separation technique that, combined with MS, provides a most powerful tool. ESI interfaced with LC has been shown to be the method of choice for direct, sensitive analysis of protein digests.[38] Postcolumn splitting enables simultaneous MS analysis and fraction collection of desired peptides for further characterization by other analytical methods. Recent significant improvements in the quality of resin packings of microbore and capillary LC columns and the reduction in flow rates increase sensitivity required for direct on-line analysis of complex mixtures in proteomics applications.[39]

CE has become a desired orthogonal separation technique to LC for peptide mapping[40,41] and intact glycoprotein analysis,[42] and is now being successfully directly coupled to ESI-MS.[43–45] The high resolution and the small sample application in CE make fast and sensitive mass detection modes like ESI-Q-TOF-MS desirable.[46,47] Very recent advancement in the high-throughput arena is made by the development of microchip devices. Here separations are carried out in etched channels in either CE, capillary electrochromatography (CEC), or pressure-assisted electrochromatography mode. Coupling to MS is also being investigated, and some success in protein digest analysis has been reported.[48]

MS Principles in Protein Analysis

Multiply-Charged Ions

ESI of a protein generates a series of multiply-charged ions proportional to the number of charged residues. In the case of an acidic solution, the basic residues — Lys, Arg, and N-terminal primary amine — contribute to an envelope of charges along the molecule. As the number of charged residues statistically varies with the size of the molecule, the envelope of ions remains in a limited m/z range (usually within 2000 Da), allowing the analysis of molecules of vastly different molecular masses, notably large biomolecules. Successive peaks in a mass spectrum represent charge states differing by one; therefore, the molecular mass M can be calculated from two consecutive peaks. In practice, the computer combines all experimental m/z values in a reiterative process and applies methods such as maximum entropy processing to give a calculated mass with the best fit to the experimental spectrum.[49] Protein-mass measurement usually gives the average mass, but isotopic mass can be obtained from Gaussian peaks even if the instrument does not resolve the isotopic envelope.[50] It has been shown that high-resolution instruments (FTMS) that resolve highly charged ions are able to give accurate mass measurements (around 10 ppm) for hundreds of proteins in a single experiment.[51] With more classical instruments, due to multiple types of heterogeneities at the primary sequence and posttranslational levels, accurate mass measurement of intact proteins is often not feasible.

Peptide Sequencing by MS/MS

Fragmentation by MS/MS is a method of choice to determine the primary sequence of peptides. A precursor peptide ion is selected in the first mass analyzer and fragmented by collision with an inert gas in the collision chamber (CID). The resulting fragment ions are separated in the second region of the instrument to give a tandem mass spectrum or MS/MS spectrum. When performed in a triple-quadrupole or ion trap instrument, this technique is a low-energy process generating fragmentation at the amide bonds along the peptide sequence. This process generates a series of ions differing by one residue. Doubly charged ions, such as those usually found in proteolytic digests with trypsin due to the charge of the primary amine at the N-terminus and the presence of basic residues, such as Lys or Arg, at the C-terminus are favored precursors in this process. Fragmentation of doubly charged ions will give rise to singly charged fragments called b-ions if the charge remains at the N-terminus, or y-ions if the charge stays at the C-terminus (Figure 10.1). Further fragmentation of b-ions generates a-ions and internal immonium ions. Nomenclature of the different fragment ions has been proposed by Roepstorff and Fohlman[52] and Biemann.[53] Isobaric residues Ile and Leu cannot be distinguished by this technique. High-energy CID that generates side-chain breakdown and complex ion spectra is required to solve this problem.[54] The fragmentation of peptides by MS/MS has been studied in detail and is generally well understood.[55–57]

Protein Identification by Database Searching

Sequence information obtained by MS/MS fragmentation of peptides can directly be used to identify the original protein. Interpretation of tandem mass spectra is a slow and difficult process if performed *de novo*, i.e., with no information about the potential sequence of the peptide. Computer programs have been written to facilitate interpretation of *de novo* MS/MS data.[58] However, with the progress of DNA sequencing, complete sequences of entire genomes became available; therefore, peptide sequencing could be simplified to a correlation with protein, expressed sequence tag (EST), or genomic databases.

Different methods for protein identification, combining MS and bioinformatics, have been described. The peptide sequence tag method[59] combines the information obtained from a short stretch of amino acid residues inside a peptide with the start mass and end mass of the series. With the addition of the precursor mass and the enzyme specificity, a comparison to theoretical fragmentation of the database is made possible, which allows a specific matching to the experimental data. Another method uses noninterpreted MS/MS spectra to compare to theoretical spectra of all peptides generated by a specified enzyme in the entire database.[60] A third method, peptide mass fingerprinting, uses single MS experiments generated from MALDI-TOF analysis of tryptic digests.[61,62] This high-throughput method relies on the mass accuracy of experimental tryptic fragments. With five peptides or even less, depending on mass accuracy, a clear identification is obtained. Peptide mass fingerprinting presents some limitation for the analysis of mixtures of molecules and is best used for high-throughput analysis of a single protein component.

The application of the improved MS techniques presented above with highly resolving separation methods, such as 2-D electrophoresis, capillary HPLC, and CE, resulted in the creation of a new science, proteomics.[63] While genomics, described by DNA databases, represents the ground stage of the cell, the study of the differential status of the cell, due to various stimuli or disease states, reflects the functional expression of protein products or proteomics. Proteomics studies are aimed at identifying the proteome, the network of proteins that define the

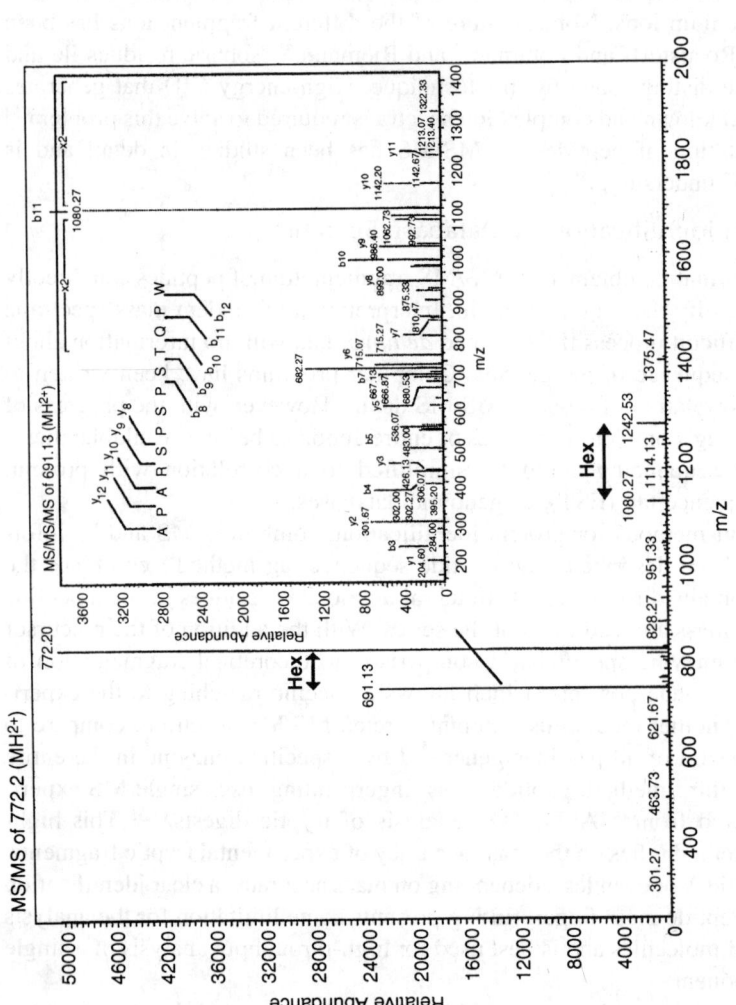

Figure 10.1 MS/MS sequencing of an O-glycosylated peptide. Application of dissociation energy on doubly charged ion 772.2 generates a tandem mass spectrum dominated by the dissociation of the glycan structure. The major ions are consistent with loss of hexose residue (mannose). The fragment ion 691.1 is then selected as a precursor ion in the next stage of fragmentation (MS³) to generate the b (N-terminal) and y (C-terminal) ions from which the amino acid sequence is reconstituted. (Data generated on ion trap LCQ Deca, Finnigan.)

active biology of the cell. Recent progress in MS techniques and instruments, combined with powerful separation methods and bioinformatics, provide discovery research with the tools to tackle these questions (reviewed in Reference 64).

USE OF MS FOR PROCESS DEVELOPMENT OF BIOPHARMACEUTICAL MOLECULES

The MS techniques described previously for characterization of the final recombinant protein product can be applied at all stages during process development. MS might be used upstream to define clone selection, processing format, and purification steps, and downstream to characterize the final product, ascertain lot-to-lot reproducibility, determine stability, and define the formulation of biopharmaceutical molecules. Presented here are some examples found either in the literature or from our own experience in which MS has been found to be a useful or necessary tool. Potential limitations of MS methods are discussed, and when appropriate, other analytical methods are mentioned that can be alternatives to MS and are also efficient tools for biopharmaceutical development.

Clone Selection and Processing Format

The first step in biopharmaceutical development is the selection of a clone in a specific cell line. Whole-mass analysis, if possible, is a fairly simple and powerful tool at this stage to verify the successful expression and translation of the desired protein. VanAdrichem et al.[65] described the use of MALDI MS to monitor protein expression in several mammalian cell lines like CHO DXB11, CHO SSF3, and hybridomas. Quantitative MALDI-TOF MS measurements of an IgG antibody and insulin during large-scale production in hybridoma cells were comparable to affinity chromatography results.

Field et al.[66] used MALDI-TOF MS to monitor the degradation of oligosaccharides during processing. Quantitative data of PNGase F-released carbohydrates from a recombinant tissue plasminogen activator mutant molecule were similar to the well-established but more time-consuming high-pH anion-exchange (AEX) chromatography method. However, ionization of sialylated species is not very efficient, and quantification in those cases is better achieved with chromatographic methods. Yet, this relatively fast mass spectrometric approach is very useful to evaluate the impact of changing cell culture conditions on the carbohydrate expression of IgG antibodies produced in CHO cells (no or very little sialylation) before scaling up the process.

In our laboratory we used whole-mass analysis or peptide mapping in combination with ES-MS to monitor the integrity of the recombinant product when comparing different processing formats such as batch vs. perfusion cell culture. We used recombinant Flt3 ligand (Flt3L) as an example of these experiments. Flt3L binds to and activates the cell surface tyrosine kinase receptor, Flt3R. This interaction initiates signaling events that regulate the proliferation and differentiation of multiple cell lineages of the hematopoietic system. The characterization of Flt3L, described in detail later in the text (see "Posttranslational Modifications"), shows that the molecule is highly heterogeneous due to a complex glycosylation profile and C-terminal processing. In the example of Flt3L, we did detect differences in the extent of glycosylation and C-terminal proteolytic processing, either when serum was used to supplement the cell culture or when the culture format was changed from perfusion to batch.

The application of MS techniques results in precise description of the bulk drug substance, the final result of process development. Information on molecule integrity, posttranslational modifications, and control of the multiple events that modify the final product quality are extremely important in the production of biopharmaceutical molecules. Feedback from analytical characterization is used to optimize clone selection and process conditions. Thanks to MS techniques, numerous cell culture parameters can be efficiently analyzed in relation to the final product. Parameters such as length of culture, lipid-linked oligosaccharide donor availability, dissolved oxygen, temperature, chemical inducers, and ammonia and glucosamine content have been shown to modify the final product quality, notably the glycosylation patterns. The influence of these factors on various therapeutic molecules, monoclonal antibodies (MAbs), glycodelin, γ-interferon, and tissue plasminogen activator have been described.[67–70] These examples show that advances in analytical techniques provide a valuable tool to optimize cell culture processes and improve the desired outcome on the bulk drug substance.

Purification Process and Impurity Identification

Efficient purification of biopharmaceutical molecules should involve a minimum of steps or unit operations. The final yield is directly related to the number of steps, and even in the case of highly efficient chromatographic methods, the yield is likely to suffer as the number of steps increases. A total of three chromatographic steps can be considered optimum. The addition of ultrafiltration/diafiltration steps in order to modify the pH and buffers for subsequent chromatography steps may also reduce the overall recovery. The addition of viral-inactivation methods — for example, acidic pH and membrane nanofiltration — increases the number of unit operations and negatively impacts the final yield. Commercial purification of MAbs represents a very favorable case, in the sense that the product of interest is captured from the crude cell medium by affinity chromatography on protein A. A highly specific chromatography step such as on protein A combines maximum yield and discrimination between product and impurities. Subsequent chromatographic steps — for example, hydrophobic interaction chromatography (HIC) and ion exchange — will play the role of polishing steps for optimized purity. Purification of other therapeutic molecules such as cytokines, clotting factors, growth factors, and soluble receptors that do not benefit from the power of affinity chromatography, is more challenging. The capture step will likely be on a high-capacity ion-exchange chromatography resin followed by a series of chromatographic steps applying to different properties of the molecule. Buffers and pH exchange by UF/DF are inserted between the chromatographic steps. Recent progress in membrane technology capitalizes on mixed modes combining ion exchange and filtration.

The development of a purification process and its subsequent scale up to commercial level involves the use of analytical techniques aimed at quantifying the recovery yield — for example, UV spectroscopy, SDS-PAGE, or RP-HPLC techniques. MS has a limited role at this stage and becomes important, rather, at a later stage in the characterization of the purification process. For example, an important factor in developing a purification process is based on the holding time of intermediates. The robustness of the process is improved by the possibility of keeping elution fractions in storage for a convenient length of time. MS is an appropriate tool to monitor product stability. In our lab, initial attempts at developing a purification process for a soluble receptor involved a series of steps in the following order: anion exchange (AEX), hydrophobic interaction chramotography (HIC), cation exchange (CEX), and hydroxyapatite (HA) chromatography. Process characterization was performed by LC-MS of Lys-C digests of intermediate fractions. As shown in Figure 10.2, degradation of the purified molecule involves C-terminal processing. Mass measurement of degraded forms lacking amino acid residues at the C-terminus indicated that the purification protocol was not optimal. LC-MS analyses showed that the degradation of the molecule was only demonstrated on CEX intermediates. Degradation of the starting material (Figure 10.2.2) resulted from exposure to acidic pH, as required by conditions used in CEX chromatography, which in turn activated contaminating proteases still present. The chromatography flowchart was modified by swapping the chromatographic step order and placing the HA column before the CEX. LC-MS of the new intermediates demonstrated that the product was no longer sensitive to CEX acidic conditions, due to efficient removal of contaminating proteolytic activities by HA chromatography (Figure 10.2-2 and Figure 10.2-3).

The goal of a purification process is to optimize the purity level of the final bulk drug substance. This is achieved by minimizing two different types of impurities: product-related and process-related. Process consistency of successive manufacturing runs is demonstrated by reproducible yields and constant product quality. Therefore, these two types of impurities have to be monitored and quantified. MS is a tool of choice for the former. Identities of product-related components are often demonstrated by MS. MALDI-TOF has been used to monitor the stability of an antibody in response to altering hybridoma cell culture conditions, notably the formation of molecular forms lacking the light chain.[71] ESI-MS has been used for rapid in-process assessment of known product-related variants, taking advantage of selective ion monitoring and minimum sample preparation. Monitoring antibody glycoforms distribution by MS is a fast and sensitive method.[72] A great deal of product-related impurities result from chemical, physical, or enzymatic degradation during the production process or, even,

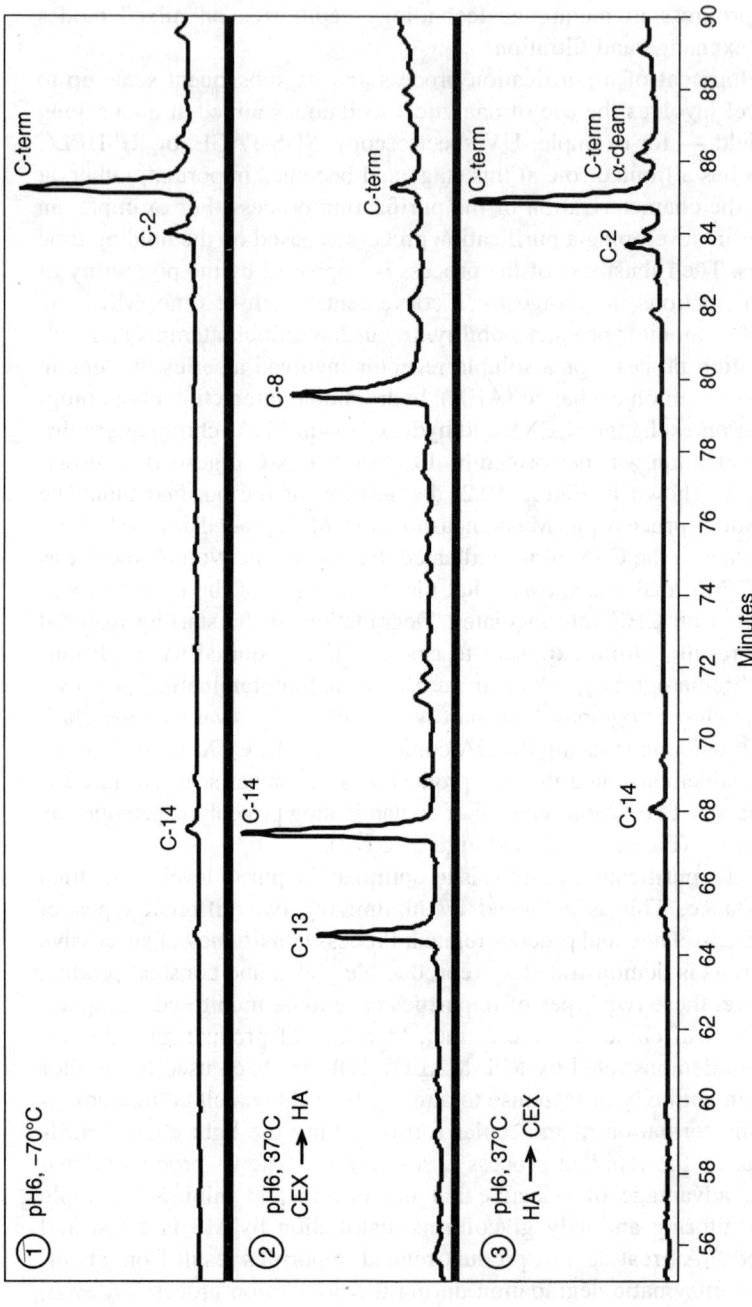

Figure 10.2 Optimization of a purification protocol by LC-MS analysis. (1) LC-MS analysis of C-terminal fragment from reference molecule purified by cation-exchange chromatography (CEX) and hydroxyapatite chromatography (HA). (2) Incubation at 37°C under acidic conditions shows degradation of the purified molecule if CEX precedes HA. (3) The molecule is stabilized when the sequence of the two purification steps is swapped.

as described above, during the purification process. MS is the essential tool to qualitatively describe these contaminants at the molecular level. Quantification of these impurities that may remain in the final bulk drug substance is better achieved by analytical chromatographic methods. For example, SEC, CE, and RP-HPLC are routinely used as quantitative assays in QC labs to measure high-molecular weight and low-molecular weight contaminants.

The limitations of MS are apparent in the analysis of the second type of contaminants, the process-related impurities. Some success can be obtained by applying proteomics techniques for the detection of these contaminants. In-gel digest of contaminating bands followed by identification of tryptic peptides against an appropriate database is a good method to identify major contaminants. Problems linked to low peptide recovery from in-gel digests can be overcome by direct analysis of protein mixtures using 2-D chromatography — for example, CEX and RP-HPLC.[73] However, these methods are not fully appropriate for the characterization of the problem at hand. The first reason is that the contaminants are not always proteins. Validation of biopharmaceutical processes calls for the quantification of potential contaminants such as viruses and DNA. In addition, contaminants can be related to carbohydrate or lipid structures. A second reason is the lack of a relevant protein database for the host cell, let alone the sequence of its genome. As a result, the positive identification of minute contaminants from MS/MS data becomes much more difficult. A third reason would be the presence of the bulk drug substance in large excess. Yates et al.[73] indicate that, typically, a protein mixture of 30 components within a 30-fold molar ratio of the main component can be identified by direct MS/MS analysis. This level is several orders of magnitude higher than potential contaminants in a commercial process of a therapeutic protein, which, compared to the major product, are quantified at the part-per-million level (ppm). Consequently, MS plays a limited role in those questions, better answered by the use of more sensitive ELISA techniques. Lot-release testing of biopharmaceuticals includes ELISA assays, such as protein A assays for MAb products, and host cells protein assays (HCP). HCP assays are developed using proteins produced in host cell transformed with a null expression vector where the gene of interest is absent. A mixture of these proteins is used as antigens for animal immunization. When the HCP ELISA has been developed, purification processes can be validated by demonstration of a significant reduction in HCP response along the purification steps. Although these tests suffer from a lack of definition of the antigens, they are widely accepted and routinely used to demonstrate purity for lot release of biopharmaceuticals.

Characterization of the Bulk Drug Substance

Whole-Mass Analysis

In several publications the mass spectral analysis of intact MAbs has been described.[74,75] MAbs are fairly big and complex, mostly recombinant biomolecules, constructed of two heavy-chain and two light-chain polypeptides that are covalently

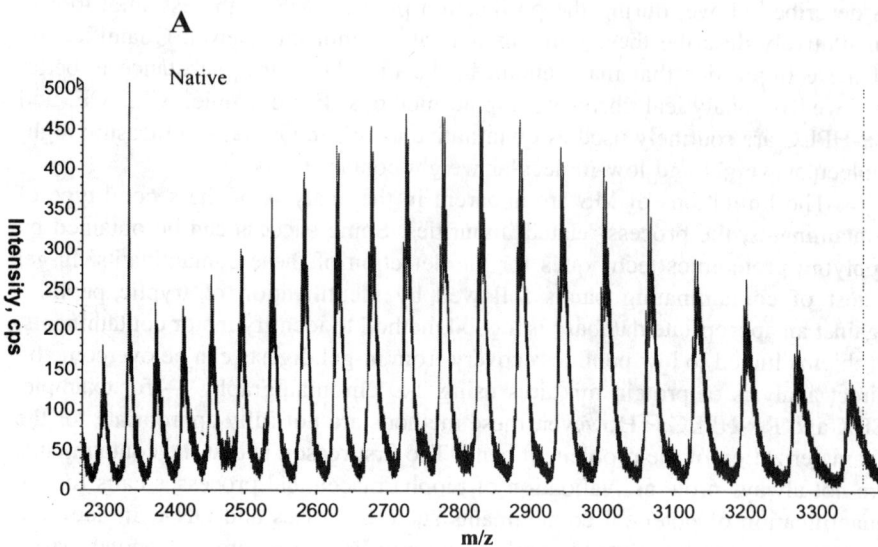

Figure 10.3   Whole-mass analysis of a monoclonal antibody. (A) Direct infusion of the antibody generates an envelope of high m/z ions ranging from 2000 to 3500. Deconvolution of the ion current signal gives the mass of the complete native molecule (147, 100.97 Da) and resolves some heterogeneity linked to the *N*-glycan structures. The major forms are consistent with molecules carrying biantennary structures capped with 0, 1, or 2 hexose (G = galactose) residues. (Data generated on an ESI-Q-Star instrument, Sciex-Applied Biosystems.)

linked via interdisulfide bonds. MAbs contain numerous posttranslational modifications such as glycosylation, cyclization of N-terminal glutamine to pyroglutamic acid on heavy and/or light chain, and carboxyl-terminal processing of heavy-chain lysine. Prior to whole-mass analysis by MS, reduction in this heterogeneity is advised. This can be accomplished by enzymatic treatment with either the endoglycosidase PNGase F or with carboxypeptidase B to remove specifically C-terminal lysine. Single-quadrupole instruments are appropriate to analyze these large biomolecules, although they suffer from the detection limits at high m/z values. Best results were achieved in our laboratories by infusing the antibody into an ESI-TOF mass spectrometer. Figure 10.3A shows a mass spectrum of an MAb, obtained by direct infusion into the Q-Star and the Mariner ESI-TOF instrument (Sciex-Applied Biosystems). Ionization of the high-mass MAb yields numerous m/z peaks in the range of 2000 to 3500. Figure 10.3B shows the same antibody spectrum after deglycosylation. After deconvolution of this m/z envelope of ions, the experimental mass accuracy obtained is 10 ppm or better. The relatively high accuracy and simplicity of this whole-mass detection gives this

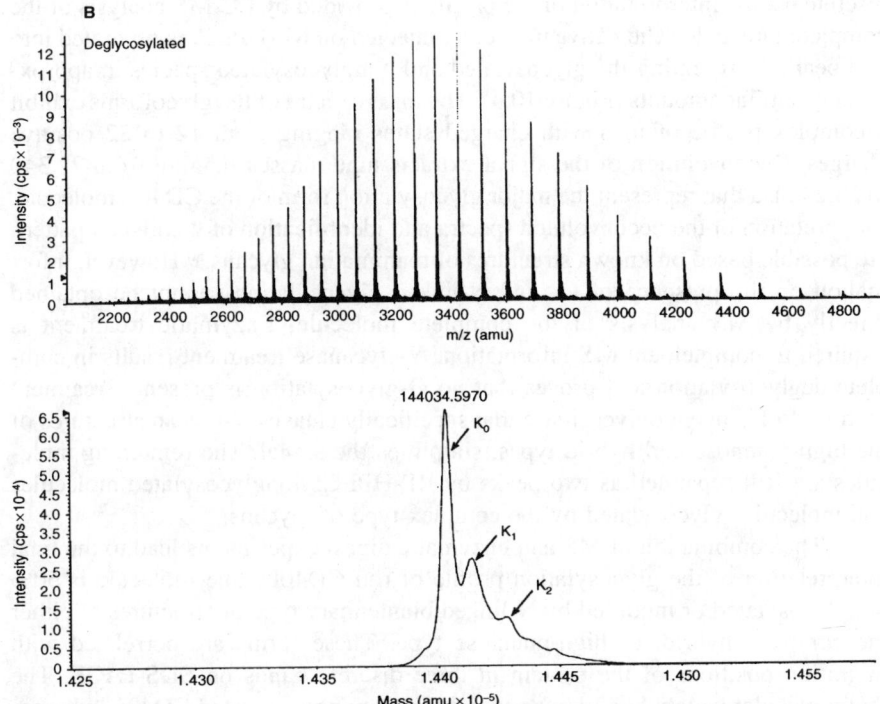

Figure 10.3 (continued) (B) Removal of the carbohydrates by *N*-glycanase treatment decreases the number of protonation sites, shifting the envelope toward higher m/z ions (2400 to 4800). Deconvolution of the signal of the deglycosylated molecule gives the mass of the antibody (144,034.59) at better than 10-ppm accuracy. The calculated major forms are explained by the presence of 0, 1, or 2 lysine residues (K) at the C-terminus of the heavy chain. (Data generated on an ESI-TOF Mariner instrument, Sciex-Applied Biosystems.)

method potential as a very fast identity method for lot release. Nedved et al.[76] described the usefulness of this method for a QC lab.

Recombinant human CD40 ligand (rhu CD40L) is used here as an example to illustrate the type of detailed information that can be obtained by direct analysis of a whole glycoprotein by MS. CD40L, a protein required for isotype switching and maturation of B cells,[77,78] has been produced from Chinese hamster ovary (CHO) cells as a fusion protein with an isoleucine zipper motif. Noncovalent association via the isoleucine zipper region results in the formation of a trimerized glycoprotein, the active form of the molecule.[79] The isoleucine zipper (amino acid residues 1–39) is fused to the rhu CD40L sequence (amino acid residues 40–189). The resulting monomer contains 189 residues with a theoretical peptide molecular weight of 21,031 Da. Rhu CD40L is separated by SDS-PAGE in three discrete bands. Interpretation of the profile is provided by LC-MS analysis of the complete molecule. The native molecule, injected on RP-HPLC, is separated into two peaks representing the glycosylated and nonglycosylated species in approximately similar amounts (Figure 10.4). The mass spectra of the glycoforms exhibit a complex profile of ions with charged states ranging from 12 to 22 positive charges. Deconvolution of the signal extracts nine masses ranging from 22,340 to 23,240 Da that represent the major glycosylation form of the CD40L molecule. Interpretation of the deconvoluted spectra and identification of the glycan pattern are possible based on known structures of mammalian glycans.[80] However, information on the presence of O- and N-linked glycosylation cannot be obtained directly by MS analysis of the complete molecule. Enzymatic treatment is required to complement MS information. N-glycanase treatment results in complete deglycosylation and proves that no O-glycosylation is present. Treatment with Endo-H, an endoglycosidase that specifically cleaves N-glycan structures of the high-mannose and hybrid types, simplifies the signal. The remaining molecules are still separated as two peaks by RP-HPLC: nonglycosylated molecules and molecules glycosylated by the complex type of glycans.

The combination of MS and enzymatic digest experiments lead to the final interpretation of the glycosylation profile of rhu CD40L. The molecule is 50% nonglycosylated or modified by N-linked biantennary type of structures of either the complex, hybrid, or high-mannose type. These forms are correlated with migration positions of the protein in three discrete bands on SDS-PAGE. The low-molecular weight band 1 corresponds to the nongycosylated CD40L, whereas the glycosylated molecules migrate at higher molecular weights in bands 2 and 3. Band 2 corresponds to CD40L modified by high-mannose or hybrid structures, whereas band 3 represents modification by glycan structures of the complex type. The major high-mannose form, $(N\text{-acetylglucosamine})_2\text{-(mannose)}_8$, corresponds to a processing of the N-glycan chain stopped at the endoplasmic reticulum stage,[81] whereas the other calculated forms correspond to an almost complete distribution of the various stages of N-glycan maturation, up to completely sialylated complex structures. The masses of the major complex forms can be interpreted as biantennary structures without sialic acid or capped by one or two neuraminic acid residues. MS of the complete molecule offers a clear evaluation of the glycosylation heterogeneity. Monitoring the distribution of glycoforms is important due to its impact on bioavailability and half-life of a protein therapeutic.

When such investigation is technically possible, as shown in the examples of MAbs and CD40L, MS analysis of complete molecules is fast, powerful, and the method of choice to obtain detailed characterization information. Such a technique should be among the first to be applied for the study of a therapeutic protein. Limitations of the technique include lack of resolution of heterogeneous mixtures and lack of interpretation of isobaric structures, identical in mass but molecularly distinct. The first point can be overcome by reduction of heterogeneity with prior enzymatic treatment or applying MS methods to protein fragments, as described in greater detail in the following text; the second point is inherent to MS and cannot be solved by the technique alone. Distinction of isobaric motifs — for example, triantennary structures with lactosamine extension vs. tetra-antennary structures — is best resolved by a combination of MS and glycosidase treatment of released glycans in combination with separation methods. MS analysis of fluorescently labeled glycans separated by LC or CE techniques is an efficient solution.[82]

Peptide Mapping

The primary goal of peptide mapping is the verification of the amino acid sequence deduced from the genetic code of the recombinant protein. The protein backbone gets cleaved by typically two or three different endoproteinases like Lys-C, trypsin, and Glu-C to achieve maps with sequence-overlapping peptide fragments. These peptide mixtures can then be separated by LC or CE and analyzed on-line by MS to obtain sequence information. Often simple mass analysis matches the predicted primary sequence of the protein. However, sometimes mutations can lead to isobaric masses of peptides that can be overseen, if no further sequence analysis like N-terminal Edman sequencing and MS/MS is carried out.

Recombinant proteins expressed in eukaryotic cells usually carry a high level of heterogeneity such as multiple N- and C-termini, modified amino acids, and glycosylation. This heterogeneity leads to the dispersion of ionization signal between very close molecular species. As a result, the mass spectrum of an intact mammalian- or yeast-expressed recombinant protein is often a broad, unresolved envelope with no practical value for characterization. To overcome this problem, complete characterization of the molecule, which is a crucial goal during the development of a production process, is approached using peptide-mapping procedures. The molecule is cleaved with proteolytic enzymes to generate smaller fragments that are separated by RP-HPLC and analyzed by MS. The information obtained on each fragment is summed up for all peptides to reconstruct the complete information on the whole molecule. ESI-MS allows peptide identification through LC-MS and LC-MS/MS techniques. Furthermore, the technique is well suited to distinguishing coeluting peptides. A method of choice for peptide identification is the use of ion trap instruments with an automatic data acquisition procedure. This method is composed of three components: (1) full-scan MS across the mass range to find the mass of the peptide, (2) zoom-scan of selected ions to determine the charge state, and (3) data-dependent MS/MS for identification by database searching. During process development, where the protein sequence is known, speed in the data interpretation is enhanced by the use of computer algorithms (ProMass, Sherpa, or Protein Prospector, http://prospector.ucsf.edu) that create peptide mass fingerprints with all possible

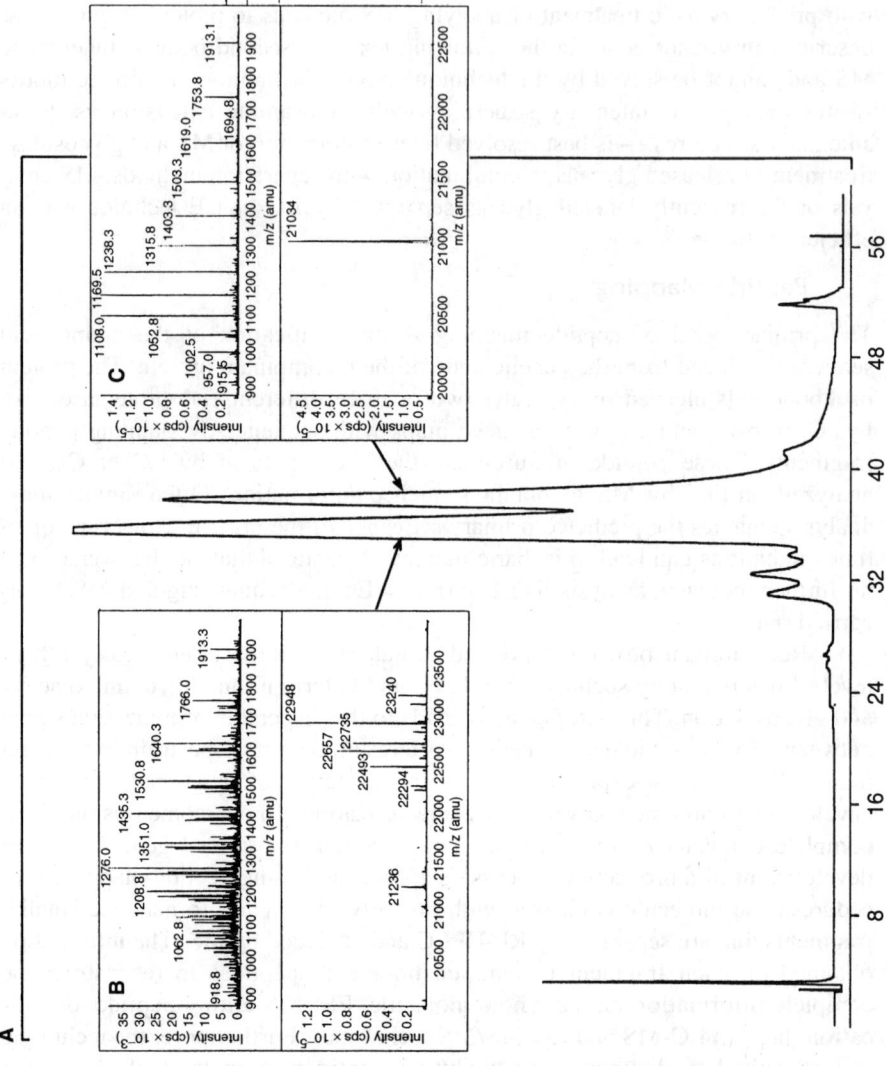

Figure 10.4 LC-MS of CD40L. Reversed-phase HPLC separates a CD40L preparation into several components. (A) Mass measurement of the complete molecule shows that the major peak (B) contains a complex mixture of glycosylated molecules (detailed in the text). The second major peak gives a simpler signal (C), deconvoluted to a mass of 21,034 Da consistent with the nonglycosylated molecule (50 ppm accuracy). (Data generated on a triple-quadrupole API350 instrument, Sciex.)

proteolytic enzymes or chemical reagents. MALDI-MS is useful as an off-line characterization procedure. The enzymatic digest mixture can be spotted on the MALDI plate and analyzed in a single MS experiment to evaluate the number of peptides generated, i.e., the sequence coverage. The high resolution of the MALDI technique is helpful to differentiate closely associated peptides, such as those generated by deamidation.

Lastly, peptide mapping is a useful tool to monitor the genetic stability of recombinant proteins. Besman and Shiba[83] demonstrated genetic stability by comparing the tryptic LC-MS maps generated at the beginning, middle, and end of a production campaign of recombinant human factor VII.

N-Terminal Heterogeneity

A common problem in process development is linked to the potential heterogeneity in the primary sequence. Although solving the underlying causes of such heterogeneity may not be always possible, it is important to recognize the problem early. MS is a valuable tool for this assessment. The goal of process development will be to define conditions to characterize heterogeneity and to ensure lot-to-lot reproducibility.

PIXY321, a human cytokine analog genetically engineered by the fusion of granulocyte-macrophage colony-stimulating factor (GM-CSF) and interleukin-3 (IL-3), offers an interesting case of these problems with heterogeneity at both N- and C-termini.[84] The molecule, developed as a treatment of chemotherapy-induced myelosuppression, was expressed in yeast under the control of various expression cassettes including the alcohol dehydrogenase 2 (ADH2) promoter and the α-mating factor expression system. Analysis of the final product showed a mixed population of proteins with four different N-terminal residues: Ala1, Ala3, Arg4, and Ser5; however, forms beginning at Pro2 or shorter than Ser5 were never obtained. The study showed that N-terminal heterogeneity depended both on the expression system and the specific sequence of the molecule. We deduced that three different proteolytic activities, required for the maturation of the natural α-factor in yeast, were responsible for the N-terminal variation. First, the action of the endoprotease ysc F (kex 2 gene product) generated Ala1, the mature sequence. Then, ysc IV activity (ste 13 gene product) removed the Ala1-Pro2 dipeptide to generate the Ala3 form. Third, an exoprotease started removing residues to generate Arg4 and Ser5 forms. The mature PIXY321 and the truncated form Ser5 were not substrates for this enzyme (Figure 10.5). The exopeptidase activity was evident following stability studies where the Ala3 form was converted to Arg4 then Ser5 form in a few weeks while the mature form stayed intact. The N-terminal variants were the result of the combined action of metal-dependent proteases and processing enzymes found in the secretion pathway. Control of the N-terminal heterogeneity required either site-directed mutagenesis of the N-terminal sequence to engineer a protease-resistant variant or the use of inhibitors to stop the action of the exoprotease and limit the heterogeneity to two sequences, Ala1 and Ala3. The stability of the process was obtained by inclusion of a metal-chelating

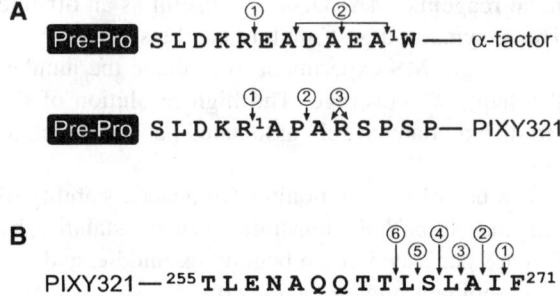

A

① ②

Pre-Pro S L D K R E A D A E A¹ W —— α-factor

① ② ③

Pre-Pro S L D K R¹ A P A R S P S P— PIXY321

B

⑥ ④ ②
 ⑤ ③ ①

PIXY321 — ²⁵⁵ T L E N A Q Q T T L S L A I F²⁷¹

Figure 10.5 PIXY321 N- and C-terminal heterogeneities. A series of enzymatic activities required for the normal processing of yeast α-factor explains the degradation of recombinant PIXY321. (A) At the N-terminus, the enzyme yscF (1) generates Ala1, then ysc IV (2) generates Ala3 and finally a metallo exoprotease (3) degrades the molecule to the final stage forms Arg4 and Ser5 (detailed in text). (B) At the C-terminus, a stretch of hydrophobic residues is a favorable substrate for a carboxypeptidase B-like activity resulting in a mixture of molecules lacking from 1 to 6 residues.

reagent that limited the heterogeneity to two forms, Ala1 and Ala3, within a specific ratio. The fact that the N-terminus of PIXY321 and GM-CSF, Ala-Pro, is a preferred substrate for the ysc IV activity makes the presence of two sequences at the N-terminus, the mature and Ala3 forms, unavoidable.

An engineered PIXY variant, Ala-Pro-Pro, a sequence similar to natural kex 2 protease, which was expressed in the α-factor expression environment, proved to be 100% homogeneous.[84] Analytical characterization of the N-terminal heterogeneity was important to define solutions for its control and to improve the process. MS of the complete molecule was not an option, as a very complex envelope of peaks that could not be deconvoluted was obtained, suggesting that numerous levels of microheterogeneities were present. Lys-C peptide mapping and LC-MS were not totally appropriate monitoring tools, as the N-terminal peptide L1 showed a progressive elution of glycopeptides carrying 1–10 mannose residues. In this case, O-linked glycosylation heterogeneity, present on 75% of the molecules, added complexity to the N-terminal sequence heterogeneity (Figure 10.6). Two analytical methods were found suitable for monitoring of the clinical lots: N-terminal sequencing and LC-MS of the peptide map of the deglycosylated molecule. Precise quantification by N-terminal sequencing of a complex mixture of four sequences is a difficult proposition. Deglycosylation by treatment with α-mannosidase followed by Lys-C peptide mapping and LC-MS is a good approach to analyze the mixture. When the process limits the complexity to two sequences, Ala1 and Ala3, in consistent ratio, N-terminal sequencing was found to be both precise and a simpler method for lot release.

C-Terminal Heterogeneity

MS was a decisive method in the interpretation of PIXY321 C-terminal heterogeneity.[84] Analysis of the peptide map of the PIXY 321 molecule digested with Lys-C revealed that the expected full-length C-terminal fragment was not present. Instead, several smaller C-terminal fragments were detected in the early portion of the chromatogram. Moreover, analysis of the molecule was hindered by the presence of a glycosylation site near the C-terminus. Quantitative analysis was obtained by preparing a deglycosylated molecule using α-mannosidase 1–2,3,6, an enzyme with a broad specificity toward various types of α-linked mannose residues. LC-MS identified six different C-termini that exhibited the following relative frequencies: Ile270, 35% > Ala269, 29% > Thr265, 15% > Ser267, 11% > Leu 268, and Leu 266, 5%. LC-MS techniques were valuable to obtain additional information on the native, glycosylated molecule. Both nonglycosylated and glycosylated C-terminal fragments were identified by monitoring of the doubly charged ions. Six C-terminal forms were present on PIXY321. The three longest, 270, 269, and 268, were found to be glycosylated on residue Ser267. The three smallest, 267, 266, and 265, were present as nonglycosylated species. MS techniques were instrumental in interpreting the complexity of the C-terminal region. The peptide map was validated as a QC method to quantify the relative distribution of each C-terminal species. Based on consecutive fermentation lots, a relative standard deviation of 12% of the areas of the C-terminal peaks in the peptide map was obtained.

MS, namely LC-MS of proteolytic fragments, is the tool of choice to describe C-terminal heterogeneity. The only alternative technique would be C-terminal sequencing using chemical methods.[85] However, C-terminal sequencing is not extremely sensitive (10 to 100 pmol required) and suffers some limitation with specific residues (proline not determined). Recent developments are aimed at overcoming those limitations.[86,87] Although chemical methods can be successfully applied, especially if the sequence is unique, in the case of rugged C-termini, where numerous related sequences are present, automatic C-terminal sequencing cannot be the answer. Our experience led us to conclude that a complex problem such as PIXY321 was beyond the capacity of the C-terminal sequencing method. On the other hand, once the interpretation of the C-terminal heterogeneity is achieved by MS techniques, monitoring the consistency of the process is usually performed by other methods. Chromatographic methods without on-line MS are appropriate. As described above, monitoring of C-terminal fragments by peptide mapping is often chosen. Even better are methods that can give information on the C-terminus by separation of the complete molecules. MAbs are well-known cases where the C-terminal heterogeneity on the heavy chain involves a charged residue, lysine. Taking advantage of this fact, separation techniques based on charge, CEX chromatography, and capillary isoelectric focusing (C-IEF) can be applied on intact antibody molecules to quantify their C-terminal heterogeneity.[88–90]

A

¹APARSPSPSTQPWEHVNAIQEALRLLDLSRDTAAEMNEEVEVISEMFDLQEPTCLQTRLELYK
L1
17, 18

⁶⁴QGLRGSLTK
L2
2

⁷³LK
L3

⁷⁵GPLTMMASHYK
L4
4

⁸⁶QHCPPTPETSCATQIITFESFK
L5
7, 8

¹⁰⁸ENLK
L6
1

¹¹²DFLLVIPFDCWEPVQEGGGGSGGGGGSAPMTQTTPLK
L7
13, 14,15

¹⁴⁹TSWVDCSNMIDEIITHLK
L8
16

¹⁶⁷QPPLPLLDFNNLNGEDQDILMENNLRRPNLEAFNRAVK
L9
10, 11,12

²⁰⁵SLQDASAIESILK
L10
6

²¹⁸NLLPCLPLATAAPTRHPIHIK
L11
7, 9

²³⁹DGDWNEFRRK
L12
3

²⁴⁹LTFYLK
L13
5

²⁵⁵TLENAQAQQTTLSLAIF
L14
early eluting peaks
(25 min)

Figure 10.6 Peptide mapping of PIXY321. (A) Cleavage of the primary sequence with Lys-C endoprotease results in 14 theoretical fragments. (B) As detailed in the text, LC-MS analysis of the Lys-C digest shows that a few peptides, notably L1, L7, and L14, elute at several retention times due to heterogeneity of both their glycan structures and amino acid sequence. The insets illustrate the complexity of the mass spectra of glycopeptide L1 eluting under peaks 17 (inset C) and 18 (inset D).

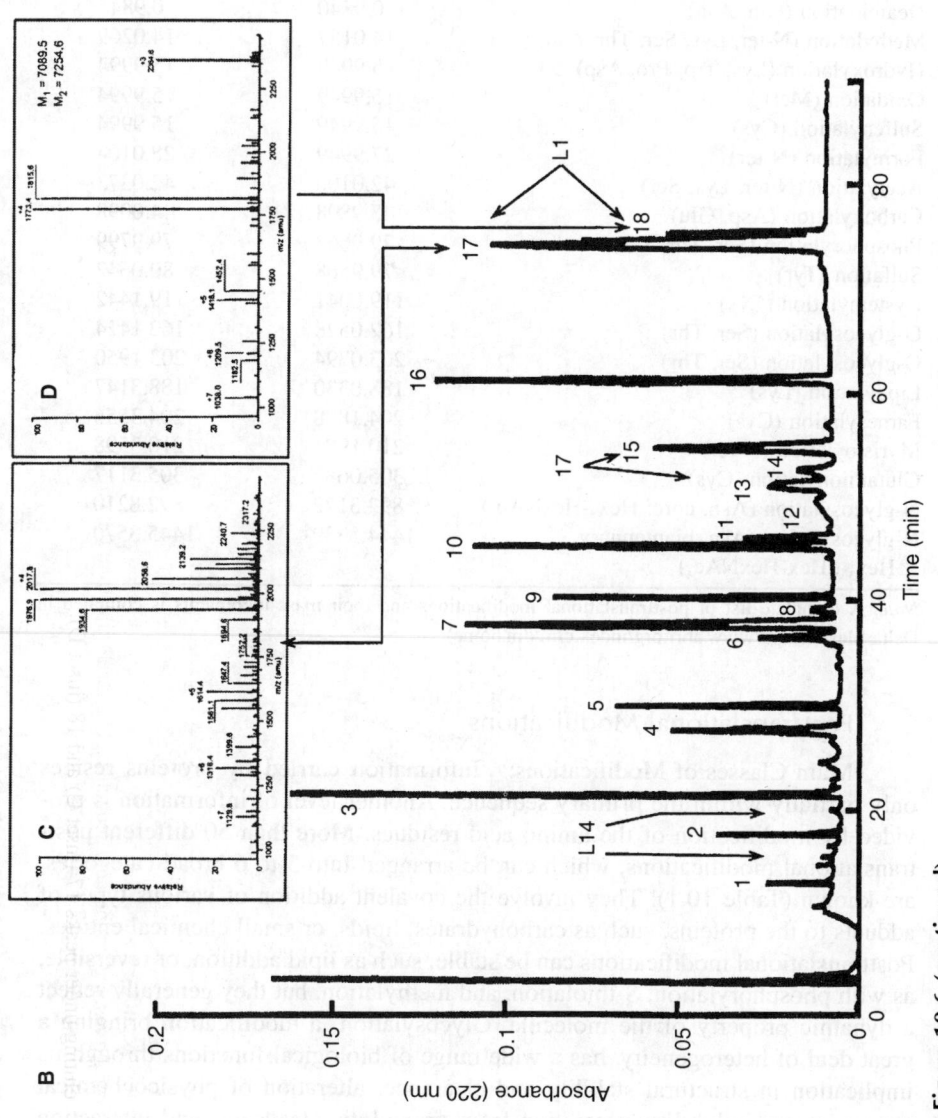

Figure 10.6 (continued)

Table 10.1 Most Common Posttranslational Modifications on Proteins

Modification (Amino Acid Residue)	Mass Change (iso)	Mass Change (avg)
Pyroglutamic acid (Gln)	17.0265	17.0306
Disulfide bond (Cys)	2.0157	2.0159
Deamidation (Gln, Asn)	0.9840	0.9847
Methylation (N-ter, Lys, Ser, Thr, Asn)	14.0157	14.0269
Hydroxylation (Lys, Trp, Pro, Asp)	15.9949	15.9994
Oxidation (Met)	15.9949	15.9994
Sulfenylation (Cys)	15.9949	15.9994
Formylation (N-ter)	27.9949	28.0104
Acetylation (N-ter, Lys, Ser)	42.0106	42.0373
Carboxylation (Asp, Glu)	43.9898	44.0098
Phosphorylation (Ser, Thr, Tyr)	79.9663	79.9799
Sulfation (Tyr)	79.9568	80.0642
Cysteinylation (Cys)	119.0041	119.1442
O-glycosylation (Ser, Thr)	162.0528	162.1424
O-glycosylation (Ser, Thr)	203.0794	203.1950
Lipoylation (Lys)	188.0330	188.3147
Farnesylation (Cys)	204.1878	204.3556
Myristoylation (N-ter)	210.1984	210.3598
Glutathionylation (Cys)	305.0682	305.3117
N-glycosylation (Asn, core: Hex_3-$HexNAc_2$)	892.3172	892.8210
N-glycosylation (Asn, biantennary, Hex_3-$dHex$-$HexNAc_4$)	1444.5339	1445.3570

Note: A complete list of posttranslational modifications and their mass increments is compiled in Delta Mass (http://www.abrf.org/index.cfm/dm.home).

Posttranslational Modifications

Main Classes of Modifications: Information carried by proteins resides only partially within the primary sequence. Another level of information is provided by modification of the amino acid residues. More than 50 different posttranslational modifications, which can be arranged into 5 to 6 broad categories, are known (Table 10.1). They involve the covalent addition of various types of adducts to the proteins, such as carbohydrates, lipids, or small chemical entities. Posttranslational modifications can be stable, such as lipid addition, or reversible, as with phosphorylation, S-thiolation, and methylation, but they generally reflect a dynamic property of the molecule. Glycosylation, a modification bringing a great deal of heterogeneity, has a wide range of biological functions through its implication in structural stability and clearance, alteration of physicochemical properties and solubility, protection from proteolytic cleavages, and interaction with receptors involved in cell traffic. Some posttranslational modifications act as signaling markers for subsequent biological events. For example, lipid modification is important for localization of proteins in membranes, and differential phosphorylation by cellular kinases is a main component of signal transduction. Precise chemical modifications of specific residues will alter their properties and result in new classes of amino acids with specific functions. A good example of this type of modification, important for biopharmaceutical development, is γ-carboxylation of glutamic acids, key to membrane interaction and bioactivity of clotting factors.[91] The posttranslational modifications of this class of proteins is especially complex and, in addition to γ-carboxylation, include glycosylation, β-hydroxylation of aspartic acid residues, sulfation of tyrosine residues, and phosphorylation of serine residues. Developing a process for the expression of fully active clotting factors by recombinant DNA technology is a difficult challenge.[92]

Disulfide formation is a crucial component of the biological pathway, required for proper protein folding, dimerization, or oligomerization into active complexes. Mispairing of disulfide bridges is sometimes observed in the case of overexpressed recombinant molecules and can result in reduced or abolished activity. Therefore, understanding the disulfide structure of recombinant molecules and eliminating incorrectly folded species is an important aspect of clinical manufacturing of protein therapeutics. Detailed information on posttranslational modifications, including comparison of natural and recombinant molecules, can be found in Krishna and Wold.[93]

In contrast to their positive role in determining the full biological activity of numerous proteins, posttranslational modifications may have a negative impact in the onset of autoimmune diseases or other pathological conditions. Several examples of the generation of autoepitopes linked to posttranslational modifications have been described.[94] Systemic lupus erythematosus (SLE) involves adverse response to proteins with phophorylation and β-aspartylation.[95] Arthritis results from autoantigenicity of certain protein motifs, degradation products of type II collagen modified by hydroxylation, certain types of glycan structures, and citrullination, or the deimination of arginine residues.[96]

Pathological conditions are also linked to posttranslational modifications such as oxidized histidine residues found in β-amyloid protein of Alzheimer's patients, or conformational variants in the case of prion-induced encephalopathies. The development of sensitive MS tools and proteomics techniques is playing an active role in the precise description of these mechanisms.[97,98]

O-Linked and N-Linked Glycosylation: The carbohydrate modifications of recombinant proteins can affect several factors *in vivo*, such as half-life, bioactivity, immunogenicity, and stability.[99–101] Hence, characterization, lot-to-lot consistency, and stability are requirements that regulatory authorities require from the manufacturer of therapeutic products. MS in combination with separation techniques has been a powerful tool to characterize the heterogeneity of glycoproteins. Information on protein glycosylation can be obtained by analyzing either the glycoproteins or the isolated glycan structures released from the protein backbone by chemical treatment or enzymatic reaction. Various analytical methods are used for the separation of glycan structures or complete glycoproteins, including gel electrophoresis, HPLC, and CE (for recent reviews see References 102 and 103). Recently, changes in glycosylation of proteins were investigated using a method based on specific lectins and surface plasmon resonance.[104]

We will focus here on MS methods for the analysis of recombinant glycoproteins using human Flt3L, introduced earlier ("Clone Selection and Processing Format"), as an example. CHO cell-derived Flt3L consists of 156 amino acids with 3 intrachain disulfide bonds. Its molecular weight when determined by whole-mass analysis varies from 18 to 29 kDa, indicating a variety of glycoforms processed to different extent. Proteins expressed in eukaryotic systems are mainly associated with N- or O-linked glycans with N-glycosidic bonds to the side chain of an Asn residue (consensus sequence Asn-X-Ser/Thr) or O-glycosidic bonds to the side group of Ser or Thr (often in a Ser-, Thr-, Pro-rich region). Flt3L contains two N-linked glycan sites (Asn100 and Asn123) and five O-linked sites, all residing in the C-terminal region (Ser136, Ser137, Thr138, Ser144, and Thr151). Reversed-phase chromatography could separate species with different extent of N-glycosylation occupancy. On-line ES-MS indicated heterogeneity within each chromatographic peak (Figure 10.7). To identify the glycosylation sites, tryptic peptides were generated and analyzed by LC-MS. Peptides T11 (contains Asn100) and T13 (contains Asn123) show masses that are increased in increments, which correspond to CHO typical biantennary core-fucosylated glycostructures with 0, 1, or 2 *N*-acetylneuraminic acid residues capping the terminal galactose residues. Because the negative charge of the neuraminic acid residue reduces the ionization efficiency in the typically applied positive ion mode, the same analysis was carried out after enzymatic treatment of Flt3L with neuraminidase to eliminate this negative charge and increase sensitivity. Several more branched and nonfucosylated structures could be identified this way. Interestingly, during development of different fermentation conditions, varying amounts of certain N-linked glycostructures were found. These forms could also be seen by whole-mass analysis of desialylated Flt3L. Sequential exoglycosidase digestion with neuraminidase, β-galactosidase, β-hexosaminidase, α-mannosidase, β-mannosidase, and α-fucosidase followed by mass spectral analysis can verify these oligosaccharide structures unambiguously.[105]

Analysis of the Flt3L O-glycosylation was more complex. The C-terminal tryptic peptide (T14) contained all O-linked glycosylation sites. LC separated at least six peptides that, when analyzed by N-terminal Edman degradation, contained the sequence of this C-terminal peptide. Mass analysis could only identify three peptides as T14 with three *O*-glycans containing the *N*-acetylgalactosamine-galactose structure with one or two neuraminic acid residues on each glycan structure. The other T14 peptide masses did not yield results that could easily be interpreted. Only when the MS/MS spectra of these unknown peptides were compared to the three identified T14 glycopeptides did it become clear that

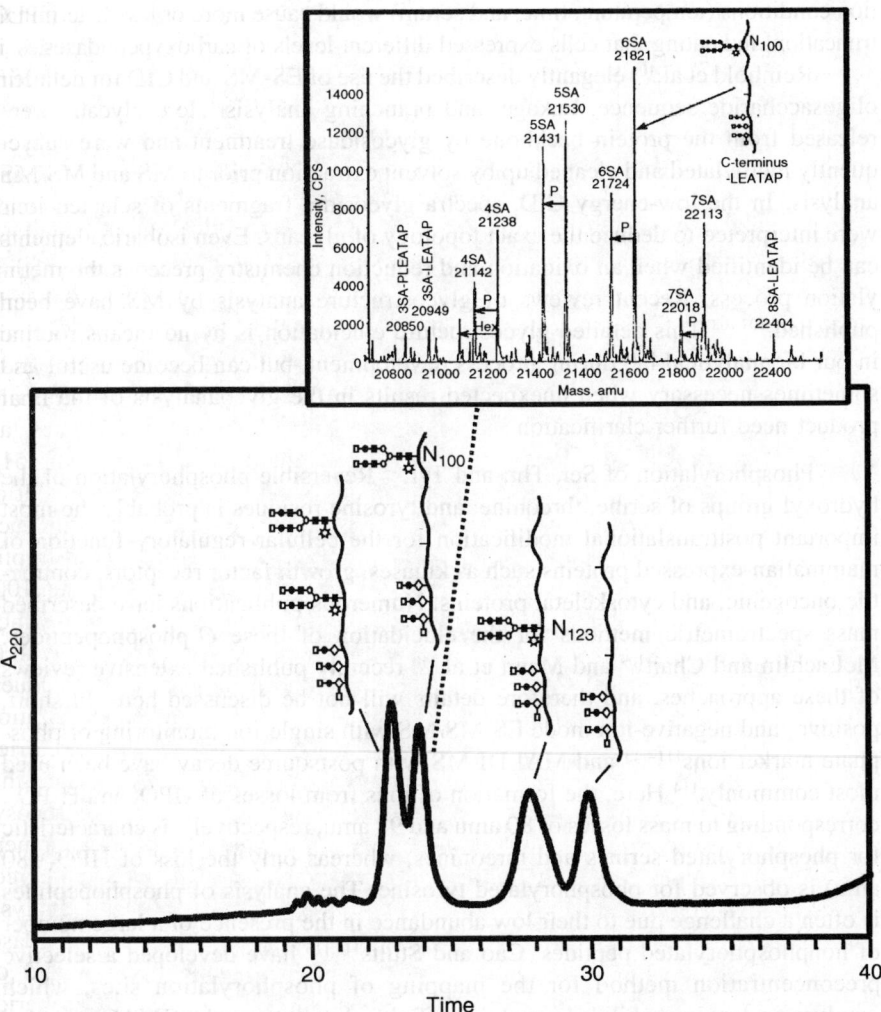

Figure 10.7 Separation of rhuFlt3L glycoforms by RP-HPLC. The recombinant molecule is separated by RP-HPLC and analyzed on-line by electrospray mass spectrometry. RhuFlt3L is separated into four N-linked glycoforms. Peak 1 is glycosylated at both N_{100} and N_{123}; peak 2 is glycosylated only on N_{100}; peak 3 is occupied only on N_{123}; peak 4 has no N-linked glycosylation. The inset illustrates the reconstructed mass spectrum from peak 2. C-terminally clipped rhuFlt3L with 3-O-glycosylation sites occupied and one biantennary core-fucosylated N-linked structure is observed with different amounts of sialic acid.

C-terminal truncation of four amino acids had occurred. Again, different fermentation conditions (temperature, time, and serum) would cause more or less C-terminal truncation, indicating that cells expressed different levels of carboxypeptidases.

Reinhold et al.[106] elegantly described the use of ES-MS and CID for detailed oligosaccharide sequence, linkage, and branching analysis. Here glycans were released from the protein backbone by glycosidase treatment and were subsequently methylated and cleaned up by solvent extraction prior to MS and MS/MS analysis. In the low-energy CID, spectra glycosidic fragments of selected ions were interpreted to deduce the exact topology of glycans. Even isobaric elements can be identified when an oxidation and reduction chemistry precedes the methylation process. Recent reviews of glycostructure analysis by MS have been published.[107,108] This detailed glycostructure elucidation is by no means routine in our bioanalytical lab during process development, but can become useful and sometimes necessary when unexpected results in the glycoanalysis of the final product need further clarification.

Phosphorylation of Ser, Thr, and Tyr: Reversible phosphorylation of the hydroxyl groups of serine, threonine, and tyrosine residues is probably the most important posttranslational modification for the cellular regulatory function of mammalian-expressed proteins such as kinases, growth factor receptors, contractile oncogenic, and cytoskeletal proteins. Numerous publications have described mass spectrometric methods for the elucidation of these O-phosphopeptides. McLachlin and Chait[109] and Mann et al.[110] recently published extensive reviews of these approaches, and therefore details will not be discussed here. In short, positive- and negative-ion mode ES-MS/MS with single-ion monitoring of phosphate marker ions[111–113] and MALDI-MS with postsource decay have been used most commonly.[114] Here, the formation of ions from losses of HPO_3 and H_3PO_4, corresponding to mass losses of 80 amu and 98 amu, respectively, is characteristic for phosphorylated serines and threonines, whereas only the loss of HPO_3 (80 amu) is observed for phosphorylated tyrosine. The analysis of phosphopeptides is often a challenge due to their low abundance in the presence of a large number of nonphosphorylated peptides. Cao and Stults[115,116] have developed a selective preconcentration method for the mapping of phosphorylation sites, which involves on-line immobilized metal–ion affinity chromatography (IMAC) coupled to CE-ES-MS and CE-ES-MS/MS^n. Sequencing the exact phosphorylation sites seems to remain challenging.

Alternative methods to MS for the analysis of protein phosphorylation would involve immunoaffinity techniques using antibodies specific for phosphorylated residues. Surface plasmon resonance technology has been recently used for the determination of active concentration of phosphotyrosine-containing proteins at picomolar levels.[117]

Phosphorylation of Glycan Residues: Inductively coupled plasma-mass spectrometry (ICP-MS) is the technique used for analysis of the metal content of complete molecules (for a review see Reference 118). Metal analysis, applied to a CHO-expressed recombinant molecule in development as anticancer therapy, indicated that approximately 0.3 mol of phosphate were present per mole of protein.[119] Interestingly, the modification was shown to occur on high-mannose-type glycans. An analytical method based on fragmentation of glycopeptides by MS/MS was developed to analyze the presence and location of this modification. In this procedure, Asn-linked glycopeptides are isolated by RP-HPLC and analyzed by nanospray ion trap MS before and after treatment with alkaline phosphatase and α-mannosidase.[119] MS/MS fragmentation patterns of glycan structures are used to identify the phosphorylated carbohydrates and their position in the glycan. Combining information from enzymatic digests and mass spectral data was necessary for complete interpretation. LC-MS analysis of a Glu-C digest of the molecule identifies, among other ions, three triply charged ions at m/z 1474.8, 1555.5, and 1636.6, which can be assigned to high-mannose-type glycan structures. MS/MS fragmentation of one example, the triply charged ion 1555.5, is shown in Figure 10.8. When treated with *E.coli* alkaline phosphatase, 1474.8 remains intact, whereas the triply charged ions 1636.6 and 1555.5 are replaced by smaller triply charged ions. No change in the ions corresponding to the protein portion of the molecule is observed, a result eliminating a potential modification located on Ser, Thr, or Tyr residues. When treated by α (1,2, 3,6) mannosidase, the triply charged ion 1636.6 remains intact, whereas 1555.5 and 1474.8 are replaced by smaller triply charged ions. Nanospray MS/MS of isolated glycopeptides shows that fragmentation along the oligosaccharide backbone is observed more readily than that of the polypeptide (Figure 10.8). Monosaccharide motifs are sequentially fragmented down to the proximal GlcNAc. These results show that nanospray tandem MS is an efficient method to identify and localize modifications on mammalian glycan structures. The presence and position of the phosphate in high-mannose glycan can be identified, based on fragmentation. We concluded that the triply charged ion 1474.8 is a glycopeptide with an N-linked site only occupied by high-mannose structure (5 mannose residues), 1555.5 is modified by an additional mannose-6-phosphate adduct, and 1636.6 is modified by two mannose-6-phosphate adducts.

Interestingly, the phospho-glycopeptide fragmentation does not generate a significant loss of the phosphate moiety. This is in contrast to the process that is observed during ESI-MS of phosphopeptides where a significant loss of HPO_3 and H_3PO_4 is observed from the parent ion (see previous subsection "ESI–Ion Trap Instruments"). It is likely that mannose-6-phosphate remains intact due to conformational constraints of the carbohydrate ring. In each glycan, loss of mannose was observed before loss of mannose-6-phosphate, even when they were equally peripheral. This indicates that the presence of phosphate on mannose results in slower glycosidic cleavage of the modified sugar. Differential sequencing of carbohydrates and peptides makes MS/MS an efficient method for the characterization of posttranslational modification of glycoproteins.

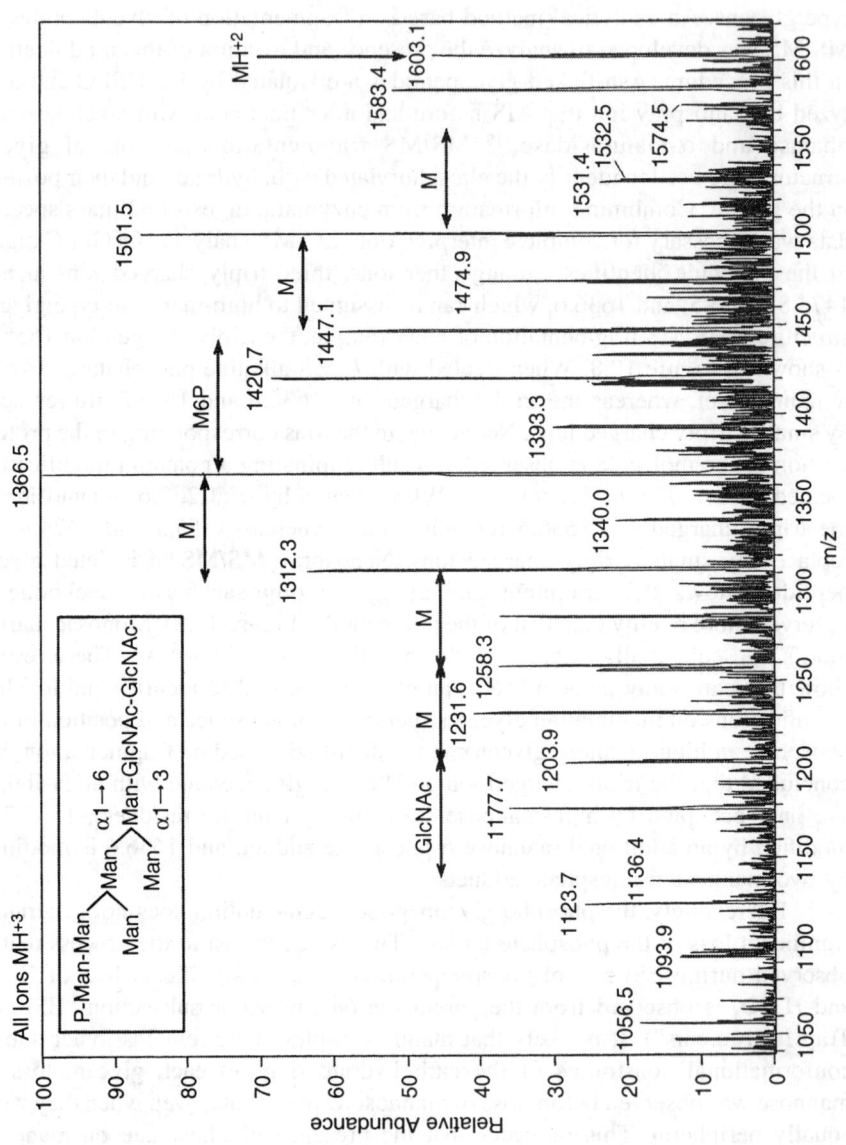

Figure 10.8 Characterization of posttranslational modifications by MS/MS fragmentation. An isolated glycopeptide (4663.3 Da) is fragmented using nanospray ionization on a Finnigan LCQ Deca ion trap mass spectrometer. The triply charged ion 1555.5 is selected as the parent ion. The presence and position of a phosphate adduct on high-mannose structure-type oligosaccharide can be identified due to the efficient fragmentation along the oligosaccharide backbone. The sequential fragmentation along the glycan structure is favored over the fragmentation of the polypeptide backbone. However, a minor ion series, similarly fragmented along the glycan, but resulting from an internal cleavage of the peptide at a Pro residue, is also observed. The glycan structure deduced from the MS/MS pattern is shown in the inset.

Disulfide Linkages, Free Thiol, and Thiol Adducts: The final piece in the recombinant protein characterization puzzle is always the location of disulfide bridges, free thiol, and sometimes thiol adducts. It is not uncommon in bacterial expression systems to find recombinant protein in an improperly folded state, which can have an impact on its biological activity. Then, exposure to redox chemistry is often implemented as part of the purification process. The presence of free thiol can also facilitate disulfide rearrangement; hence the amount of free cysteines in the protein backbone should be determined during product characterization. An efficient procedure for the location of free cysteine residues is the addition of a mass tag at acidic pH where disulfide scrambling is minimized. Sulfhydryl addition on the double bond of *N*-ethyl maleimide (NEM) results in 125-Da peptide mass increase. Subsequent modification of specific peptides can be detected by single ion monitoring of the 125-Da mass excess. MS/MS sequencing of the detected peptides is finally performed to determine the precise location of the tag.[120] Other detection methods as alternatives to MS for the detection of free cysteines are usually done spectroscopically after derivatization of the free thiol with UV-absorbing or fluorescent tags. Cysteinylation and glutathionylation of nonbridged cysteines are sometimes identified during peptide mapping. A mass increase of 119 amu and 305 amu, respectively, is then found on these peptides.[121] The most common approach to disulfide linkage analysis uses endoprotease digestion of the native form, followed by reduction–alkylation and mass spectral analysis of the peptides to determine which cysteines are involved in the disulfide linkage. Proteolysis with pepsin, chemical cleavage using partial acid hydrolysis, or cyanogen bromide cleavage have been used because these reactions are carried out at low pH where disulfide rearrangement is unlikely to occur. Dormady et al.[122] described an elegant approach to reduce the disulfide exchange during glutamyl endopeptidase digestion by using immobilized enzyme cartridges operated in the stopped-flow mode. The digestion time is orders of magnitude shorter in comparison to the lengthy free-solution reaction, and the possibility of artifacts due to disulfide exchange is significantly reduced. However, proteins are often resistant to proteolysis without prior reduction and alkylation. Furthermore, some cysteines are located adjacent to each other and enzymatic cleavage is not possible between them; hence the disulfide assignments remain ambiguous. In this case, manual Edman degradation can be performed until the first cysteine is reached. Qi et al. have recently published the disulfide structure identification of a highly knotted, cysteine-rich peptide by cyanylation/cleavage mass mapping.[123] This method uses partial reduction with tris(2-carboxyethyl) phosphine hydrochloride at pH 3 followed by immediate blocking of the nascent sulfhydryl groups by cyanylation. The cyanylated species were then separated by LC, collected, and treated with aqueous ammonia, which cleaved the peptide chain on the N-terminal side of cyanylated cysteine residues. MALDI-TOF analysis before and after complete reduction then identified the exact linkages. Other methods for disulfide bond determination involve pepsin digestion at acidic pH in the presence of ^{18}O

water. The disulfide bonded peptides have distinct isotopic profiles that facilitate their identification.[124]

Lastly, it has been shown that in rare cases trisulfide bridges (Cys-SSS-Cys) can be found as hydrophobic variants of recombinant proteins. The experimental mass of such a trisulfide-bridged peptide would be increased by 32 amu. Andersson et al.[125] have described this phenomenon for recombinant human growth hormone during expression in *E. coli*. However, the receptor binding properties were not affected by this trisulfide modification.

Stability and Degradation Pathways

An important part of the development of biopharmaceuticals involves the design of optimized formulations. MS has an essential role in monitoring product integrity during stability studies. The use of accelerated stability studies such as exposure to extreme temperature, shear force, and pH reveals the potential weaknesses of the molecule and potentially modifies specific residues along the primary sequence. The purpose of these experiments is to forecast what could happen to the molecule during long-term storage. Accelerated stability studies are designed to rapidly identify buffer conditions and additives that minimize protein modification. These studies must always be confirmed with real-time stability studies. Proteins kept for months and years on storage in liquid formulation undergo slow modifications. Most common chemical modifications involve oxidation and deamidation of specific residues. MS is the best tool to monitor these modifications.

Oxidation

Activated oxygen species can react with protein and alter their biological properties, notably in physiological events related to aging.[126] Sulfur-containing residues are primary oxidation targets in proteins. Numerous proteins are inactivated by oxidation of their methionine residues with significant biological consequences.[127] For example, inactivation of α1-proteinase inhibitor (α1-antitrypsin) due to oxidation of specific methionine residues by cigarette smoke results in the pathology of pulmonary emphysema in smokers.[128,129] *In vivo*, methionine oxidation can be reversed by a specific enzyme, methionine sulfoxide reductase, whose activity requires the thioredoxin regeneration system. Site-directed mutagenesis experiments and MS measurements show that the catalytic mechanism involves oxidoreduction of three cysteine residues and a sulfenic acid intermediate.[130] The presence of such an enzyme suggests that reversible oxidation of methionine residues is involved in regulation of biological activity.[131] *In vitro*, such chemical modification of protein therapeutics is usually detrimental and therefore is a primary concern for biopharmaceutical companies in their quest for a stable formulation. MS is the appropriate tool to measure oxidation of protein molecules, either from LC-MS of enzymatic digests or directly from analysis of the complete molecule. Mass measurements show a variation of 16 mass units, consistent with modification of methionine residues into methionine sulfoxide derivatives, whereas more complete oxidation to methionine sulfone requires drastic conditions and is typically not observed. Oxidation can be observed naturally on proteins in storage from reaction with dissolved oxygen or can be artificially induced by the use of chemical reagents such as *N*-chlorosuccinimide or hydrogen peroxide. Time-course experiments classify the different methionine residues for their susceptibility. Bulky peroxides such as *tert*-butyl hydrogen peroxide are good probes to target the surface-exposed residues.

Artificial oxidation of GM-CSF performed with *N*-chlorosuccinimide or hydrogen peroxide generates transient oxidized intermediates that eventually evolve to a completely oxidized but intact molecule. RP-HPLC of the complete molecule is a good technique to separate discrete molecules identified by MS as carrying one, two, three, or four oxidized methionine residues.[132] Enzymatic digestion — trypsin, in this case — is required to locate the modifications on specific fragments. When a fragment carries several oxidized residues, the location of the modifications can be determined by MS/MS. Time-course studies and MS/MS experiments show that one to four methionine residues can be modified with the following order of susceptibility: M46 = M79 > M36 > M80. This result is in agreement with the three-dimensional structure of the molecule showing that residue M80 is buried, M46 and M79 are exposed, and residue M36 has its sulfur atom partially buried (Figure 10.9). Interestingly, the three residues can be independently oxidized with *N*-chlorosuccinimide, but another oxidizing reagent, hydrogen peroxide, attacks partially buried M36 more easily than exposed M79, which is left intact. The small size of the reagent and specific structure constraints could explain the results.

Even if LC-MS and MS/MS are the methods of choice to localize oxidized residues on proteins, they suffer some limitations. In an example such as CHO-expressed TNFR:Fc, a large dimerized 125-kDa molecule with complex glycosylation heterogeneities, quantification of artificially oxidized residues is only partially resolved by MS. From one to seven methionine residues can be sequentially oxidized and quantified with the following order of susceptibility: M174 > M448 > M30 = M272. Three methionine residues, M187, M378, and M448, located on tryptic fragments carrying heterogeneous glycosylation patterns, are not quantified by MS. The solution to this problem was obtained with an alternative characterization method, using a combination of cleavage by cyanogen bromide and N-terminal sequencing.[133] Proteins are specifically cleaved at methionine residues by treatment with cyanogen bromide, whereas oxidized methionines do not undergo this cleavage reaction. Chemical treatment at various concentrations of *N*-chlorosuccinimide, from 0 to 5000 μM, was performed to generate a series of partially oxidized molecules. Repetitive yields graphs of CNBr peptides were determined, and the Y-intercept was taken as the recovery value for each peptide. The amount of methionine oxidation was determined by calculating the percent difference in recoveries between the 0-μM reference and

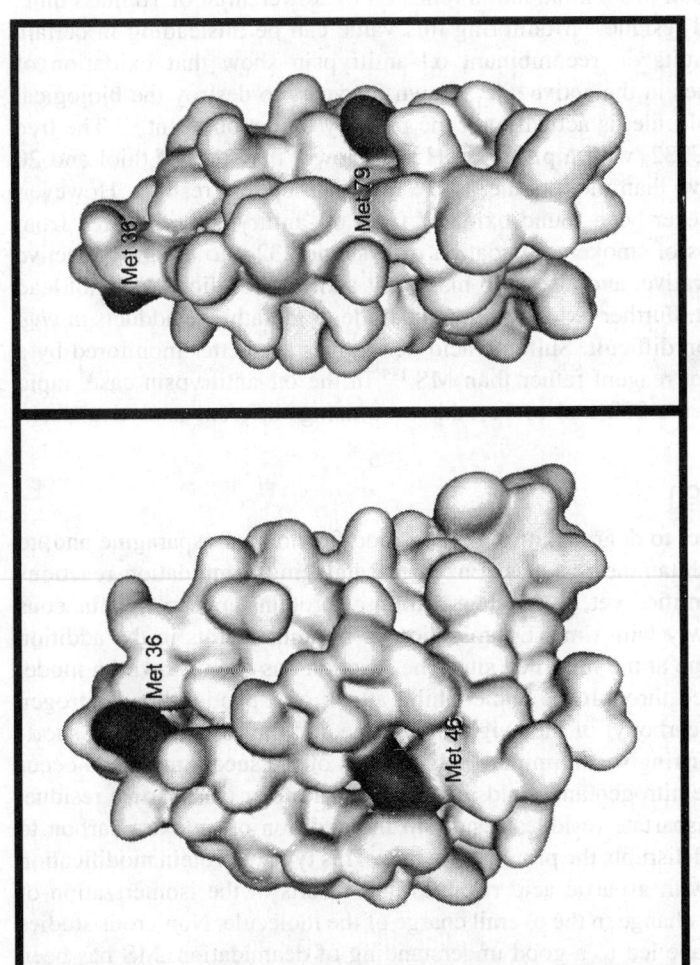

Figure 10.9 Tridimensional structure of GM-CSF.[148] The tridimensional structure shows that the four methionine residues present on the molecule have different degrees of solvent exposure. The sulfur atoms are either fully exposed (residues M46 and M79), partially exposed (residue M36), or totally buried (residue M80). Forced oxidation experiments described in the text show that residue M80 is unaffected, whereas local structural constraints make M79 less susceptible to oxidation than predicted by the model.

the various treatments. Increased oxidation results in fewer CNBr peptides being generated and therefore N-terminally sequenced. The data showed that M187 and M223 were the most readily oxidized residues, while M378 was the most resistant to oxidation.[133] This CNBr-NTS method offers a sensitive and direct evaluation of methionine susceptibility to oxidation. It can also assess peptides that are not easily quantified by LC-MS mapping.

MS analysis of protein oxidation relies on the differential of 16 mass units to define oxidized residues. Monitoring this value can be misleading in certain cases. Recent results on recombinant α1-antitrypsin show that oxidation of methionine residues in the active site, known for years to destroy the biological activity of the molecule, is actually not the primary oxidation event.[134] The free cysteine residue C232, with a pKa 1.5 pH units lower than that of thiol and 20 times more reactive than methionine, is the most susceptible residue. However, this residue has never been found oxidized from α1-antitrypsin recovered from the lung secretions of smokers. Oxidation of cysteine 232 into a highly reactive sulfenic acid derivative, and also a 16 mass unit addition reaction, does not lead to a stable product. Further oxidation into disulfide or glutathione adducts *in vivo* makes its detection difficult. Sulfenic acid derivatives are better monitored by a specific fluorescent reagent rather than MS.[135] In the α1-antitrypsin case, rapid oxidation of cysteine 232 had to be addressed using different analytical techniques.

Deamidation

Proteins are subject to degradation linked to modifications of asparagine and, to a lesser degree, glutamine residues. Enzymes catalyzing deamidation reactions have not been identified yet, so the degradation of proteins appears spontaneous and increases slowly with time. Deamidation of proteins results in the addition of a negative charge at the modified site. The reaction has been shown on model peptides to proceed through the nucleophilic attack of a peptide bond nitrogen on the side-chain carbonyl of the neighbor residue and the formation of a short-lived, five-member ring succinimide.[136] Hydrolysis of the succinimide can occur on each side of the nitrogen and yields either an aspartate or isoaspartate residue. Formation of isoaspartate residues results in the addition of an extra carbon to the main chain and disrupts the protein structure. This type of protein modification occurs similarly with aspartic acid residues and results in the isomerization of the residue, but no change in the overall charge of the molecule. Numerous studies since the 1970s have led to a good understanding of deamidation. MS has been used as a fast tool to measure the deamidation rates of 400 possible near-neighbor combinations in model pentapeptides.[137] Deamidation half-times were found to vary from 1 day to over 30 years. The deamidation rate is controlled primarily by the carboxyl-side residue with little contribution of the amino side residue. The data confirm the well-documented rapid deamidation rates of sequences involving carboxyl-side residues of small sizes, most notably Asn-Gly sequences.[138] It has been proposed that deamidation plays a role as a molecular timer of biological events.[139] Because of variable rates of deamidation along protein sequences, timed mechanisms such as protein turnover and aging could be influenced by deamidation.[140] Site-directed mutagenesis of sensitive sites could improve protein half-lives.

A challenge in the process development of biopharmaceuticals is to stabilize the molecule and limit its degradation. Deamidation events often result in progressive loss of activity of the molecule. Several analytical methods, including MS, RP-HPLC, N-terminal sequencing, and specific enzymatic reaction, can be used to analyze deamidation and therefore control it. MS used in complement to RP-HPLC is the preferred method. Modification of asparagine residues into aspartic acid residues in a peptide is identified by LC-MS of an enzymatic digest as a change in retention times and an increase of one mass unit. Modification with retention of configuration of the peptide bond (asparagine to α-aspartic acid) results in modified peptides eluting later than the asparagine-containing counterparts. Modification with change of configuration of the peptide bond (asparagine to β-aspartic acid) results in retention times that are difficult to predict. The addition of an extra carbon in the main chain modifies the structure of the peptide extensively and the resulting peptide may elute before or after the original molecule. A typical example of the presence of modified peptides is shown in Figure 10.10. The precise localization of the modification, in the case of multiple potential deamidation sites within a peptide, can be assessed by N-terminal sequencing. Aspartic acid and asparagine residues are positively identified, whereas β-aspartic acid residues are negatively confirmed, as the presence of a β-linkage stops the Edman degradation cycle.

The physiological significance of β-aspartate is confirmed by the presence of a repair enzyme, L-isoaspartyl methyltransferase.[141] Methylation of the α-carbonyl group facilitates reformation of the succinimide ring and, after hydrolysis, repair of the original asparagine deamidation sites to aspartate residues. The repair mechanism is relatively inefficient as it only partially removes β-aspartate. Moreover, the original asparagine residue is not really restored, but rather mutated into α-aspartate. L-isoaspartyl methyltransferase is the basis of a test for the presence of isoaspartate residues in the Isoquant isoaspartate detection kit commercially available from Promega (Madison, WI). This method offers an alternative to MS for the detection of deamidation. The reaction is standardized using a β-aspartate-containing peptide reference. Incorporation of a ^{3}H-labeled methyl group in the analyzed protein and its subsequent release as radioactive methanol can be quantified in a radioactivity counter and converted to moles of β-aspartate per mole of protein.[142] A drawback of this enzymatic method is the use of radioactive precursors. This can be overcome, although at the expense of sensitivity, by focusing, rather, on the analysis of cold reaction by-products by RP-HPLC. An advantage of the method would be the possibility of analyzing an unpurified mixture of proteins. During process development, the Isoquant kit can

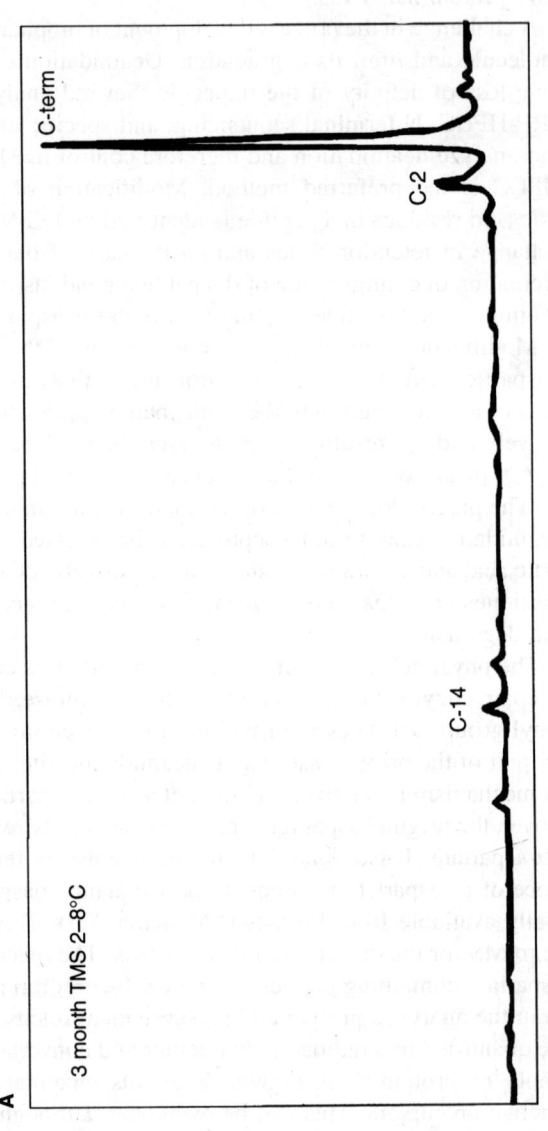

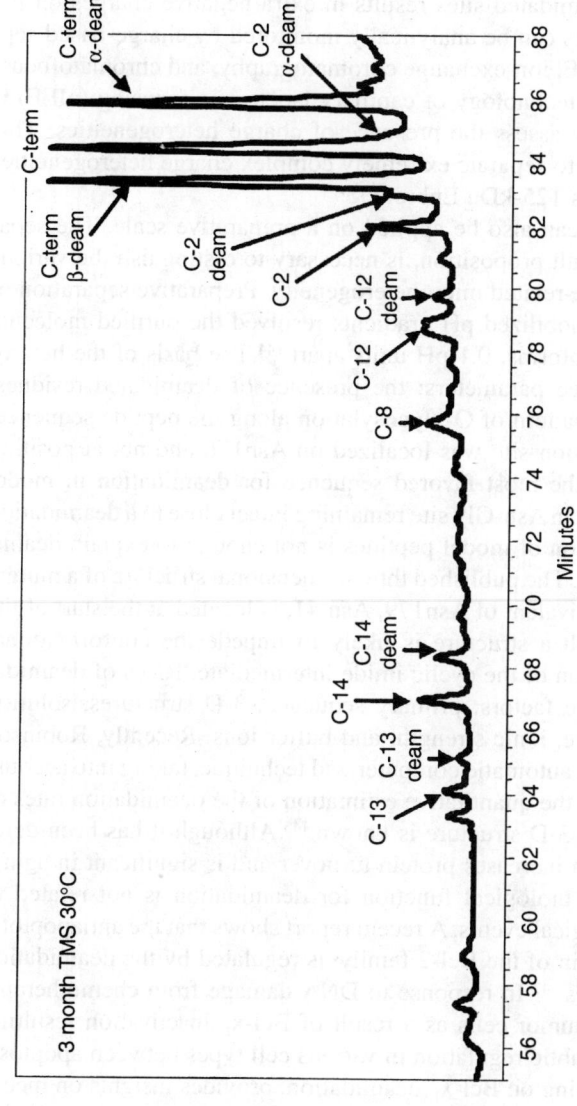

Figure 10.10 Stability studies analysis by LC-MS. Long-term stability studies (3 months, 30°C) are evaluated by LC-MS analysis of a C-terminal peptide fragment. Various degradation mechanisms are visualized, removal of C-terminal residues due to proteolytic activities, isomerization and deamidation of specific asparagine residues. Future development efforts will allow the use of this methodology to assess progress toward a stable formulation.

be very informative for the in-process assay of intermediates. During stability studies, the Isoquant kit can be used as a fast assay to assess deamidation events in a large series of sample conditions without the need of extra handling, digestion, and LC-MS. However, the conversion of asparagines to α-aspartate residues cannot be monitored by this method.

The presence of deamidated sites results in extra negative charges on the protein. Deamidation events can be analytically monitored by charge-based separation methods, such as CE, ion-exchange chromatography, and chromatofocusing. The newly developed technology of capillary isoelectric focusing (cIEF) is a good tool to analytically assess the presence of charge heterogeneities. This technique has the capacity to separate extremely complex charge heterogeneities in large molecules, such as 125-kDa Enbrel®.[143]

Isoelectric focusing can also be applied on a preparative scale. The separation of isoforms, a difficult proposition, is necessary to distinguish the various molecular causes of charge-related microheterogeneity. Preparative separation of PIXY321 isoforms in immobilized pH gradients resolved the purified molecule into a series of discrete isoforms, 0.1 pH units apart.[144] The basis of the heterogeneity was linked to three parameters: the presence of deamidated residues, charged glycans, and the pattern of O-glycosylation along the peptide sequence. Interestingly, the deamidation site was localized on Asn176 and not at position 179 on an Asn-Gly site, the most favored sequence for deamidation in model peptides.[84] This rare event, an Asn-Gly site remaining intact close to a deamidation site, proves that deamidation of model peptides is not enough to explain deamidation in complex proteins. The published three-dimensional structure of a mutant of IL-3 shows that the equivalent of Asn179, Asn 41, is located at the start of the fifth short α-helix.[145] Such a structure is likely to impede the conformational flexibility and the formation of the cyclic imide intermediate. Rates of deamidation depend upon multiple factors: primary sequence, 3-D structures, solution chemistry, pH, temperature, ionic strength, and buffer ions. Recently, Robinson and Robinson proposed an automatic computerized technique, taking into account these multiple factors, for the quantitative estimation of the deamidation rates of any protein for which the 3-D structure is known.[146] Although it has been demonstrated that deamidation increases protein turnover and is significant in aging, the first clear proof of a biological function for deamidation is not related to molecular timing of biological events. A recent report shows that the antiapoptotic activity of Bcl-x_L, a protein of the Bcl-2 family, is regulated by the deamidation of two asparagine residues.[147] In response to DNA damage from chemotherapy, apoptosis is activated in tumor cells as a result of Bcl-x_L inactivation resulting from deamidation. This subtle regulation in various cell types between apoptosis and antiapoptosis, depending on Bcl-x_L deamidation, provides insights on mechanisms of apoptosis and the effectiveness of chemotherapy in cancer treatment. It is the first report that correlates deamidation with a precise impact on cell physiology.

MS IN QC OF BIOPHARMACEUTICALS

Although MS plays a crucial role in process development, it has not yet found its way as a routine QC tool for lot release and stability monitoring. We believe the main reason for this phenomenon is the nonquantitative aspect for protein analysis by MS. Ionization of proteins would need to be very reproducible. In our experience, this is the case if the protein or peptide is homogeneous. However, in most cases, protein pharmaceuticals show heterogeneity either due to glycosylation, aggregation, or proteolytic degradation. The results for heterogeneous proteins can be only semiquantitative with no assurance of accurate data for the different species. This being said, we did develop a semiquantitative method for the consistency monitoring of the N-terminal heterogeneity of GM-CSF.[149] The formulated protein was chromatographed over a reversed-phase column before it was analyzed on-line by electrospray MS. Ratios of ion intensities of the different N-terminal species were plotted, and we could show very good reproducibility from day to day and operator to operator. This method was used in the development of a new formulation that would prevent N-terminal degradation. The chromatographic mass spectrometric approach (with autosampling) was much faster and more convenient than the alternative method of Edman degradation and was much preferred in the QC lab.

Hunt et al.[150] proposed the incorporation of LC-MS techniques to eliminate other identity assays required in the QC lab. A LCQ ion trap mass spectrometer was added downstream of the UV spectrometer. In addition to a UV trace, a total ion current chromatogram was generated. Comparison to a reference chromatogram was performed to define identity without extensive interpretation of the mass spectral data. This LC-MS method eliminated the need for other identity assays such as N-terminal sequencing and was validated as a release assay for three proteins.[150]

Nedved et al.[76] discussed the implementation of MS in a QC lab for lot release of therapeutic MAbs. In the QC environment, the transfer of methods must be documented, the methods validated, the personnel trained, the instrument calibrated, release specification set, and data handling and report generation made uniform and routine. Electrospray analysis of native MAbs was validated for the parameters of accuracy, intermediate precision, and intra-assay precision using a Sciex API 150 instrument. The accuracy of mass measurements allowed the authors to use the method as an identity assay. A similar approach for the validation of a peptide-mapping procedure using the LC-MS technique was found more challenging. An algorithm had to be created for the automated analysis of peptide maps defining comparability of a sample to the reference. Set limits of m/z peaks were validated and the degree of overall pattern match of m/z intensities was specified using the R-factor, a parameter scoring the overall agreement between LC-MS experiments. Some validation issues remain, such as method automation and software development and validation, limiting fast progress.[76] However, the high accuracy of MS and its capacity to analyze biopharmaceuticals makes the technique desirable in a QC environment, and its use is likely to increase.

CONCLUSION

Recent instrumentation improvements have resulted in the explosive growth of MS. The technique is crucial for the understanding of many different questions of protein chemistry and structure. Originally applied as a research tool, MS is moving into process development and is heading toward later stages of manufacturing and QC. Some limitations still exist, notably related to instrument costs, even if the development of the technique has a favorable impact. We have presented an overall view of different problems where MS has been efficiently applied, but we have also offered a discussion about alternative methodologies to answer process development questions. We have shown that other methods — chromatography and immunochemistry techniques, for example — sometimes offer a better choice for development purposes. The successful development of biopharmaceuticals to the commercialization stage requires a battery of analytical tools to answer questions and monitor progress. However, it is clear that MS is the primary method for the analysis of these drugs. New generations of instruments with higher resolving power are emerging that will ensure continuous development of MS applications in process development.

ACKNOWLEDGMENTS

We are indebted to many colleagues in the process development department of Immunex/Amgen for the work they performed on molecules illustrating this review. We would especially like to thank Alison Wallace, Theresa Martinez, Nancy Nightlinger, Shawn Novick, and Amy Guo for sharing some experimental results. We are grateful to Randy Ketchem for images of tridimensional structures and Gary Carlton for artwork. We thank Dean Pettit and Jim Thomas for a critical review of the manuscript.

REFERENCES

1. Thompson J.J. (1913), *Rays of Positive Electricity and Their Applications to Chemical Analysis*, Longmans Green, London.
2. Aston F.W. (1933), *Mass Spectra and Isotopes*, Edward Arnold, London.
3. Griffiths I.W. (1997), J.J. Thompson — the centenary of his discovery of the electron and his invention of mass spectrometry, *Rapid Commun. Mass Spectrom.* 11, 2–16.
4. Dempster A.J. (1918), A new method of positive ray analysis, *Phys. Rev.* 11, 316–324.
5. Munson M.S.B. and Field F.H. (1966), Chemical ionization mass spectrometry. General Introduction, *J. Am. Chem. Soc.* 88(12), 2621–2630.
6. Barber M., Bordoli R.S., Elliott G.L., Sedgwick R.N., and Tyler A.N. (1981), Fast atom bombardment of solids (FAB): a new ion source for mass spectrometry, *J. Chem. Soc. Chem. Commun.*, 325–327.
7. Barber M., Bartoldi R.S., Elliott G.J., Sedgwick R.N., and Tyler A.N. (1982), Fast atom bombardment mass spectrometry, *Anal. Chem.* 54, 645A–657A.

8. Dole M., Mach L.L., Hines R.L., Mobley R.C., Ferguson L.D., and Alice M.B. (1968), Molecular beams of macroions, *J. Chem. Phys.* **49**, 2240–2247.

9. Fenn J.B., Mann M., Meng C.K., Wong S.F., and Whitehouse C.M. (1989), Electrospray ionization for mass spectrometry of large biomolecules, *Science* **246**, 64–71.

10. Iribarne J.V. and Thompson B.A. (1976), On the evaporation of small ions from charged droplets, *J. Chem. Phys.* **64**, 2287–2294.

11. Cole R.B. (Ed.) (1997), *Electrospray Ionization Mass Spectrometry: Fundamentals, Instrumentation, and Applications*, John Wiley & Sons, New York.

12. Cole R.B. (2000), Some tenets pertaining to electrospray ionization mass spectrometry, *J. Mass Spectrom.* **35**, 763–772.

13. Kebarle J. (2000), A brief overview of the present status of the mechanisms involved in electrospray mass spectrometry, *J. Mass Spectrom.* **35**, 804–817.

14. Amad D.H., Cech N.B., Jackson G.S., and Enke C.G. (2000), Importance of gas-phase proton affinities in determining the electrospray ionization response for analytes and solvents, *J. Mass Spectrom.* **35**, 784–789.

15. Cech N.B. and Enke C.G. (2001), Practical implications of some recent studies in electrospray ionization fundamentals, *Mass Spectrom. Rev.* **20**, 362–387.

16. Tanaka K., Waki H., Ido Y., Akita S., Yoshida Y., and Yoshida T. (1988), Protein and polymer analyses up to m/z 100,000 by laser ionization time-of-flight mass spectrometry, *Rapid Commun. Mass Spectrom.* **2**, 151–153.

17. Karas M., Bachman D., Bahr U., and Hillenkamp F. (1987), Matrix-assisted ultraviolet laser desorption of non-volatile compounds, *Int. J. Mass Spectrom. Ion. Process.* **78**, 53–68.

18. Hillenkamp F. and Karas M. (1990), Mass spectrometry of peptides and proteins by matrix assisted ultraviolet laser desorption/ionization, *Meth. Enzymol.* **193**, 280–294.

19. Jarrold M.F. (2000), Peptides and proteins in the vapor phase, *Annu. Rev. Physical Chem.* **51**, 179–207.

20. Karas M., Gluckmann M., and Schafer J. (2000), Ionization in matrix-assisted laser desorption/ionization: singly charged molecular ions are the lucky survivors, *J. Mass Spectrom.* **35**, 1–12.

21. Paul W., Reinhard H.P., and von Zahn U. (1958), *Z. Phys.* **152**, 143.

22. Jennings K.R. and Dolnikwski G.G. (1990), Mass analyzers, *Method Enzymol.* **193**, 37–61.

23. Jonscher K.R. and Yates J.R. (1997), The quadrupole ion trap mass spectrometer — a small solution to a big challenge, *Anal. Biochem.* **244**, 1–15.

24. Colby S.M., King T.B., and Reilly J.P. (1994), Improving the resolution of matrix assisted laser desorption ionization time of flight mass spectrometry by exploiting the correlation between ion position and velocity, *Rapid Comm. Mass Spectrom.* **8**, 865–868.

25. Vestal M.L., Juhasz P., and Martin S.A. (1995), Delayed extraction matrix assisted laser desorption/ionization time of flight mass spectrometry, *Rapid Commun. Mass Spectrom.* **9**, 1044–1050.

26. Brown R.S. and Lennon J.J. (1995), Mass resolution improvement by incorporation of pulsed ion extraction in a matrix assisted laser desorption/ionization linear time of flight mass spectrometry, *Anal. Chem.* **67**, 1998–2003.

27. Dodonov A.F., Chernushevich I.V., and Laiko V.V. (1994), In time-of-flight mass spectrometry, Cotter R.J. (Ed.), *American Chemical Society Symposium Series*, 549, 108. American Chemical Society, Washington, D.C.

28. Verentchikov A.N., Ens W., and Standing K.G. (1994), Reflecting time-of-flight mass spectrometer with an electrospray in source and orthogonal extraction, *Anal. Chem.* **66**(1), 126.

29. Von Broke A., Nicholson G., and Bayer E. (2001), Recent advances in capillary electrophoresis/electrospray-mass spectrometry, *Electrophoresis* **22**, 1251–1266.

30. Morris H.R., Paxton T., Dell A., Langhorne J., Berg M., Bordoli R.S., Hoyes J., and Bateman R.H. (1996), High sensitivity collisionally-activated decomposition tandem mass spectrometry on a novel quadrupole/orthogonal-acceleration time-of-flight mass spectrometer, *Rapid Commun. Mass Spectrom.* **10**, 889–896.

31. Shevchenko A., Wilm M., Vorm O., and Mann M. (1996), Mass spectrometric sequencing of proteins from silver stained polyacrylamide gels, *Anal. Chem.* **68**, 850–858.

32. Loboda A.V., Krutchinsky A.N., Bromisrski M., Ens W., and Standing K.G. (2000), A tandem quadrupole/time of flight mass spectrometer with a matrix-assisted laser desorption/ionization source: design and performance, *Rapid Commun. Mass Spectrom.* **14**, 1047–1057.

33. Medzihradszky K.F., Campbell J.M., Baldwin M.A., Falick A.M., Juhasz P., Vestal M.L., and Burlingame A.L. (2000), The characteristics of peptide collision-induced dissociation using a high-performance MALDI-TOF/TOF tandem mass spectrometer, *Anal. Chem.* **72**(3): 552–558.

34. He F., Hendrickson C.L., and Marshall A.G. (2001), Baseline mass resolution of peptide isobars: a record for molecular mass resolution, *Anal. Chem.* **73**, 647–650.

35. Ge Y., Lawhorn B.G., El Naggar M., Strauss E., Park J.H., Begley T.P., and McLafferty F.W. (2002), Top down characterization of larger proteins (45 kDa) by electron capture dissociation mass spectrometry, *J. Am. Chem. Soc.* 124(4): 672–678.

36. Smith R.D., Anderson G.A., Lipton M.S., Masselon C., Pasa-Tolic L., Shen Y., and Udseth H.R. (2002), The use of accurate mass tags for high-throughput microbial proteomics, *Omics* **6**, 61–90.

37. Chase D.H. (2001), Mass spectrometry in the clinical laboratory, *Chem. Rev.*, **101**(2), 445–478.

38. Ling V., Guzzetta A.W., Canova-Davis E., Stults J.T., Hancock W.S., Covey T.R., and Shushan B.I. (1991), Characterization of the tryptic map of recombinant DNA derived tissue plasminogen activator by high-performance liquid chromatography-electrospray ionization mass spectrometry, *Anal. Chem.* **63**, 2909–2915.

39. Mann M., Hendrickson R.C., and Pandey A. (2001), Analysis of proteins and proteomes by mass spectrometry, *Annu. Rev. Biochem.* **70**, 437–473.

40. He Y., Zhong W., and Yeung E.S. (2002), Multiplexed on-column protein digestion and capillary electrophoresis for high-throughput comprehensive peptide mapping, *J. Chromatogr. B* **782**, 331–341.

41. Boss H.J., Watson D.B., and Rush R.S. (1998), Peptide capillary zone electrophoresis mass spectrometry of recombinant human erythropoietin: an evaluation of the analytical method, *Electrophoresis* **19**(15), 2654–2664.

42. Apffel A., Chakel J., Udiavar S., Swedberg S., Hancock W.S., Sounders C., and Pungor Jr. E. (1998), Application of new technology to the production of a "well-characterized biological," *Dev. Biol. Stand.* **96**, 11–25.

43. Hsieh F.Y., Cai J., and Henion J. (1994), Identification of trace impurities of peptides and alkaloids by capillary electrophoresis-ionspray mass spectrometry, *J. Chromatogr. A* **679**, 206–211.

44. Severs J.C., Hofstadler S.A., Zhao Z., Senh R.T., and Smith R.D. (1996), The interface of capillary electrophoresis with high performance Fourier transform ion cyclotron resonance mass spectrometry for biomolecule characterization, *Electrophoresis* **17**(12): 1808–1817.

45. Cao P. and Moini M. (1997), A novel sheathless interface for capillary electrophoresis/electrospray ionization mass spectrometry using an in-capillary electrode, *J. Am. Soc. Mass Spectrom.* **8**, 561–564.

46. Muddiman D.C., Rockwood A.L., Gao Q., Severs J.C., Udseth H.R., and Smith R.D. (1995), Application of sequential paired covariance to capillary electrophoresis electrospray ionization time-of-flight mass spectrometry: unraveling the signal from the noise in the electropherogram, *Anal. Chem.* **67**, 4371.

47. Cao P. and Moini M. (1998), Capillary electrophoresis/electrospray ionization high mass accuracy time-of-flight mass spectrometry for protein identification using peptide mapping, *Rapid Commun. Mass Spectrom.* **12**, 864–870.

48. Wang C., Oleschuk R., Ouchen F., Li J., Thibault P., and Harrison D.J. (2000), Integration of immobilized trypsin bead beds for protein digestion within a microfluidic chip incorporating capillary electrophoresis separations and an electrospray mass spectrometry interface, *Rapid Commun. Mass Spectrom.* **14**(15), 1377–1383.

49. Green B.N., Hutton T., and Vinogradov S.N. (1996), Analysis of complex protein and glycoprotein mixtures by electrospray ionization mass spectrometry with maximum entropy processing, *Method. Mol. Biol.* **61**, 279–294.

50. Yerguei J., Heller D., Hansen G., Cotter R.J., and Fenselau C. (1983), Isotopic distribution in mass spectra of large molecules, *Anal. Chem.* **55**, 353–356.

51. Jensen P.K., Pasa-Tolic L., Anderson G.A., Horner J.A., Lipton M.S., and Smith R.D. (1999), Probing proteomes using capillary isoelectric focusing-electrospray ionization fourier transform ion cyclotron resonance mass spectrometry, *Anal. Chem.* **71**, 2076–2084.

52. Roepstorff P. and Fohlman J. (1984), Proposal for a common nomenclature of sequence ions in mass spectra of peptides, *Biomed. Mass Spectrom.* **11**, 601.

53. Biemann K. (1990), Nomenclature for peptide fragments ions, *Meth. Enzymol.* **193**, 886–887.

54. Johnson R.S., Martin S.A., Biemann K., Stulz J.T., and Watson J.T. (1987), Novel fragmentation process of peptides by collision-induced decomposition in a tandem mass spectrometer: differentiation of leucine and isoleucine, *Anal. Chem.* **59**, 2621–2625.

55. Papayannopoulos I.A. (1995), The interpretation of collision-induced dissociation tandem mass spectra of peptides, *Mass Spectrom. Rev.* **14**, 49–73.

56. Dongré A.R., Jones J.L., Somogyi A., and Wysocki G.H. (1996), Influence of peptide composition, gas-phase basicity, and chemical modification on fragmentation efficiency: evidence for the mobile proton model, *J. Am. Soc. Mass Spectrom.* **118**, 8465–9374.

57. Tsaprailis G., Hair H., Somogyi A., Wysocki V.H., Zhong W., and Futrell J.H. (1999), Influence of secondary structure on the fragmentation of protonated peptides, *J. Am. Chem. Soc.* **121**, 5142–5154.

58. Johnson R.S. and Taylor J.A. (2000), Searching sequence databases via *de novo* peptide sequencing by tandem mass spectrometry, in *Methods in Molecular Biology*, Vol. **146**, *Mass Spectrometry of Proteins and Peptides*, pp. 41–61, Chapman J.R., Ed., Humana Press, Totowa, NJ.

59. Mann M. and Wilm M.S. (1994), Error-tolerant identification of peptides in sequence databases by peptide sequence tags, *Anal. Chem.* **66**, 4390–4399.

60. Eng J.K., McCormack A.L., and Yates J.R. (1994), An approach to correlate tandem mass spectral data of peptides with amino acid sequences in protein database, *J. Am. Soc. Mass Spectrom.* **5**, 976–989.

61. Perkins D.N., Pappin D.J., Creasy D.M., and Cottrell J.S. (1999), Probability-based protein identification by searching sequence databases using mass spectrometry data, *Electrophoresis* **20**, 3551–3567.

62. Clauser K.R., Baker P., and Burlingame A.L. (1999), Role of accurate mass measurement (± 10ppm) in protein identification strategies employing MS or MS/MS and database searching, *Anal. Chem.* **71**, 2871–2882.

63. Yates J.R. (1998), Mass spectrometry and the age of the proteome, *J. Mass Spectrom.* **33**, 1–19.

64. Mann M., Hendrickson R.C., and Pandey A. (2001), Analysis of proteins and proteomes by mass spectrometry, *Annu. Rev. Biochem.* **70**, 437–473.

65. VanAdrichem J.H.M., Bornson K.O., Conzelmann H., Gass M.A.S., Eppenberger H., Kresbach G.M., Ehrat M., and Leist C.H. (1998), Investigation of protein patterns in mammalian cells and culture supernatants by matrix assisted laser desorption/ionization mass spectrometry, *Anal. Chem.* **70**, 923–930.

66. Field M., Papac D., and Jones A. (1996), The use of high-performance anion-exchange chromatography and matrix-assisted laser desorption/ionization time-of-flight mass spectrometry to monitor and identify oligosaccharide degradation, *Anal. Biochem.* **239**, 92–98.

67. Yuk H. and Wang D.I. (2002), Changes in the overall extent of protein glycosylation by CHO cells over the course of batch culture, *Biotechnol. Appl. Biochem.* **36**, 133–140.

68. Andersen D.C., Bridges T., Gawlitzek M., and Hoy C. (2000), Multiple cell culture factors can affect the glycosylation of Asn-184 in CHO-produced tissue-type plasminogen activator, *Biotech. Bioeng.* **70**, 25–31.

69. Van den Nieuwenhof I.M., Koistinen H., Easton R.L., Koistinene R., Kamarainen M., Morris H.R., Van Die I., Seppala M., Dell A., and Van den Eijnden D.H. (2000), Recombinant glycodelin carrying the same type of glycan structures as contraceptive glycodelin-A can be produced in human kidney 293 cells but not in CHO cells, *Eur. J. Biochem.* **267**, 4753–4762.

70. Kunkel J.P., Jan D.C., Butler M., and Jamieson J.C. (2000), Comparisons of the glycosylation of a monoclonal antibody produced under nominally identical cell culture conditions in two different bioreactors, *Biotechnol. Prog.* **16**, 462–470.

71. Kyuchnicchenko V., Rodenbrock A., Thommes J., Kula M., Heine H., and Biselli M. (1998), Analysis of hybridoma cell culture processes by CE-SDS and MALDI-MS, *Biotechnol. Appl. Biochem.* **27**, 181–188.

72. Wan H.Z., Kaneshiro S., Frenz J., and Cacia J. (2001), Rapid method for monitoring galactosylation levels during recombinant antibody production by electrospray mass spectrometry with selective-ion monitoring, *J. Chromatogr. A.* **913**, 437–446.

73. Yates J.R., Link A.L., and Schieltz D. (2000), Direct analysis of protein in mixtures, in *Mass Spectrometry of Proteins and Peptides* (Chápman J.R., Ed.), Humana Press, Totowa, NJ.

74. Roberts G.D., Johnson W.P., Burman S., Anumula K.R., and Carr S.A. (1995), An integrated strategy for structural characterization of the protein and carbohydrate components of monoclonal antibodies: application to anti-respiratory syncytial virus MAb, *Anal. Chem.* **67**, 3613–3625.

75. Matamoros Fernandez L.E., Kalume D.E., Calvo L., Fernandez Mallo M., Vallin A., and Roepstorff P. (2001), Characterization of a recombinant monoclonal antibody by mass spectrometry combined with liquid chromatography. *J. Chromatogr. B, Biomed. Sci. Appl.* **752**, 247–261.

76. Nedved M.L., Kretschmer M., and Sunday B.R. (2002), Mass Spectrometry in the QC Lab: Are We Ready?, Poster communication, Well Characterized Biologicals Conference, Washington, D.C., 2002.

77. Spriggs M.K. (1994), The role of CD40 ligand in human disease. *Adv. Exp. Med. Biol.* **365**, 233–244.

78. Fanslow W.C., Srinivasan S., Paxton R., Gibson M.G., Spriggs M.K., and Armitage R.J. (1994), Structural characteristics of CD40 ligand that determine biological function, *Semin. Immunol.* **6**, 267–278.

79. Morris A.E., Remmele R.L. Jr., Klinke R., Macduff B.M., Fanslow W.C., and Armitage R.J. (1999), Incorporation of an isoleucine zipper motif enhances the biological activity of soluble CD40L (CD154), *J. Biol. Chem.* **27**, 418–423.

80. Kobata A. (1992), Structures and functions of the sugar chains of glycoproteins, *Eur. J. Biochem.* **209**, 483–501.

81. Varki A., Cummings R., Esko J., Freeze H., Hart G., and Marth J. (1999), Biosynthesis, metabolism, and function of N-glycans, in *Essentials of Glycobiology*, pp. 85–100. Cold Spring Harbor Laboratory Press, Cold Spring Harbor, NY.

82. Harvey D.J. (2001), Identification of protein-bound carbohydrates by mass spectrometry, *Proteomics* **1**, 311–328.

83. Besman M.J. and Shiba D. (1997), Evaluation of genetic stability of recombinant Human Factor VIII by peptide mapping and on-line mass spectrometric analysis. *Pharm. Res.* **14**(8), 1092–1098.

84. Balland A., Krasts D.A., Koch K.L., Gerhart M.J., Stremler K.E., and Waugh S.M. (1998), Characterization of the microheterogeneities of PIXY321, a genetically engineered GM-CSF/IL-1 fusion protein expressed in yeast, *Eur. J. Biochem.* **251**, 812–820.

85. Hardeman K., Samyn B., Van der Eycken J., and Van Beeumen J. (1998), An improved chemical approach toward the C-terminal sequence analysis of proteins containing all natural amino acids, *Protein Sci.* **7**, 1593–1602.

86. Samyn B., Hardeman K., Van der Eycken J., and Van Beeumen J. (2000), Applicability of the alkylation chemistry for chemical C-terminal protein sequence analysis, *Anal. Chem.* **72**, 1389–1399.

87. Li J. and Liang S. (2002), C-terminal sequence analysis of peptides using triphenylgermanyl isothiocyanate, *Anal. Biochem.* **302**, 108–113.

88. Ma S. and Nashabeh W. (2001), Analysis of protein therapeutics by capillary electrophoresis, *Chromatographia Suppl.* **53**, 75–98.

89. Harris R.J. (1995), Processing of C-terminal lysine and arginine residues of proteins isolated from mammalian cell culture, *J. Chromatogr. A* **705**(1): 129–134.

90. Moorehouse K.G., Nashabeh W., Deveney J., Bjork N.S., Mulkerrin M.G., and Ryskamp T. (1997), Validation of an HPLC method for the analysis of the charge heterogeneity of the recombinant monoclonal antibody IDEC-C2B8 after papain digestion, *J. Pharm. Biomed. Anal.* **16**, 593–603.

91. Furie B., Bouchard B.A., and Furie B.C. (1999), Vitamin K-dependent biosynthesis of gamma-carboxyglutamic acid, *Blood* **93**, 1798–1808.

92. White G.C., Pickens E.M., Liles D.K., and Roberts H.R. (1998), Mammalian recombinant coagulation proteins: structure and function, *Transfusion Sci.* **19**, 177–189.

93. Krishna R.G. and Wold F. (1998), Posttranslational modifications, in *Proteins — Analysis and Design*, Angeletti R.H. (Ed.), 121–126, Academic Press, San Diego.

94. Doyle H.A. and Mamula M.J. (2001), Post-translational protein modifications in antigen recognition and autoimmunity, *Trend. Immunol.* **22**(8), 443–449.

95. Neugebauer K.M., Merrill J.T., Wener M.H., Lahita R.G., and Roth M.B. (2000), SR proteins are autoantigens in patients with systemic lupus erythematosus. Importance of phosphoepitopes, *Arthritis Rheum.* **43**, 1768–1778.

96. Zhou Z. and Menard H.A. (2002), Autoantigenic posttranslational modifications of proteins: does it apply to rheumatoid arthritis?, *Curr. Opin. Rheumatol.* **14**, 250–253.

97. Meri S. and Baumann M. (2001), Proteomics: posttranslational modifications, immune responses, and current analytical tools, *Biomol. Eng.* **18**, 213–230.

98. Mann M. and Jensen O.N. (2003), Proteomic analysis of post-translational modifications, *Nature Biotechnol.* **21**, 255–261.

99. Goochee C.F., Gramer M.J., Andersen D.C., Bahr J.B., and Rasmussen J.R. (1991) The oligosaccharides of glycoproteins: bioprocess factors affecting oligosaccharide structure and their effect on glycoprotein properties, *Biotechnology* (New York) **9**(12): 1347–1355.

100. Jenkins N. and Curling E.M. (1994), Glycosylation of recombinant proteins: problems and prospects, *Enzyme Microb. Technol.* **16**, 354–364.

101. Jenkins N., Parekh R.B., and James D.C. (1996), Getting the glycosylation right: implications for the biotechnology industry, *Nature Biotechnol.* **14**, 975–981.

102. Kuster B., Krogh T.N., Mortz E., and Harvey D.J. (2001), Glycosylation analysis of gel-separated proteins, *Proteomics.* **1**, 350–361.

103. Kakehi K., Kinoshita M., and Nakano M. (2002), Analysis of glycoproteins and the oligosaccharides thereof by high-performance capillary electrophoresis — significance in regulatory studies on biopharmaceutical products, *Biomed. Chromatogr.* **16**, 103–115.

104. Liljeblad M., Lundblad A., and Pahlsson P. (2002), Analysis of glycoproteins in cell culture supernatants using a lectin immunosensor technique, *Biosens. Bioelectron.* **17**, 883–891.

105. James D.C. (1996), Analysis of recombinant glycoproteins by mass spectrometry, *Cytotechnology* **22**, 17–24.

106. Reinhold V.N., Reinhold B.B., and Costello C.E. (1995), Carbohydrate molecular weight profiling, sequence, linkage, and branching data: ES-MS and CID, *Anal. Chem.* **67**, 1772–1784.

107. Dell A. and Morris H.R. (2001), Glycoprotein structure determination by mass spectrometry, *Science* **291**, 2351–2356.

108. Harvey D.J. (2003), Matrix-assisted laser desorption/ionization mass spectrometry of carbohydrates and glycoconjugates, *Int. J. Mass Spectrom.* **226**, 1–35.

109. McLachlin D.T. and Chait B.T. (2001), Analysis of phosphorylated proteins and peptides by mass spectrometry, *Curr. Opin. Chem. Biol.* **5**(5), 591–602.

110. Mann M., Ong S.E., Gronborg M., Steen H., Jensen O.N., and Pandey A. (2002), Analysis of protein phosphorylation using mass spectrometry: deciphering the phosphoproteome, *Trend. Biotechnol.* **20**, 261–268.

111. Ding J., Burkhart W., and Kassel D.B. (1994), Identification of phosphorylated peptides from complex mixtures using negative-ion orifice-potential stepping and capillary liquid chromatography/electrospray ionization mass spectrometry, *Rapid Commun. Mass Spectrom.* **8**: 94–98.

112. Busman M., Schey K.L., Oatis Jr. J.E., and Knapp D.R.J. (1996), Identification of phosphorylation sites in phosphopeptides by positive and negative mode electrospray ionization tandem mass spectrometry, *J. Am. Soc. Mass Spectrom.* **7**, 243–249.

113. Annan R.S. and Carr S.A. (1996), Phosphopeptide analysis by matrix-assisted laser desorption time-of-flight mass spectrometry, *Anal. Chem.* **68**(19), 3413–3421.

114. Annan R.S. and Carr S.A. (1997), The essential role of mass spectrometry in characterizing protein structure: mapping posttranslational modifications, *J. Protein Chem.* **16**(5), 391–402.

115. Cao P. and Stults J.T. (1999), Phosphopeptide analysis by on-line immobilized metal-ion affinity chromatography-capillary electrophoresis-electrospray ionization mass spectrometry, *J. Chromatogr.* **853**(1–2), 225–235.

116. Cao P. and Stults J.T. (2000), Mapping the phosphorylation sites of proteins using on-line immobilized metal-ion affinity chromatography/capillary electrophoresis/electrospray ionization multiple stage tandem mass spectrometry, *Rapid Commun. Mass Spectrom.* **14**(17), 1600–1606.

117. Sigmundsson K., Masson G., Rice R., Beauchemin N., and Obrink B. (2002), Determination of active concentrations and association and dissociation rate constants of interacting biomolecules: an analytical solution to the theory for kinetic and mass transport limitations in biosensor technology and its experimental verification, *Biochemistry* **41**, 8263–8276.

118. Beauchemin D. (2002), Inductively coupled plasma mass spectrometry, *Anal. Chem.* **74**, 2873–2893.

119. Wallace A., Boyce J., and Balland A. (2002), Structural Characterization of Mannose-6-phosphate-containing glycans by tandem mass spectrometry, IBC's Conference on "The Impact of Post-Translational and Chemical Modifications on Protein Therapeutics," San Diego, CA.

120. Young Y., Zeni L., Rosenfeld R.D., Stark K.L., Rohde M.F., and Haniu M. (1999), Disulfide assignment of the C-terminal cysteine knot of agouti-related protein (AGRP) by direct sequencing analysis, *J. Peptide Res.* **54**, 514–521.

121. Shinina M.E., Carlini P., Polticelli F., Zappacosta F., Bossa F., and Calabrese L. (1996), Amino acid sequence of chicken Cu, Zn-containing superoxide dismutase and identification of glutathionyl adducts at exposed cysteine residues, *Eur. J. Biochem.* **237**(2), 433–439.

122. Dormady S.J., Lei J.M., and Regnier F.E. (1999), Eliminating disulfide exchange during glutamyl endopeptidase digestion of native protein, *J. Chromatogr.* **864**(2), 237–245.

123. Qi J., Wu J., Somkuti G.A., and Watson J.T. (2001), Determination of the disulfide structure of sillucin, a highly knotted, cysteine-rich peptide, by cyanylation/cleavage mass mapping, *Biochemistry* **40**, 4531–4538.

124. Gorman J.J., Wallis T.P., and Pitt J.J. (2002), Protein disulfide bond determination by mass spectrometry, *Mass Spectrom. Rev.* **21**, 183–216.

125. Andersson C., Edlund P.O., Gellerfors P., Hansson Y., Holmberg E., Hult C., Johansson S., Kordel J., Lundin R., Mendel-Hartvig I., Noren B., Wehler T., Widmalm G., and Ohman J. (1996), Isolation and characterization of a trisulfide variant of recombinant human growth hormone formed during expression in *Escherichia coli, Int. J. Peptide Protein Res.* **47**(4), 311–321.

126. Schoneich C. (1999), Reactive oxygen species and biological aging: a mechanistic approach, *Exp. Gerontol.* **34**, 19–34.

127. Vogt W. (1995), Oxidation of methionyl residues in proteins: tools, target and reversal, *Free Rad. Biol. Med.* **18**, 93–105.

128. Beatty K., Robertie P., Senior R.M., and Travis J. (1982), Determination of oxidized alpha-1-proteinase inhibitor in serum, *J. Lab. Clin. Med.* **100**(2), 186–192.

129. Evans M.D., Church D.F., and Pryor W.A. (1991), Aqueous cigarette tar extracts damage human a1-proteinase inhibitor, *Chem. Bio. Int.* **79**, 151–164.

130. Boschi-Mueller S., Azza S., Sanglier-Cianferani S., Talfournier F., Van Dorsselear A., and Barnalnt G. (2000), A sulfenic acid enzyme intermediate is involved in the catalytic mechanism of methionine sulfoxide reductase from *E.coli, J. Biol. Chem.* **275**, 35908–35913.

131. Moskovitz J., Flescher E., Berlett B.S., Azare J., Poston J.M., and Stadtman E.R. (1998), Overexpression of peptide-methionine sulfoxide reductase in *Saccharomyces cerevisiae* and human T cells provides them with high resistance to oxidative stress, *Proc. Nat. Acad. Sci. U.S.A.* **95**, 14071–14075.

132. Balland A., Stremler K., Paxton R., Krakover J., Klinke R., and Sassenfeld H. (1994), Oxidation of human GM-CSF and analysis of the modified residues by LC-MS and MS/MS, Proceedings of the 42nd ASMS Conference on Mass Spectrometry.

133. Gerhart M.J., Balland A., and Paxton R.J. (1997), Evaluation of CNBr digestion followed by Edman sequencing as a tool for assessing methionine susceptibility to oxidation in native molecules, *ABRF '97: Techniques at the Genome-Proteome Interface*, Baltimore, MD.

134. Griffiths S.W., King J., and Cooney C.L. (2002), The reactivity and oxidation pathway of cysteine 232 in recombinant human alpha 1-antitrypsin, *J. Biol. Chem.* **277**, 25486–25492.

135. Griffiths S.W. and Cooney C.L. (2002), Relationship between protein structure and methionine oxidation in recombinant human alpha 1-antitrypsin, *Biochemistry* **41**, 6245–6252.

136. Geiger T. and Clarke S. (1987), Deamidation, isomerization and racemization at asparaginyl and aspartyl residues in peptides. Succinimide-linked reactions that contribute to protein degradation, *J. Biol. Chem.* **262**, 785–794.

137. Robinson N.E., Robinson A.B., and Merrifield R.B. (2001), Mass spectrometric evaluation of synthetic peptides as primary structure models for peptide and protein deamidation, *J. Peptide Res.* **57**, 483–493.

138. Wright H.T. (1991), Nonenzymatic deamidation of asparaginyl and glutaminyl residues in proteins, *Crit. Rev. Biochem. Mol. Biol.* **26**, 1–52.

139. Robinson A.B., McKerrow J.H., and Cary P. (1970), Controlled deamidation of peptides and proteins: an experimental hazard and a possible biological timer, *Proc. Nat. Acad. Sci. U.S.A.* **66**, 753–757.

140. Robinson N.E. (2002), Protein deamidation, *Proc. Natl. Acad. Sci. U.S.A.* **99**, 5283–5288.

141. Johnson B.A., Langmark E.L., and Aswad D.W. (1987), Partial repair of deamidation damaged calmodulin by protein carboxyl methyltransferase, *J. Biol. Chem.* **262**, 12283–12287.

142. Aswad D.W., Parandandi M.W., and Schurter B.T. (2000), Isoaspartate in peptides and proteins: formation, significance and analysis, *J. Pharm. Biomed. Anal.* **21**, 1129–1136.

143. Jochheim C., Novick S., Balland A., Mahan-Boyce J., Wang W.C., Goetze A., and Gombotz W.R. (2001), Separation of Enbrel (rhu TNFR:Fc) isoforms by capillary isoelectric focusing, *Chromatographia Suppl.* **53**, S59-S65.

144. Balland A., Mahan-Boyce J.A., Krasts D.A., Daniels M., Wang W., and Gombotz W.R. (1999), Characterization of the isoforms of PIXY321, a GM-CSF/IL-3 fusion protein, separated by preparative isoelectric focusing on immobilized pH gradients, *J. Chromatogr. A* **846**, 143–156.

145. Feng Y., Klein B.K., and McWherter C.A. (1996), Three-dimensional solution structure and backbone dynamics of a variant of human interleukin-3, *J. Mol. Biol.* **259**, 524–541.

146. Robinson N.E. and Robinson A.B. (2001), Prediction of protein deamidation rates from primary and three-dimensional structure, *Proc. Nat. Acad. Sci. U.S.A.* **98**, 4367–4372.

147. Deverman B.E., Cook B.L., Manson S.R., Niederhoff R.A., Langer E.M., Rosava I., Kulans L.A., Fu X., Weinberg J.S., Heinecke J.W., Roth K.A., and Weintraub S.J. (2002), Bcl-xL deamidation is a critical switch in the regulation of the response to DNA damage, *Cell* **111**, 51–56.

148. Walter M.R., Cook W.J., and Ealick S.E. (1992), Three-dimensional structure of recombinant human granulocyte-macrophage colony-stimulating factor, *J. Mol. Biol.* **224**, 1075–1085.

149. Pettit D.K., Jochheim C.M., Nighlinger N.S., Willans F.L., Leach K.J., and Gombotz W.R. (2001), EDTA prevents N-terminal proteolysis in a pre-filled syringe formulation of Leukine™, ACS National Meeting, San Diego, CA.

150. Hunt J., Meng H., Welk P., Austin F., and Buckel S. (2001), A Simple Mass Spectrometry Method for a Lot Release Assay, Poster communication. Well Characterized Biologicals Conference, 2001.

11

Analytical Techniques
for Biopharmaceutical
Development — ELISA

Joanne Rose Layshock

INTRODUCTION

Enzyme-linked immunosorbent assay (ELISA) is based on the specific reaction between an antibody and an antigen. One of the reagents in the reaction is labeled with an enzyme that generates a colorimetric product that can be measured with a spectrophotometric device. The color intensity correlates with the concentration of specific antibody and the respective antigen. The reaction can be formatted in various ways in a multiwell plate (microtiter plate) with the common formats being the sandwich assay, the competitive assay, and the direct assay. (See Figure 11.1.)

The sandwich assay is the format used most often to quantitate a target antigen or analyte. In the sandwich assay, two antibodies are used that bind to different parts of the antigen. One of the antibodies is bound to, or coated on, the solid surface (mictotiter plate wells), whereas the other has a label attached to it (Figure 11.1a). Alternatively, a secondary conjugated antibody can be used to detect the bound primary antibody (Figure 11.1b). If the antigen is present in the sample solution, it links the two antibodies. Therefore, the label is retained on the plate where it can be detected by use of a colorimetric substrate.

The competitive assay is another format used to quantitate an analyte. An unlabeled analyte competes with a labeled analyte (enzyme-conjugated molecule) for binding to a specific capture antibody (Figure 11.1c).

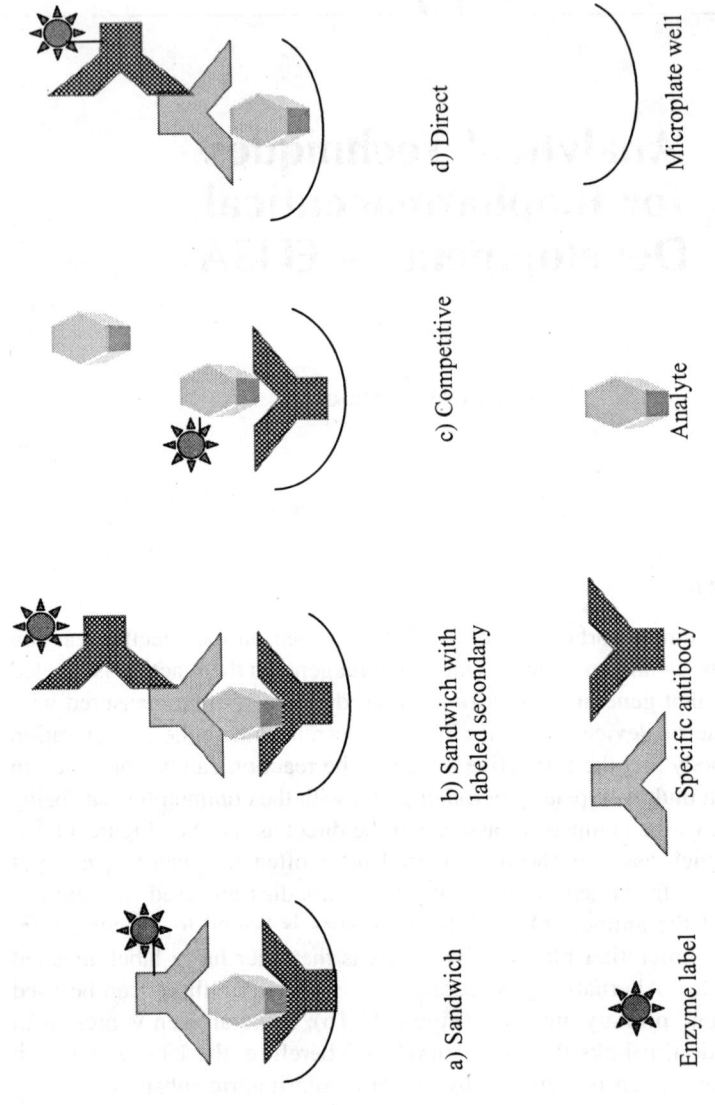

Figure 11.1 (a) Sandwich assay: analyte is captured between an antibody coated on the microplate well and an enzyme-labeled antibody. (b) Sandwich assay: analyte is captured between two specific antibodies and detected with secondary enzyme-labeled antibody. (c) Competitive assay: an enzyme-labeled analyte competes with unlabeled analyte for binding to the antibody. (d) Direct assay: antibodies to the analyte coated directly on the plate are detected with an enzyme-labeled antibody.

The direct assay format is the one used most often to detect antibodies to an antigen, with the antigen coated directly onto the solid phase (Figure 11.1d). This is used mostly for diagnostic purposes. The direct assay can detect whether or not a patient has reacted to an infectious organism by producing an antibody response to an antigenic protein. This format can be applied to biopharmaceutical product development when the drug product itself is an antibody, and the specific antigen is coated onto the plate.

The ELISA can be used as one component of a battery of analyses. Rarely is only one method used in isolation. Other tests include chromatographic methods such as reversed-phase high-performance liquid chromatography (HPLC), size exclusion chromatography, and physical structure analytical methods such as UV spectral analysis, mass spectroscopy, etc.

The ELISA is a versatile method that can be used throughout the biopharmaceutical product development process, from small-scale research to cell-line selection, to monitoring fermentation and downstream processing, and to product release testing. The use of the microtiter plates allows for high sample throughput and various degrees of automation. ELISAs satisfy the biopharmaceutical production requirements for specific, accurate, precise, and reproducible assays.

APPLICABILITY OF THE ELISA METHOD FOR ITS INTENDED USE

Characterization: Characterizing the Protein's Physical–Chemical and Biological Properties

Identification and Quantitation

At an early stage in product development, there must be some material made available for use as ELISA reagents. ELISA requires a "standard" for quantitation and specific antibodies to the drug protein of interest. If these are not already available, either commercially or internally, then time, effort, and expense must be reserved for reagent development as well as assay development activities. The specific antibody is the key reagent in an ELISA. The antibody defines the specificity and sensitivity of the assay. To date there is no successful substitution for the routine production of specific antibodies by immunization of an animal with the antigen target of interest.

The ELISA can be used for identification and quantitation of the protein product (biopharmaceutical) of interest throughout the development, production, and manufacturing process. For example, in the initial development phase, ELISAs can aid in the selection of the best cell line. In the early manufacturing steps, it can be used to identify the appropriate product-containing pools or fractions in process to be subjected to further purification. Because of the selectivity of ELISA, it is a suitable tool to select out the protein of interest from complex protein mixtures, such as cell culture fermentation media or product pools in early steps of protein recovery as well as downstream processing. Even complex mixtures do not require much sample preparation. It is important to determine

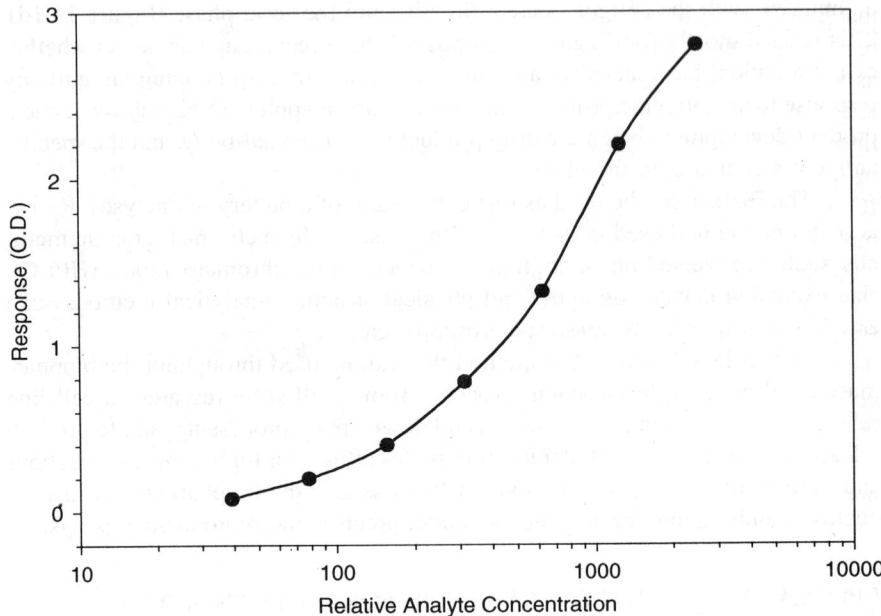

Figure 11.2   An example of a typical ELISA standard curve.

that the sample matrix does not contain any buffer or salt components, surfactants, or protein materials that interfere with the antibody–antigen reactions of ELISA. Often, simple sample dilution is effective in removing or reducing interference. If there is interference, this must be accounted for in the final calculations of sample recovery. The ELISA can still be used, with appropriate compensation.

The most commonly used format for quantitation assays is the sandwich assay format. Typically, a monoclonal antibody (MAb) is used to capture the product. It is then detected by another antibody, usually enzyme-labeled. A reference standard is used from which to compare the response of an unknown test sample. There is a relative increase in measured response (optical density) with increasing analyte concentration. Figure 11.2 is an example of a typical ELISA standard curve.

A competitive assay could also be used for quantitation. In a competitive assay, unlabeled antigen competes for labeled antigen. Examples include ELISAs for vaccine product antigens, such as recombinant proteins from viruses, or nonvaccine antigens such as growth factors or cytokines.

In addition to its use in quantitating the product, ELISA can also be used as a readout tool for cell-based bioassays or other assays that measure a biological activity conferred by the drug product. The end point of the bioassay may be a particular induced cellular protein (e.g., cytokine) or a reduction of a viral protein that can be quantitated by an ELISA. There are many commercially available reagents as well as complete ELISA kits for various cytokines and cell markers as well as common viral proteins. Sadick et al.[1] developed an assay designated KIRA, for kinase receptor activation. The KIRA assay is capable of quantifying ligand bioactivity by measuring ligand-induced receptor tyrosine kinase activation in terms of receptor phosphorylation. Assays using this method for the ligands IGF-1 and NGF showed excellent correlation with the more classical end point bioassays. Figure 11.3 is a schematic diagram of the IGF-I KIRA.

Product Variants

Product-related substances are molecular variants of the desired product formed during manufacture and storage, which are active and have no deleterious effect on the safety and efficacy of the drug product.[2] Many recombinant protein products are inherently heterogeneous, mixtures of closely related structures or product variants. These variants possess properties comparable to the desired product and are not considered impurities. It is only when they do *not* have properties of the desired product that their presence is problematic.

Product-related impurities are molecular variants of the desired product (e.g., precursors, certain degradation products arising during manufacture and storage) which do not have properties comparable to those of the desired product with respect to activity, efficacy, and safety.[2] Variants may exert different biological effects (potentially uncontrolled or hazardous) and specifically lead to antibody formation in the patient.[3]

Product variants are generated by a number of genetic, chemical, and physical changes:

- Genetic changes can give rise to product variants by mutations in the product-encoding gene itself.
- Chemical changes include posttranslational modifications including glycosylation, phosphorylation, disulfide bond formation and exchange (scrambling), proteolysis or hydrolysis, and deamidation or oxidation of amino acids.[4]
- Physical changes include denaturation, precipitation, adsorption, and aggregation (with like molecules or with excipients).[4]

Product variants can also be the result of action of cellular proteins on the drug molecule, such as proteolytic processing inside and outside cells. Enzymes from the host cell can make modifications of the product protein.

Product variants can also be generated by in-process procedures, such as those used for viral inactivation, for example. These procedures could alter the protein structure, forming new epitopes. These types of changes could potentially be detected by ELISA because of the specificity of the antigen–antibody interaction. In the case of vaccine production, an ELISA could be used to monitor viral inactivation. For this, a panel of MAbs, if available, could be used.

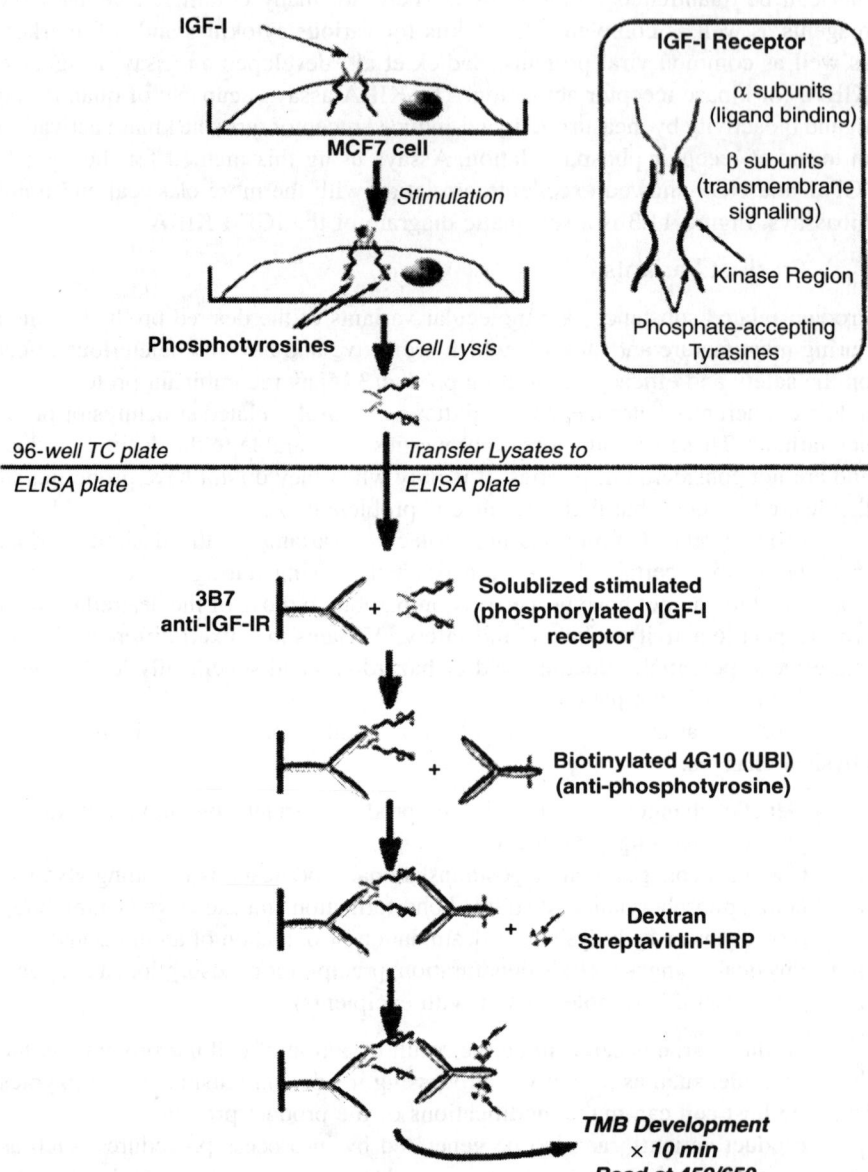

IGF-I

MCF7 cell

Stimulation

Phosphotyrosines   *Cell Lysis*

IGF-I Receptor

α subunits
(ligand binding)

β subunits
(transmembrane
signaling)

Kinase Region

Phosphate-accepting
Tyrasines

96-well TC plate

ELISA plate

Transfer Lysates to

ELISA plate

**3B7
anti-IGF-IR**   +   **Soublized stimulated
(phosphorylated) IGF-I
receptor**

+   **Biotinylated 4G10 (UBI)
(anti-phosphotyrosine)**

+   **Dextran
Streptavidin-HRP**

**TMB Development
× 10 min
Read at 450/650**

Figure 11.3   Schematic diagram of the IGF-I KIRA.

The most frequently encountered product variants are:[2]

- Truncated forms: hydrolytic enzymes or chemicals may catalyze the cleavage of peptide bonds.
- Other modified forms: deamidated, isomerized, mismatched S-S linked, oxidized, or altered conjugated forms (e.g., glycosylation, phosphorylation).
- Aggregates: include dimers and higher multiples of the desired product.

The ELISA is an appropriate method to detect some, but not all, of these product variants. The reactivity of antibodies is not affected much by glycosylation of the antigen. An ELISA would not be the most appropriate method to analyze product variants due to differences in glycosylation. ELISA would be appropriate for analysis of variants, such as aggregates, that contain a different protein structure that can be specifically recognized by an antibody.

Aggregates: For proper biological activity, proteins must maintain their three-dimensional (native) conformation. Protein aggregation almost invariably leads to severely reduced biological activity.[5] It is desirable to detect protein aggregates because they are likely to be more immunogenic than the monomer. This was shown for aggregated human growth hormone, insulin, and IgG.[6] In general, protein aggregates can be induced by stress conditions, such as exposure to extremes in temperature and pH, introduction of a high air/water or solid/water interface, or addition of certain pharmaceutical additives.[6]

Braun and Alsenz[6] used an ELISA to detect aggregates in interferon-alpha (IFN-α) formulations. They analyzed IFN-α formulations for possible aggregate formation because all marketed interferons are reported to induce antibodies to some extent. Because of its stabilizing effects, human serum albumin (HSA) is used in the formulation of marketed IFN-α at a great excess over IFN-α itself. HSA can also interact with other proteins. Braun and Alsenz developed an ELISA for the detection of both IFN-α–IFN-α and HSA–IFN-α aggregates. A MAb was used for the capture and detection of the IFN-α and a polyclonal for the detection of HSA. The assay is shown schematically in Figure 11.4.

The aggregate-specific ELISAs could be used to monitor the aggregate-inducing processes during IFN-α formulation and storage in an early phase and the development of aggregate-free IFN-α formulations. The ELISAs were highly sensitive, needed low protein concentrations, worked in the presence of excipients, and required no pretreatment.[6]

Proper Protein Refolding: The final protein structure of the recombinant protein is often obtained by the promotion of proper refolding. When the production of recombinant proteins results in inclusion bodies, an ELISA can be used to assure that the final protein product has been properly refolded and to monitor the refolding process. To do this requires antibodies that can only recognize and bind to a particular conformation that is only present upon proper refolding.

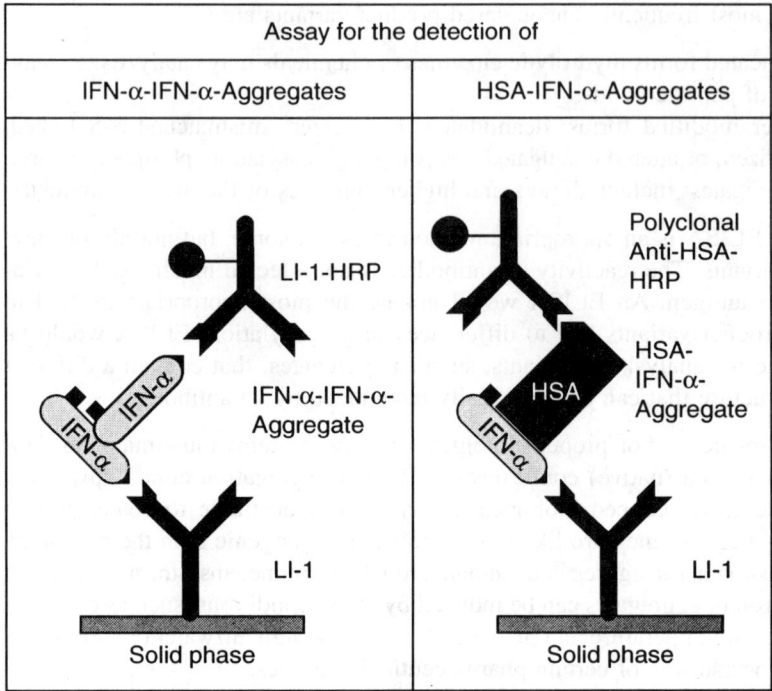

Figure 11.4 Schematic representation of ELISAs for the detection of IFN-α–IFN-α and HSA–IFN-α aggregates. IFN-α–IFN-α aggregates are detected using the same monoclonal anti-IFN-α antibody (LI-1) as capture and detection antibodies. HSA–IFN-α aggregates are captured by the anti-IFN-α antibody LI-1, and IFN-α-bound HSA is identified by a polyclonal and anti-HSA antibody. For simplicity, aggregates are illustrated at a 1:1 molar ratio (HRP = horseradish peroxidase).

Vandenbroeck et al.[7] used an ELISA to determine the recovery of immunoreactive porcine interferon-gamma (IFN-γ) from *E. coli* inclusion bodies. The ELISA used a polyclonal coating antibody with detection by a MAb. The inclusion bodies were solubilized in diluted 6 *M* guanidine/HCl and IFN subsequently refolded by its removal. The antiviral activity of the interferon was measured with a bioassay using the cytopathic effect (CPE) of vesicular stomatitis virus (VSV) on bovine kidney cells. The results of this study showed that the immunoreactivity measured by ELISA matched the biological activity measured by bioassay.

Tsouloufis et al.[8] used an ELISA to assess the refolding of a recombinant subunit of the extracellular domain of the human muscle acetylcholine receptor expressed in *E. coli*. The plates were coated with refolded or unfolded protein and then reacted with conformationally dependent MAbs. The use of specific

MAbs enabled identification of the protein segments required for native conformation.

Merli et al.[9] used an ELISA to optimize refolding of soluble tumor necrosis factor receptor type I (sTNF-RI) in *E. coli*. Native conformation of the molecule is maintained by 12 disulfide bridges. At different time intervals, aliquots of each well in the refolding microplate were diluted and tested by ELISA. The ELISA format used was a double-antibody sandwich assay and was based on the use of MAbs directed, respectively, against a neutralizing and a nonneutralizing epitope on the sTNF-RI molecule. The protein was characterized by biological and functional assays, and a good correlation was observed between all the data (biological assay, ligand-directed ELISA, and double-determinant sandwich ELISA).

Purity: Detecting Impurities and Contaminants

Biopharmaceuticals are subjected to strict regulations to monitor and quantitate impurities to maintain product safety, quality, integrity, and efficacy.[4] The following are definitions that relate to purity:

Impurity: Any component present in the drug substance or drug product that is not the desired product, a product-related substance, or excipient including buffer components. It may be either process- or product-related. According to ICH Guidelines on Impurities in New Drug Substances,[10] impurities at a level greater than 0.1% should be identified. So should degradation products observed in stability studies (for stability identification, the threshold is 1% for a maximum daily dose of < 1 mg to 0.1% for a maximum daily dose of > 2 g[10]). Identification below the 0.1% level is usually not considered necessary.[10]

Process-related impurities: Impurities that are derived from the manufacturing process. They may be derived from cell substrates (e.g., HCPs, host cell DNA), cell culture (e.g., inducers, antibiotics, or media components), or downstream processing (e.g., processing reagents or column leachables).

Product-related impurities: Molecular variants of the desired product (e.g., precursors, certain degradation products arising during manufacture or storage) that do not have properties comparable to those of the desired product with respect to activity, efficacy, and safety.

Contaminants:[2] Any adventitiously introduced materials (e.g., chemical, biochemical, or microbial species) not intended to be part of the manufacturing process of the drug substance or drug product.[2]

Degradation products:[2] Molecular variants resulting from changes in the desired product or product-related substance brought about over time or by the action of, e.g., light, temperature, pH, water, or by reaction with an excipient or the immediate container/closure system. Such changes may occur as a result of manufacture or storage (e.g., deamidation, oxidation, aggregation, proteolysis). Degradation products may be either product-related substances or product-related impurities.

Of all the possible contaminants and impurities of a biopharmaceutical product, organisms (bacteria, virus, mycoplasma) and their products (DNA, endotoxin, host protein), media components, and raw materials, it is most appropriate to use an ELISA for the HCP impurities and some of the process residuals (media components and raw materials). Impurities from media components are known or expected unlike those from the host cell.

Host Cell Impurities: Various organisms have been used to produce recombinant proteins: yeast, bacteria (e.g., *E. coli*), insect cells, and mammalian cells such as Chinese hamster ovary (CHO) cells. During the purification process, some HCPs can copurify with the protein product. Because of the specificity of the antigen–antibody interaction, an ELISA can be used to detect and quantitate the contaminating HCPs. Detecting host impurities is important for quality process control as well as for product safety issues. The intent is to avoid "unsafe" levels of residual HCPs which might lead to adverse reactions.[11]

The final purity required depends on final use of the product: e.g., vaccine with one immunization vs. hormone with chronic use. A detection range of 1 to 100 ppm of residual HCPs has been quoted as a regulatory (and analytical) benchmark for therapeutic proteins.[11] Many biotech companies have limited the range to 1 to 10 ppm. The sensitivity and specificity of any unique HCP assay that is used to support such a target should be demonstrated accordingly.[11]

Because ELISA is based on the specificity of antibody recognition of the antigen, the assay is only as good as the antibodies used. Proprietary reagents and assays must be developed and validated for the quantitation of host cell-specific proteins. Prolonged time frames must be anticipated when planning for the manufacture and characterization of complex HCP immunogens, the elicitation, purification, and qualification of anti-HCP immunoreagents, and the development and validation of multianalyte HCP immunoassays.[11] A commitment of time and expense must be made as it may take 3 to 6 months to elicit, purify, and characterize an acceptable polyclonal anti-HCP immunoreagent.[11] There are some commercially available antibodies to some of the most commonly used host cell expression systems. The commercial antibodies should be extensively evaluated before use and would be acceptable if they detected most of the specified host cell impurities. The evaluation should be done by another immunological-based assay, such as Western blot analysis.

There are two approaches to developing a host cell impurity assay, each with advantages and disadvantages. Both approaches require the production of polyclonal antibodies for use in an ELISA, but the immunogen used to elicit those antibodies is what differentiates the two approaches. The first approach is to use the host cell homogenate preparation as the immunogen. The host cell homogenate is produced from null host cells or the same host cell and vector systems as the production cell line, only lacking the cDNA of the product. All other characteristics (i.e., fermentation conditions) should be identical to the production process.[12] This is referred to as a "generic" assay and the goal is to measure all the HCP impurities that could be present.[13] The second approach is to use a "purified" host cell preparation or only those HCPs that typically copurify with the product as the immunogen. This is referred to as a "process-specific" assay.[13] It also uses host cell homogenate from null host cells, but the homogenate is then subjected to the same final purification process or to some of the purification steps as the recombinant DNA product.

The primary disadvantage of the generic assay is that the antibody may not detect all HCPs because some proteins are not sufficiently immunogenic or present in high enough concentrations to be detected by antibodies. The detection of HCPs should be differentiated from the detection of process impurities such as media components. Some proteins are at a low level, and there could be an immunodominant protein (such as albumin, a common media component). This can be avoided by preparing the host cell homogenate in serum-free media. To increase the response to proteins that are at a low concentration, or to weakly immunogenic proteins, there are methods available, such as chemical modification, passive immunization, and cascade immunization.[11,13] Passive immunization[11] involves immunization of antigen combined with purified IgG from previous postimmunization bleeds. Cascade immunization uses immune IgG to adsorb major antigens from the immunogen preparation. This facilitates the antibody response to weaker or minor antigens.

The host antigen preparation is used in several capacities of assay development as well as routine analysis:

- As an immunogen to generate specific antibodies
- For immunoaffinity matrix, to affinity-purify the antibodies generated
- As an assay standard

The antigen preparation used for immunogen and the antibody reagent generated should be evaluated using a combination of other techniques such as acrylamide gel and Western blot. The Western blot can identify those proteins seen on an acrylamide gel that are reactive with the antisera. This method can be used to evaluate the bleeds obtained over time during the immunization process. The analysis of the antibodies can also yield the specific identity of HCP impurities, especially particularly troublesome or recurring impurities, and differentiate among other process impurities such as media components.

The primary disadvantage of the process-specific assay is that it is specific for a particular purification process, and if any changes are made to this process, the antibodies originally developed are obsolete. Many factors can alter the final population of HCP, expressed, including a change in fermentation that affects host cell viability or a failure in one of the purification steps during a production run. If any unanticipated changes occur and change the host cell impurity load, this could result in a contaminated final product in which the contaminants cannot be detected by the process-specific assay. The impurity could be from a different population of HCPs than the population from a process that did not "fail".[13]

Hoffman advises, "Relying solely on a process-specific assay is ill advised and can result in failure to detect atypical process contaminants. In cases with a defined, persistent, and problematic host cell protein impurity, a down-stream process-specific assay may be justified. It is critical that the immunoassay be capable of detecting every possible host cell protein contaminant."[13]

In the development of an ELISA for host cell impurities, you also have to consider copurification of a HCP that is homologous to the product; the host species version of the recombinant protein, e.g., urokinase, is known to be present in many continuous cell lines.[3] Due to the similarity in structure, it is possible that an endogenous homologous protein molecule could copurify with the desired product.[3]

Because the quality of the impurity assay depends on the quality of the antibodies, there are several important points to consider regarding the immunization. The animals may have to undergo substantial boosting, especially for a "generic" assay. This is to elicit antibodies to proteins of lower concentration or to those that may be less immunogenic. The animal species chosen for antisera production is not critical. Rabbits and goats are often used as the species of choice. The antisera generated from the host immunogen should be pooled from several animals and subjected to immunoaffinity (ideally to the same antigen preparation used as the immunogen) or purified before assay development. The resultant antibody can be used as a capture antibody to coat the plate, and as a detecting antibody by conjugating it with avidin or biotin.[12]

There are many examples of ELISAs used for detecting host cell impurities in the literature. Pauly et al.[12] developed an ELISA to detect impurities in erythropoietin that had a detection limit of around 0.05 ng/ml. SDS polyacrylamide gel and Western blot analysis were used to confirm the spectrum of proteins detected and to demonstrate the specificity of the antibody preparation. Anicetti et al.[14] describe an assay for the detection of *E. coli* proteins in recombinant DNA-derived human growth hormone. Whitmire and Eaton[15] report on an immunoligand assay for quantitation of process-specific *E. coli* host cell contaminant proteins in a recombinant bovine somatotropin.

The host cell impurity assay can be used as a tool for detecting impurities in the various in-process purification steps as well as the final product. When using the assay for the in-process steps, it is important to test the samples in the appropriate buffers or matrix. Any sample recovery problems from buffers need to be determined and accounted for in the analysis.

Process Residuals: A source of potential impurities can be process residuals as well as host cells. Process residuals include fermentation media components as well as raw materials used in chromatography (immunoglobulin affinity ligands), etc. According to the ICH Guidelines,[2] process-related impurities are impurities that are derived from the manufacturing process. They may be derived from cell substrates, cell culture, or downstream processing.

Cell Culture-Derived: Media-Derived Protein Impurities. Immunoassays can detect low impurity levels (<1 ppm).[4] The ELISA is probably one of the most sensitive analytical methods. If bovine serum is used as a media component, then testing should include ELISAs for bovine serum albumin (BSA), bovine transferrin, bovine fetuin, and bovine IgG. Often hormones and growth factors, such as insulin or insulinlike growth factor, are used as media components. ELISAs should be used to detect and quantitate these residuals in the various production steps as well as in the final product. There are commercially available antibodies to most commonly used media components. If proprietary media components are used, then the same investment in time and effort is required for the production of specific antibodies, as described above for host cell impurities.

If the product is an antibody, then it is essential to distinguish the immunoglobulin product, e.g., mouse IgG, from any media immunoglobulin components, e.g., bovine IgG. Lucas et al.[16] developed an immunoassay to measure nanogram quantities of bovine IgG in the presence of a large excess of a structurally homologous protein, mouse MAb. The bovine IgG was a contaminant that copurified with the product from a protein A column. For the bovine IgG assay, whole IgG and protein A-purified IgG reacted differently in the assay. It is important to evaluate these types of assays for cross-reactivity. For other media components, such as chemicals or antibiotics, ELISA is probably not the most appropriate method due to the low immunogenicity of chemicals. Techniques such as HPLC would be better to detect these chemical components.

Downstream-Derived: Column Leachates. If a chromatography medium used in purification contains lectins, an ELISA can be used to detect contaminating lectin. For example, for a downstream column using a lectin such as the *Galanthus nivalis* (snowdrop) lectin, the concentration of unbound *G. nivalis* lectin in bulk solutions can be determined with an ELISA. Reagents for lectins are commercially available. A quantitative sandwich ELISA could be used to quantitate the lectin: a capture antibody (goat anti-*Galanthus nivalis* lectin, Vector Labs, Burlingame, CA) is adsorbed to the surface of a microtiter plate. Biotinylated goat anti-*Galanthus nivalis* lectin (Vector Labs) is used to detect bound samples and standards, followed by a streptavidin-HRP conjugate for detection.

Other common impurities, such as immunoglobulins and protein A, result from the immunoaffinity purification of recombinant proteins or MAbs.[16] If affinity chromatography is used to purify an antigen, then an ELISA can be used to detect contaminating levels of MAbs leached from the column. An assay for the antibody needs to detect the antibody in the presence and absence of its specific antigen.

Lucas developed an assay for protein A contamination.[16] Protein A from *S. aureus* is commonly used for purifying immunoglobulin because of its specificity for binding immunoglobulins of several species. For an assay to accurately quantitate amounts of contaminating protein A, it must be able to measure it in the presence of a large excess of another protein with which it interacts (because bound antibody might reduce the exposed regions of protein A available for recognition in the immunoassay). For a protein A assay, it was necessary to use F(ab)2 fragments for the capture and detection of antibodies because of the known reactivity of protein A with the Fc portion of antibodies.[16] MAb was added in the diluent for the standard curve of protein A. A standard ELISA can quantitate 50 pg/ml protein A in samples and is excellent for monitoring protein A impurities in process intermediates and final product.[4]

Stability

A variety of analytical methods, such as ELISA and HPLC, can be used to evaluate the effect of excipients or lyophilization on the stability of the biopharmaceutical product. Some parameters the analytical methods should evaluate are degradation, chemical and physical changes, aggregation, adsorption, and loss of biological activity.

Identifying Stable Formulations: Detecting the Effect of Excipients and Lyophilization

Proteins must be stable during processing, storage, and reconstitution (if lyophilized). Although liquid formulation is preferred for protein biopharmaceuticals, it may not always be the most stable presentation. The biopharmaceutical protein may need to be lyophilized to maintain stability. Lyophilization involves the removal of water from a frozen substance by sublimation and water vaporization under vacuum.[17] But in some cases, this process may itself cause protein instability.

Product-specific formulations should be developed, taking account of specific degradation pathways to prevent chemical and physical changes, aggregation, adsorption and loss of biological activity, and to provide long-term stability on storage.[18]

Protein instability is caused by both chemical degradation reactions and physical processes, and usually results in loss of potency.[19] High temperatures increase the flexibility and collision frequency of proteins in solution, which can result in aggregation and/or precipitation.[19] The process may adversely affect the tertiary structure of protein, and the protein then may undergo an unfolding process when it loses the surrounding water.[17] Unfolded protein molecules may have a higher tendency to aggregate.[17] The aggregation of the protein often leads to precipitation upon reconstitution.[17] After lyophilization, during storage, the proteins can undergo aggregation.[5] Aggregation can result from covalent and/or noncovalent interactions. Moisture, oxygen, and light all can induce aggregation.[5] Aggregated proteins may exhibit decreased bioactivity, altered half-life, and enhanced immunogenicity.[19]

Inclusion of a cryoprotectant in the formulation can stabilize the protein during the freezing and drying stages of lyophilization.[17] Excipients frequently used for stabilization include sugars, amino acids, surfactants, fatty acids, proteins such as HSA, and a range of salts and buffers.[18] Solvent additives can affect protein stability by direct interaction with the protein (binding to the protein), by indirect action through effects on the solvent (increasing solvent surface tension or viscosity), or by a combination of both of these mechanisms.[19] Water-soluble synthetic polymers, such as polyethylene glycol (PEG) or poly(vinylpyrrolidone) (PVP),[19] have been used as protein stabilizing agents. The protein–solvent interaction determines the effect of the cosolutes on the stability of proteins, rather than a direct interaction of solutes with proteins.[17]

Vemuri et al.[17] looked at the effects of various cryoprotectants, freezing rates, and buffer systems on the shelf-life of lyophilized recombinant alpha$_1$-antitrypsin (rAAT). Alpha$_1$-antitrypsin (AAT) is labile in solution; therefore, a more stable presentation was required. A competitive ELISA was used to measure total AAT in a sample. The AAT in the sample competed with HRP-labeled AAT for binding to the specific antibody. A stable formulation containing lactose as a cryoprotectant was found that maintained the protein's specific activity.

Recombinant human IL-11 was under evaluation in human clinical trials for use as a thrombopoietin product for chemotherapy support.[18] An ELISA was developed to measure adsorption (residual concentration in various containers over time). The ELISA was formatted with an anti-IL-11 MAb for capture and a biotinylated antibody for detection.[18] A combination of HSA and Tween-20 was required to address both the adsorption of IL-11 to glass and retention of biological activity postlyophilization.

Lyoprotectants are added to the formulation to stabilize the protein both during the freeze-drying process and on storage.[20] Ressing et al. investigated the effect of freeze drying on the stability of a mouse IgG monoclonal (MN12).[20] MN12 is directed against the class 1 outer membrane protein of meningococcal strain H44/76. ELISA was one method used to determine changes in the physicochemical properties of the monoclonal, by measuring antigen-binding capacity.[20] Aggregation of proteins during the freeze-drying process, often leading to insoluble protein, is a common observation.[20] Degradation may occur during storage of the product.[20]

When the product itself is an antibody, ELISA can be used to measure the binding to its specific target antigen. Gombotz et al.[19] studied stabilization of IgM class human MAb used to provide therapeutic protection in animals infected with group B streptococci (GBS).[19] The ELISA was used to study the nature of the interactions between the cryoprotectant (PVP) and the antibody product (4B9). The ELISA for antigen binding measured the ability of the antibody to bind to the GBS group polysaccharide and to quantitate human IgM, which measures the amount of 4B9 antibody bound by an antihuman immunoglobulin. An ELISA was used to determine if the cryoprotectant (PVP) had an effect on the GBS antigen-binding activity of the 4B9 antibody or its ability to be quantitated by binding to an antihuman immunoglobulin.[19]

Evaluating Real-Time and Accelerated Stability

If a MAb was available that could differentiate between the native and denatured forms of a protein product, an ELISA could be used as a stability-indicating assay. The antibody could also be used for purification of the native molecule by removal of the denatured form.[21] A MAb specific for the denatured protein can detect small amounts of the denatured protein in the presence of the native form.[21]

Gu et al.[22] examined the stability of IL-1 in aqueous solution. They used HPLC for separation of degradation products, then analyzed these by ELISA and bioassay. This is a good example of using a variety of methods together to obtain the most information. The stability (of stressed samples) was analyzed at various temperature and pH conditions using HPLC, ELISA, and IEF-PAGE, and bioassay. The results indicated that the degradation mechanism of IL-1β in aqueous solution is primarily aggregation-precipitation at or about 39°C and possible deamidation at or below 30°C.

Another example of a MAb drug product is a chimeric mouse–human monoclonal IgG antibody specific for the Lewis-Y antigen found on the surface of tumor cells.[23] The antibody was labeled with ^{131}I to target tumor cells for radioimmunotherapy or cancer therapy, and an ELISA was used to measure binding activity. An antiidiotype-binding ELISA was used to quantitate the binding activity of selected stressed stability samples of chimeric L6.[23] The antiidiotype 1B antibody binds to the murine Fab portion of chimeric L6 and is therefore a potential indicator of chemical and conformational changes in the antigen-binding region of the protein.[23] ELISA can still bind lower-molecular weight fragments or aggregates that may also have binding activity and falsely elevate the remaining concentration using this assay.[23]

Determining Shelf Life

According to current FDA guidelines, an acceptable pharmaceutical product should exhibit less than 10% deterioration after 2 years.[5] Protein biopharmaceuticals usually have to be stored under refrigerated conditions or freeze dried to achieve an acceptable shelf life. ELISAs for aggregates and product variants as discussed previously would be applicable for determining shelf life.

QC and Manufacturing

Product Testing, Identification, Quantitation

ELISAs can be used for identification and quantitation of a biopharmaceutical product or for quantitation of impurities or contaminants as discussed previously. They can be used throughout the manufacturing process as well as in quality control or the product release stage just as they are used in all the other stages of product development. To be used for quality control, GMP practices must be followed. All methods need to be validated so that the assay's performance is documented. ELISAs should have internal quality controls to monitor assay acceptability and performance over time. The assay must have defined limits, so it will be known if and when an assay is not acceptable.

Monitoring In-Process Steps

ELISAs can be used for identification and quantitation of the product as well as impurities in the various purification steps (as discussed previously). They can be used to document the removal of known impurities and contaminants, and in process validation to demonstrate batch-to-batch consistency of manufacturing.

STRENGTHS AND WEAKNESSES

In general, the strengths of the ELISA are its selectivity and specificity, whereas its weaknesses are related to precision of measurement. Each assay varies — depending on the antibody, enzyme, enzyme conjugate, and measurement, as well as on assay format. Each should be validated so that its unique performance characteristics are known.

Selectivity

The selectivity of ELISA is based on the reaction between antigen and antibody and thus depends on the specificity of the antibody or antibody pairs. It provides the power and uniqueness of this method as an analytical tool. Selectivity can be viewed either as a weakness or a strength depending on the particular application. Degraded forms of products or products with an altered structure may contain immune-specific epitopes that are recognized by the antibody and not differentiated from full-length or unaltered products. In this case, selectivity would be a weakness. This emphasizes the importance of MAbs. The use of MAbs can provide specificity to the epitope of interest. If a MAb recognized an epitope that was only on intact protein and not on degraded protein, selectivity would be viewed as a strength.

Polyclonal antibodies can react with many epitopes, whereas MAbs are restricted to one epitope on proteins that do not have repeating sequences.[24] By definition, polyclonal immunoassays are generally much more sensitive but less specific than monoclonal assays. Bispecific or hybrid antibodies can be used to increase the affinity. Bispecific antibodies are formed by the fusion of two previously established hybridomas to produce antibodies displaying the binding characteristics of both of the antibodies in one molecule.[25]

The binding of the antigen and antibody can be affected by several factors, including the conjugated label chosen for detection and the method used to conjugate the label, as well as the assay format itself. The selectivity of the ELISA can be affected by the assay format. In an ELISA with a two-site sandwich format, independent epitopes are bound by different antibodies.[26] The specificity comes from multiple site recognition. Polyclonal antibodies can react with many epitopes on a complex antigen surface.[24]

Because of antibody-based selectivity, ELISAs are capable of handling samples that are impure or only semipurified. It is possible to perform ELISAs in a variety of matrices. This is in contrast to other methods such as HPLC that require relatively pure material. During the development and validation of the ELISA method, it needs to be demonstrated that the ELISA is not affected by interfering substances that could be in the test sample, such as buffers, salts, contaminating proteins, and excipients. It also needs to be demonstrated that the conjugated antibody does not bind nonspecifically to the coated solid phase.

Sensitivity

Improvements in ELISA techniques over the years have lead to increased sensitivity. Sensitivity is ultimately determined by the affinity (and avidity) of the antibody used, although the label and detection system used contribute to the overall sensitivity as well. The use of some of the recently developed labels and substrates can amplify the signal for increased sensitivity. The level of sensitivity required is dependent on the application. If ELISA is used to quantitate the product or provide identification of the product in final containers, a sensitivity of μgs may be sufficient. However, if ELISA is used to quantitate impurities, a sensitivity of 1 to 2 ng may be required. For impurity ELISAs, it is possible to have good sensitivity when detecting multiple analytes. It is even possible to get in the picomolar range or even lower using techniques such as enzyme amplification, which generates an increased signal. There are a wide variety of labels now available, including those that are fluorescent and chemiluminescent.[26] With ELISA modifications such as immuno-PCR, it is possible to detect as little as 1000 molecules.[25]

Some ways to increase sensitivity include the use of MAbs or the use of a more sensitive substrate for the enzyme. The ELISA's sensitivity is from the inherent magnification of the enzyme–substrate reaction.

Accuracy

Accuracy expresses the closeness of agreement between a measured test result and its theoretical true value.[27] Accuracy is one of the properties that must be evaluated during assay validation. It is affected by the antigen–antibody reaction as well as the error of measurement. The use of plastic for the solid surface can affect the antigen–antibody reaction. There are sometimes steric constraints and the molecule can change when bound to the surface.[24] Native epitopes could be denatured or altered just by binding to the solid phase.[24] There can also be a difference in affinity depending on whether antigen or antibody is bound to the solid phase. All of these parameters can be investigated during assay development.

Accuracy also depends on the error of measurement. During assay development, it should be demonstrated that there are no plate edge effects, that is, the response is the same in every well of the multiwell plate. The adsorbing capacity of each well in the same plate should not vary; manufacturers today can certify the homogeneity of the wells on the plate. If there are any differences, those usually occur in wells in the outer edges of the plate. If the edge effect cannot be eliminated with further development, these wells must be eliminated from the assay.

One common problem of ELISAs affecting accuracy is the hook effect. This is when the signal does not increase with increasing concentration but actually decreases. Another weakness of ELISA is that, compared to other techniques, it has a limited dynamic range. Extrapolation beyond the limits of the range of the standard curve can lead to inaccuracies.

Precision

Precision is a quantitative measure of the random variation between repeated measurements from multiple sampling of the same homogenous sample under specified conditions.[27] The weakness of the ELISA is its imprecision. The imprecision is related to the nature of the biological reaction — the reaction between antigen and antibody — and its inherent variability. Typically, the precision of an average ELISA is about 20% relative standard deviation, but can be as high as 30% in some circumstances.

Speed — Same-Day Results

Once the reagents are available, the ELISA method is fast and can be modified to accommodate high throughput. But the time to obtain the appropriate reagents must be considered. It may take considerable time to generate the appropriate antibodies by immunization, as discussed previously. Also considerable time must be allowed to develop and optimize the assays.

Over the years, in addition to developments with ELISA reagents such as labels, there have been improvements in automation. This has enabled ELISA to be utilized as a high-throughput tool. Typically, ELISAs can be performed in several hours to days. The most common practice is to precoat the microtiter plate for an overnight incubation period, with the remainder of the steps performed the following day. While ELISAs are fast when compared to other assays such as bioassays, which can take days to weeks, they might be considered slow when compared to methods like HPLC, in which the time from sample injection to chromatogram is a matter of minutes.

Simplicity

The ELISA uses equipment that is commonly accessible in immunochemical laboratories. They are usually formatted using microtiter plates, which have 96 wells. Depending on how many wells are used for standards and controls, many test samples can be run on one plate. It is common practice to run multiple plates in one assay. This number can be greatly increased if part or parts of the assay are automated, that is, at least with the use of an automatic plate washer or automatic pipetter to dispense solutions or perform serial dilutions. Most of this equipment is relatively inexpensive. Some kind of photometric detector (spectrophotometer) or a plate reader is required. More common than not, the plate reader is controlled by software of various levels of sophistication for data recording and analysis, as well as for reporting.

ELISA is a method that is simple to perform. Ferris and Fischer[28] evaluated the performance of subjects with no formal laboratory training or experience (sixth and seventh graders), and this sample of "analysts" were able to perform the ELISA with success.

COMPARISON TO OTHER METHODS

Other Immunological Methods

Western Blot

The Western blot, also known as immunoblot, is an analytical method in which proteins are separated by polyacrylamide gel electrophoresis, transferred by electroelution to a nitrocellulose membrane, reacted with antibodies, and detected by a labeled detecting antibody that precipitates substrate onto the membrane wherever the specific antigen–antibody reaction occurred. The Western blot is most useful in early biopharmaceutical product development. It is able to associate specific bands on the acrylamide gel with reactivity with a particular antibody. It provides information regarding molecular weight and subunit structure. The ELISA does not. The advantage of the Western blot is that the samples can be complex mixtures; they do not need to be purified. The disadvantage of this method is that it is not strictly quantitative. Also, some antibodies, particularly some MAbs, may not be able to react with a target protein that has been subjected to this method, whereas they may react with a target protein in solution or present as the solid phase in an ELISA. Some proteins are denatured somewhat during the separation and transfer process.

The Western blot method is often used in the analysis of host cell impurities. It can be used to identify a recurring impurity. O'Keefe et al. used a Western blot to identify an *E. coli* protein impurity in the preparation of the recombinant fibroblast growth factor (aFGF).[29] By using specific antisera to the *E. coli* host cell proteins, they were able to isolate the impurity and determine its N-terminus amino acid sequence to confirm its identity. Antibodies could be used to determine the concentration of this impurity in sample preparations.

ELISA with Other Labels

The "E" in the acronym ELISA stands for "enzyme," indicating that an enzyme is linked, or conjugated, to one of the reactants, most commonly to the antibody.

Enzyme labels are currently by far the most common labels in use. Enzymes include HRP, alkaline phosphatase (AP) and, less often, beta-galactosidase (β-gal). The effectiveness of the conjugated antibody depends on the antibody, the label, and the procedure chosen to link the two.[30]

Other immunoassays are based on the same antibody–antigen binding reaction but use a different labeling system for detection. Instead of an enzyme label, there are radioactive isotopes, and fluorescent and luminescent labels. Some important immunoassays are defined below:

RIA: Radioimmuno assay; uses a radioactive label on the antibody or antigen. These were among some of the first types of immunoassays.

FIA: Fluorescence immunoassay; uses a fluorescent tag on the antibody or antigen. Fluorescent labels absorb light of one wavelength and reemit it at another wavelength. The label is excited by UV and emits visible light. Common fluorescent labels are fluorescein, Texas red, and GFP (green fluorescent protein).

CLIA: Chemiluminescent immunoassay; uses variety of light-generating labels. Luminescent labels are widely replacing radioactive ones in some applications because they can provide the same sensitivity without the hazard. Chemiluminescence uses specific chemicals that, when reacted, give out light, or use a blocking group that, when removed, generates light. A phosphate group is a common blocking group used to detect AP. Bioluminescence uses specialized enzyme systems that can also generate light using ATP. The most common is luciferase.

Unlabeled Format: Electrochemical Immunoassays

The immunosensor is a type of immunoassay that does not depend on labels. The antibody or antigen is coupled to a reactive surface, and then responds to the binding of the complementary antigen or antibody. Indirect immunosensors use separate labeled species to facilitate detection after binding. Direct immunosensors are those in which the reaction between an antibody and its corresponding antigen give rise to changes in the reactive surface, which can then be measured. Direct immunosensors can carry out real-time monitoring of the sample.[25]

The BIAcore biosensor measures biospecific interactions in real time.[31] The principle is to immobilize the antigen on a sensorchip surface while the antibody is allowed to continuously flow over the surface or vice versa. The interaction is then detected by surface plasmon resonance (SPR), directly registered and presented as a sensorgram. Association and dissociation rate constants can be calculated from SPR data. The BIAcore has been used in antibody engineering: in screening, selection, characterization, and epitope mapping during the different steps of generating antibodies.

Another antibody-based method is immunoligand assay (ILA) technology. The Threshold system (Molecular Devices, Sunnyvale, CA) shortens assay development and assay turnaround time.[11] It adapts a sensitive detection system originally designed for quantitation of DNA at picogram levels and commercially available IgG-labeling reagents compatible with the detector technology.[11]

Ghobrial et al.[32] used ILA to measure HCPs impurities. In this method, immune complexes are formed between the HCPs, biotinylated capture antibodies, and fluorescein-labeled polyclonal antibodies. The complexes are then immobilized on a biotinylated nitrocellulose strip in the presence of streptavidin. After washing, an antifluorescein–urease conjugate is added, thereby forming a complex that is directly proportional to the amount of residual HCPs present. The urease hydrolyzes a solution of substrate urea to release ammonia, causing a localized pH change, which is then read by the Threshold electronics.[4] This was used for CHO proteins in recombinant human erythropoietin,[32] *E. coli* HCPs in recombinant human basic fibroblast growth factor, *E. coli* HCPs in recombinant human alpha interferon, and *E. coli* HCPs in recombinant bovine somatotropin.[11]

Protein pharmaceuticals are so complex that multiple methods are required to gain a complete picture of the sample. Methods are often used in combination. For example, a chromatography system may be used to analyze fluorescent reactants from an immunoassay-based method.

Another technology where two methods are combined is tandem liquid chromatography–immunoassay (LC-IA). LC-IA increases the selectivity and the sensitivity of assays by removing interfering species. The LC system provides a wide variety of new, high-sensitivity, high-speed methods for carrying out immunological assays. This is important for monitoring analytes in complex biological matrices.

Other Methods (Not Immunology Based)

At one end of the analytical spectrum is the bioassay, which can demonstrate what biological activity the biopharmaceutical molecule may possess, regardless of molecular structure. At the other end are structural methods that elucidate the molecular structure of the molecule, regardless of biological activity. Somewhere in the middle of the spectrum is the ELISA.

Structural Analytical Methods: HPLC, CE

Methods to measure the structure of biopharmaceuticals include tryptic peptide mapping, HPLC, capillary electrophoresis (CE), mass spectroscopy, and circular dichroism spectra.

Reversed-phase HPLC uses a nonpolar stationary phase and a polar mobile phase. The characteristics are operational simplicity, high efficiency, column stability, and ability to analyze simultaneously a broad spectrum of both closely related and widely different compounds. Separation is based on hydrophobicity.[33] Findlay et al. provide a comparison of chromatography methods with immunoassays (Table 11.1).[27]

Other commonly used techniques include capillary electrophoresis (CE) and mass spectrometry. These techniques are discussed in other chapters in this book.

Table 11.1 Differences between Chromatographic Assays and Immunoassays

	Chromatographic assays	Immunoassays
Basis of measurement	Physicochemical properties of analyte	Antigen–antibody reaction
Analytical reagents	Well-characterized and widely available	Unique and usually not widely available
Analytes	Small molecules	Small molecules and macromolecules
Detection method	Direct	Indirect
Sample pretreatment	Yes	Usually no
Calibration model	Linear	Nonlinear
Assay environment	Contains organic solvents	Aqueous
Time required for development	Weeks	Months (due to time needed for Ab generation)
Intermediate (interassay) imprecision	Low (<10%)	Moderate (<20%)
Source of imprecision	Intraassay	Interassay
Assay working range	Broad	Limited
Cost of equipment	Expensive	Inexpensive
Analysis mode	Series, batch	Batch
Assay throughput	Good	Excellent

Bioassay

Bioassays are used together with other methods to test biopharmaceutical products for potency, identity, purity, and stability. However, only a bioassay can assess the potency or biological effects of a product. Bioassays can be *in vivo* or *in vitro*. *In vivo*, or animal potency assays, measure the drug product's effect on the whole organism. The trend is away from *in vivo* and to *in vitro*, when *in vitro* methods are available. *In vitro* bioassays are typically cell culture-based assays that measure the drug product's effect on the cell. Bioassays monitor biologically active sites, whereas ELISAs monitor immunoreactive sites. These sites may or may not be the same. Bioassays can be used to monitor the desired relevant biological effects of a product and also to detect any undesirable effects of a product for safety concerns. They may be able to distinguish the activity or inactivity of product variants, whereas ELISA may not. One disadvantage of the bioassay is its imprecision. Bioassays may have percent residual standard deviation (%RSD) of 25 to 30%. Another disadvantage is that it often can take days or weeks to complete. They are useful in all phases of product development, from research to release testing for quality control.

Bioassays can be used to assess the effect of molecular heterogeneity due to glycosylation, whereas ELISA cannot. The sensitivity of the bioassay and ELISA are often comparable. Both methods can be used as stability-indicating test methods.

References

1. Sadick, M.D., A. Intintoli, V. Quarmby, A. McCoy, E. Canova-Davis, and V. Ling (1999). Kinase receptor activation (KIRA): a rapid and accurate alternative to endpoint bioassays. *J Pharm Biomed Anal* **19**(6): 883–891.
2. FDA (1999). ICH harmonised tripartite guideline. Specifications: Test procedures and acceptance criteria for biotechnological/biological products. *Fed Regist* **64**: 44928.
3. Berthold, W. and J. Walter (1994). Protein purification: aspects of processes for pharmaceutical products. *Biologicals* **22**(2): 135–50.
4. DiPaolo, B., A. Pennetti, L. Nugent, and K. Venkat (1999). Monitoring impurities in biopharmaceuticals produced by recombinant technology. *PSTT* **2**(2): 70–82.
5. Costantino, H.R., R. Langer, and A.M. Klibanov (1994). Solid-phase aggregation of proteins under pharmaceutically relevant conditions. *J Pharm Sci* **83**(12): 1662–1669.
6. Braun, A. and J. Alsenz (1997). Development and use of enzyme-linked immunosorbent assays (ELISA) for the detection of protein aggregates in interferon-alpha (IFN-alpha) formulations. *Pharm Res* **14**(10): 1394–1400.
7. Vandenbroeck, K., E. Martens, S. D'Andrea, and A. Billiau (1993). Refolding and single-step purification of porcine interferon-gamma from *Escherichia coli* inclusion bodies. Conditions for reconstitution of dimeric IFN-gamma. *Eur J Biochem* **215**(2): 481–486.

8. Tsouloufis, T., A. Mamalaki, M. Remoundos, and S.J. Tzartos (2000). Reconstitution of conformationally dependent epitopes on the N-terminal extracellular domain of the human muscle acetylcholine receptor alpha subunit expressed in *Escherichia coli*: implications for *myasthenia gravis* therapeutic approaches. *Int Immunol* **12**(9): 1255–1265.

9. Merli, S., A. Corti, and G. Cassani (1995). Production of soluble tumor necrosis factor receptor type I in *Escherichia coli:* optimization of the refolding yields by a microtiter dilution assay. *Anal Biochem* **230**(1): 85–91.

10. FDA (2000). ICH Draft Revised Guidance on Impurities in New Drug Products. *Fed Regist* **65**: 44791.

11. Eaton, L.C. (1995). Host cell contaminant protein assay development for recombinant biopharmaceuticals. *J Chromatogr A* **705**: 105–114.

12. Pauly, J.U., B. Siebold, R. Schulz, W. List, G. Luben, and F.R. Seiler. (1990). Development of an ELISA for the detection and determination of contaminating proteins in recombinant DNA derived human erythropoietin. *Behring Inst Mitt* **86**: 192–207.

13. Hoffman, K. (2000). Strategies for host cell protein analysis. *Biopharm–Appl T Bio* **13**(5): 38–45.

14. Anicetti, V.R., E.F. Fehskens, B.R. Reed, A.B. Chen, P. Moore, M.D. Geier, and A.J. Jones (1986). Immunoassay for the detection of *E. coli* proteins in recombinant DNA derived human growth hormone. *J Immunol Methods* **91**(2): 213–224.

15. Whitmire, M.L. and L.C. Eaton (1997). An immunoligand assay for quantitation of process specific *Escherichia coli* host cell contaminant proteins in a recombinant bovine somatotropin. *J Immunoassay* **18**(1): 49–65.

16. Lucas, C., C. Nelson, M.L. Peterson, S. Frie, D. Vetterlein, T. Gregory, and A.B. Chen (1988). Enzyme-linked immunosorbent assays (ELISAs) for the determination of contaminants resulting from the immunoaffinity purification of recombinant proteins. *J Immunol Methods* **113**(1): 113–122.

17. Vemuri, S., C.D. Yu, and N. Roosdorp (1994). Effect of cryoprotectants on freezing, lyophilization, and storage of lyophilized recombinant alpha 1-antitrypsin formulations. *PDA J Pharm Sci Technol* **48**(5): 241–246.

18. Page, C., P. Dawson, D. Woollacott, R. Thorpe, and A. Mire-Sluis (2000). Development of a lyophilization formulation that preserves the biological activity of the platelet-inducing cytokine interleukin-11 at low concentrations. *J Pharm Pharmacol* **52**(1): 19–26.

19. Gombotz, W.R., S.C. Pankey, D. Phan, R. Drager, K. Donaldson, K.P. Antonsen, A.S. Hoffman, and H.V. Raff (1994). The stabilization of a human IgM monoclonal antibody with poly(vinylpyrrolidone). *Pharm Res* **11**(5): 624–632.

20. Ressing, M.E., W. Jiskoot, H. Talsma, C.W. Van Ingen, E.C. Beuvery, and D.J. Crommelin (1992). The influence of sucrose, dextran, and hydroxypropyl-beta-cyclodextrin as lyoprotectants for a freeze-dried mouse IgG2a monoclonal antibody (MN12). *Pharm Res* **9**(2): 266–270.

21. Werner, R.G., W. Berthold, H. Hoffmann, J. Walter, and W. Werz (1992). Immunological techniques in biotechnology research. *Biochem Soc Trans* **20**(1): 221–226.

22. Gu, L.C., E.A. Erdos, H. Chiang, T. Calderwood, K. Tasai, G.C. Visor, J. Duffy, W. Hsu, and L.C. Foster (1991). Stability of interleukin1B (IL-1B) in aqueous solution: analytical methods, kinetics, products, and solution formulation implications. *Pharm Res* **8**(4): 485–490.

23. Paborji, M., N.L. Pochopin, W.P. Coppola, and J.B. Bogardus (1994). Chemical and physical stability of chimeric L6, a mouse-human monoclonal antibody. *Pharm Res* **11**(5): 764–771.

24. Pesce, A.J. and J.G. Michael (1992). Artifacts and limitations of enzyme immunoassay. *J Immunol Methods* **150**(1–2): 111–119.

25. Ronald, A. and W.H. Stimson (1998). The evolution of immunoassay technology. *Parasitology* **117 Suppl.**: S13–27.

26. Self, C.H. and D.B. Cook (1996). Advances in immunoassay technology. *Curr Opin Biotechnol* **7**: 60–65.

27. Findlay, J.W., W.C. Smith, J.W. Lee, G.D. Nordblom, I. Das, B.S. DeSilva, M.N. Khan, and R.R. Bowsher (2000). Validation of immunoassays for bioanalysis: a pharmaceutical industry perspective. *J Pharm Biomed Anal* **21**(6): 1249–1273.

28. Ferris, D.G. and P.M. Fischer (1992). Elementary school students' performance with two ELISA test systems. *JAMA* **268**(6): 766–770.

29. O'Keefe, D.O., P. DePhillips, and M.L. Will (1993). Identification of an *Escherichia coli* protein impurity in preparations of a recombinant pharmaceutical. *Pharm Res* **10**(7): 975–979.

30. Avrameas, S. (1983). Enzyme immunoassays and related techniques: development and limitations. *Curr Top Microbiol Immunol* **104**: 93–9.

31. Malmborg, A.C. and C.A. Borrebaeck (1995). BIAcore as a tool in antibody engineering. *J Immunol Methods* **183**(1): 7–13.

32. Ghobrial, I.A., D.T. Wong, and B.G. Sharma (1997). An immuno-ligand assay for the detection and quantitation of contaminating proteins in recombinant human erythropoietin (r-HuEPO). *BioPharm–Technol Bus* **10**(1): 42–45.

33. Krstulovic, A.M. and P.R. Brown (1982). *Reversed-Phase High Performance Liquid Chromatography: Theory, Practice, and Biomedical Applications*. John Wiley & Sons, New York.

12

Applications of NMR Spectroscopy in Biopharmaceutical Product Development

Yung-Hsiang Kao, Ping Wong, and Martin Vanderlaan

ABBREVIATIONS

CPMG	Carr–Purcell–Meiboom–Gill
CE	Capillary electrophoresis
HPLC	High-performance liquid chromatography
HEPES	N-2-Hydroxyethylpiperazine-N″-2-ethanesulfonic acid
LC	Liquid chromatography
MES	2-(N-Morpholino)-ethanesulfonic acid
MOPS	3-(N-Morpholino)-propanesulfonic acid
^{1}H-NMR	Proton nuclear magnetic resonance
ppm	Chemical shift unit (part per million, Hz/MHz)
TEA	Tetraethylammonium
TFA	Trifluoracetic acid
TMA	Tetramethylammonium
Tris	2-Amino-2hydroxymethyl-1,3-propanediol
UF/DF	Ultrafiltration/diafiltration

NMR AND BIOPHARMACEUTICAL PRODUCT DEVELOPMENT

Ever since its discovery more than 50 years ago, nuclear magnetic resonance (NMR) spectroscopy has been an important analytical method for chemists. The introduction of FT (Fourier transform) NMR techniques in the early 1970s made it even more powerful and widely used. NMR is undoubtedly one of the most valuable tools in the discovery stage of pharmaceutical product development. For example, structural biologists often use NMR to determine the three-dimensional solution structures of biomolecules of pharmaceutical importance.[1-3] In recent years, NMR has also been employed to screen small-molecule combinatorial libraries in search of drug leads that bind to target proteins.[4-6] For a small-molecule drug candidate, NMR is an indispensable analytical technique in the development stage as well. Using NMR to identify drug metabolites, to determine degradation products of a drug, and to elucidate impurities in a drug substance are some important applications of NMR in a small-molecule drug product development program.

In contrast, NMR is fairly underutilized beyond the discovery phase of the biopharmaceutical product development, mainly because the sensitivity and resolution of NMR for macromolecules such as proteins are much lower than those for small molecules. The characterization of a protein product and its variants is a critical part of a protein-based pharmaceutical product development program. The determination of a protein structure using NMR, while providing extremely valuable information, is not trivial and not always feasible. Thus, NMR has not been used often in biopharmaceutical product characterization studies. However, there are still many analytical issues that can readily be addressed by NMR in a biopharmaceutical product development program. For example, process validation studies, as required by the regulatory agencies, must be conducted to demonstrate that impurities (often small molecules) from the cell culture fluids and process buffers used in the manufacturing process are removed from the final product. In addition, the manufacturing process must not cause objectionable levels of extractables from process equipment (filters, columns, gaskets, etc.).[7] Many of these process-related impurities can be easily detected by fairly simple and general NMR methods.

Traditionally, HPLC, GC-MS, or LC-MS methods were used to monitor the clearance of small-molecule impurities. These analytical techniques often require unique solvents, columns, methods, reagents, detectors, and buffers for each analyte to be quantified. The NMR method, albeit not the most sensitive technique, normally does not have these problems. In this chapter, some examples will be used to demonstrate that NMR is a fast, generic, and reliable analytical technique for solving analytical problems encountered in the development of biopharmaceutical products. The NMR techniques described here require minimal sample handling and use simple standard NMR methods. They can easily be implemented and used for process development and validation purposes.

NMR is generally considered an expensive and insensitive technique. Despite higher cost and lower sensitivity, the advantages that NMR offers are significant and may outweigh its weaknesses. In fact, the cost of NMR can be offset easily by the savings in time and effort to solve biopharmaceutical analytical problems. If the instrument capital cost is a concern, performing these NMR tests in a contract lab might be an alternative. The sensitivity of NMR, while lower than those of other techniques such as mass spectrometry, is often adequate for solving analytical problems in process development and validation.

NMR SPECTROSCOPY

The underlying physical principles of NMR have been established and are well understood.[8] Applications of both solid- and solution-state NMR spectroscopy can be found in many different disciplines. It is routinely used in structural elucidation of organic and inorganic compounds, polymers, and biomolecules (e.g., proteins, nucleic acids, and carbohydrates). Additionally, NMR can be used to study molecular interactions (e.g., protein–protein and protein–ligand), molecular dynamics, and chemical reactions. It has also been used extensively in medical research and imaging (magnetic resonance imaging).

NMR is basically one form of absorption spectroscopy. Interested readers can find details of NMR theory and methodology elsewhere.[8-10] Briefly, NMR signals arise from the transitions between nuclear spin states. In the presence of an external magnetic field, nuclei with nonzero spin angular momentum (for example, a proton has a spin of 1/2) will have nondegenerate nuclear spin states. An electromagnetic radiation applied to a nucleus at a frequency corresponding to the energy difference between these spin states will result in a transition between these spin states. The radiation that induces such transitions is in the radio frequency (RF) range and delicately dependent on the chemical environment of each nucleus in a molecule. A plot of absorption peak intensities vs. frequencies constitutes an NMR spectrum. All modern NMR instruments are operated in FT mode, which uses RF pulses to excite all nuclei in a sample simultaneously and then detects the nuclear free induction decay (FID). During the FID, the excited nuclei return to their equilibrium states while releasing the RF energy, which can be detected by a tuned RF coil. The FT of FID gives rise to a normal absorption spectrum. The FT method allows for much faster multiple-transient acquisition and, therefore, results in much higher sensitivities than the conventional continuous wave (CW) mode of operation.

COMMONLY USED NMR METHODS

There are a large number of one- and multidimensional NMR methods available for solving different analytical problems. Typically, only a few standard NMR experiments are necessary for solving most process-related analytical problems (Table 12.1). These methods are all straightforward and easy to execute routinely. A repertoire of more complex methods can be found elsewhere.[9]

A solvent-suppression method is normally required in NMR applications because the solvent signals, if not suppressed, may saturate the receiver and hinder

Table 12.1 Commonly Used FT-NMR Methods

Experiment	Description	Applications in Process Development and Validation
Standard one-dimensional NMR	One-dimensional NMR — simple one-pulse experiment, typically with presaturation of solvent during the recycle delay with a weak RF field	To quantify small molecules To identify some simple small molecules
CPMG	Carr–Purcell–Meiboom–Gill spin-echo method	To suppress signals with short relaxation time (e.g., protein signals) Small molecules can be detected and quantified in the presence of proteins
TOCSY	Two-dimensional total correlation spectroscopy	To elucidate structure of organic molecules To establish proton coupling network and molecular connectivity
COSY	Homonuclear correlation spectroscopy	To elucidate structure of organic molecules To establish proton coupling network within a molecule
NOESY	Homonuclear NOE (nuclear Overhauser) spectroscopy	To elucidate structure of organic · molecules To determine the spatial proximity of nuclei
HSQC HMQC	Heteronuclear single-quantum/multiple-quantum correlation spectroscopy	To elucidate structure of organic molecules To determine heteronuclear coupling connectivity
HMBC	Heteronuclear multiple-bond correlation spectroscopy	To elucidate structure of organic molecules To establish long-range (i.e., multibond) heteronuclear coupling

the observation of weak peaks, which are usually the signals of interest. The easiest solvent-suppression method uses a low-power saturation pulse applied at the solvent (H_2O in most cases) peak frequency for 1 to 2 sec prior to the 90° observation pulse to reduce the solvent peak. Other more sophisticated solvent-suppression techniques can also be used.[11–13]

In many cases, the analytical tasks are simply to detect and quantify a specific known analyte. Examples include the detection and quantification of commonly used buffer components (e.g., Tris, acetate, citrate, MES, propylene glycol, etc.). These simple tasks can readily be accomplished by using a standard one-dimensional NMR method. In other situations, the analytical tasks may involve identifying unknown compounds. This type of task usually requires homonuclear and heteronuclear two-dimensional NMR experiments, such as COSY, TOCSY, NOESY, HSQC, HMBC, etc. The identification of unknown molecules may also require additional information from other analytical methods, such as mass spectrometry, UV-Vis spectroscopy, and IR spectroscopy.[14]

One-dimensional proton NMR spectroscopy is the most straightforward method for process validation and development. It can be used as a limit test, i.e., to demonstrate that a particular analyte is below the detection limit. It can also be used to accurately quantify an analyte by comparing the NMR peak area from a test sample against a standard curve. To get accurate quantitation, it is important to keep the acquisition parameters and conditions constant for both standard and test samples. For example, the receiver gain, power level, and duration of all pulses must stay the same within an assay. In addition, the probe should remain tuned for all samples.

For biopharmaceutical process validation, it is often necessary to detect and quantify small molecules in the presence of large protein molecules. A standard CPMG spin-echo experiment[15,16] may be used for this purpose. The spin-echo method reduces the broad signals, which arise mostly from large molecules such as proteins, while it preserves the sharp signals, which arise mostly from small molecules. If spin-echo cannot satisfactorily reduce the protein signals, removing the protein from the sample by ultrafiltration is an easy alternative. However, extra care must be taken to ensure that ultrafiltration does not introduce any impurities or remove analytes of interest. The ability to analyze small molecules either in the presence of protein or after a simple filtration step to remove protein is essential in the purity test for a biopharmaceutical product.

Because the sensitivity of NMR is the highest for protons compared to other nuclei, all examples of quantitation work described in this chapter are based on proton NMR data. The signals from other NMR active nuclei such as ^{19}F or ^{13}C may also be used for quantitation. The quantification of TFA using ^{19}F NMR is a good example. However, except for ^{19}F, the sensitivities and detection limits are usually compromised in these measurements because nuclei other than ^1H and ^{19}F typically have a lower natural abundance and a lower magnetogyric ratio.

ADVANTAGES AND DISADVANTAGES OF NMR FOR BIOPHARMACEUTICAL PROCESS DEVELOPMENT AND VALIDATION

One major advantage of NMR over other types of spectroscopy is that NMR signals are highly specific and quantitative. Because the resonance frequency (chemical shift) of a nucleus is closely related to its local environment in a molecule and the structure of that molecule, each compound will have a unique set of NMR signals. In addition, the peak intensity of an NMR signal is linearly proportional to the population of the nuclei that contribute to that signal. This is true not only for the nuclei from the same molecule, but also for the nuclei from different molecules. The relative peak intensities of the NMR signals from the same molecule directly reflect the relative numbers of each type of nucleus in the molecule. In essence, NMR is a way to obtain a relative nuclei count for a molecule. It is also valid to compare intensities of NMR peaks arising from different molecules. Therefore, NMR offers a convenient and nondestructive way of obtaining relative concentrations of impurities, isomer ratios, etc. If accurate standards can be prepared, it is very straightforward to obtain absolute concentrations for a given analyte in solution. The specific and quantitative characteristics of NMR spectroscopy make it a very useful technique in quantitative analytical chemistry.

There are many other advantages of using NMR as an analytical method for process development or validation. For example, sample preparation required for NMR analysis is usually minimal. There is no need to perform buffer exchange or complex chemical reaction (e.g., chemical labeling) prior to NMR measurement. In addition, there is very little restriction on the nature of samples that can be analyzed by NMR. For proton NMR (most commonly used) analysis in aqueous solution, the only requirement is that the molecules to be analyzed by NMR must contain nonexchangeable protons. NMR measurement usually is compatible with a wide range of solvents (e.g., organic solvents, aqueous buffers, buffers containing high concentration of salt, etc.). These advantages significantly reduce the assay development time.

Like all analytical techniques, NMR spectroscopy also has its weakness. The major limitation is its sensitivity. For proton (the most sensitive nucleus) NMR, the detection limit is typically 10 to 100 μM for small organic molecules. This detection limit can be achieved fairly easily within a reasonable experimental time at 500 MHz. However, NMR is still significantly less sensitive than other techniques such as mass spectrometry. In spite of this limitation, the proton NMR is still sensitive enough to detect small organic molecules at a level near 1 to 10 $\mu g/ml$, which is sufficient for process validation purposes in most cases. The sensitivity and resolution of NMR are also limited by the size of molecules. While NMR has routinely been used in studying structures of macromolecules such as polymers, proteins, DNA, and polysaccharides, the NMR signals arising from these large molecules are normally much broader and weaker. Therefore, structural studies of these molecules by NMR often require a high sample concentration (>1 mM). In extreme cases, the NMR signals of large molecules become so broad that they cannot be observed even at a very high sample concentration. Quantitative applications of NMR are therefore primarily limited to small molecules.

CLEARANCE OF PROCESS-RELATED IMPURITIES

Trace levels of process-related impurities might have a profound effect on the quality and safety of a biopharmaceutical protein product. Therefore, the process validation study must demonstrate the removal of these impurities to an acceptable level by the recovery process. Detecting a low-level (<10 μg/ml) small-molecule impurity in the presence of active protein and excipients at much higher concentrations is a daunting analytical problem. In many cases, NMR is an excellent method for solving such challenging problems. The main advantage of using NMR is that the final formulated protein sample can be analyzed directly by NMR. The ability of detecting process buffer components in a formulated protein product by proton NMR is illustrated by a few examples described below. The same method can easily be applied to quantitation of other small proton-containing molecules.

Trace A in Figure 12.1 is a proton NMR spectrum of a control sample containing a mixture of three process buffer components (succinate, TEA, and TMA) in sodium acetate, a formulation buffer for a recombinant protein. The proton NMR spectrum in Figure 12.1 (trace A) shows that the signals of the three process buffer components are all unique and well resolved from each other. These unique signals from different molecules demonstrate that simultaneous detection of multiple analytes in a single spectrum is possible if the signals from those analytes do not overlap. Traces B and C are the proton NMR spectra of the formulated recombinant protein (2.5 mg/ml protein in sodium acetate buffer) with and without the spike of the three process buffer components (10 μg/ml each). Because the protein in this sample is a large molecule (MW > 25,000), the protein signals appear in the spectrum as very broad and weak peaks over a large spectral width. These broad and weak protein signals usually give rise to a rough or wavy baseline in the spectrum (trace B in Figure 12.1). This type of baseline makes the detection and quantification of low-level small-molecule impurities difficult. However, the protein signals can be minimized by the spin-echo method (traces D and E in Figure 12.1) to allow for better detection and quantification of small molecules. The NMR spectra in Figure 12.1 clearly indicate that TEA and TMA are not present in this formulated protein sample, whereas a small amount of succinate is present. The level of succinate in the final formulated protein sample was determined to be about 12 μg/ml based on its peak area and a standard curve.

Because NMR can be used to detect and quantify process buffer components, it is a convenient method to monitor clearance of the penultimate buffer during the final diafiltration step of a purification process. An example is shown in Figure 12.2. In this case, a protein was initially in a buffer containing several components including Tris, HEPES, and citrate (top spectrum). The buffer was exchanged during the diafiltration with 10 m*M* histidine (the diafiltration buffer). After 10 diavolumes, the Tris and HEPES are reduced below the limit of quantification, and the residual citrate level is 64 μg/ml. This example demonstrates

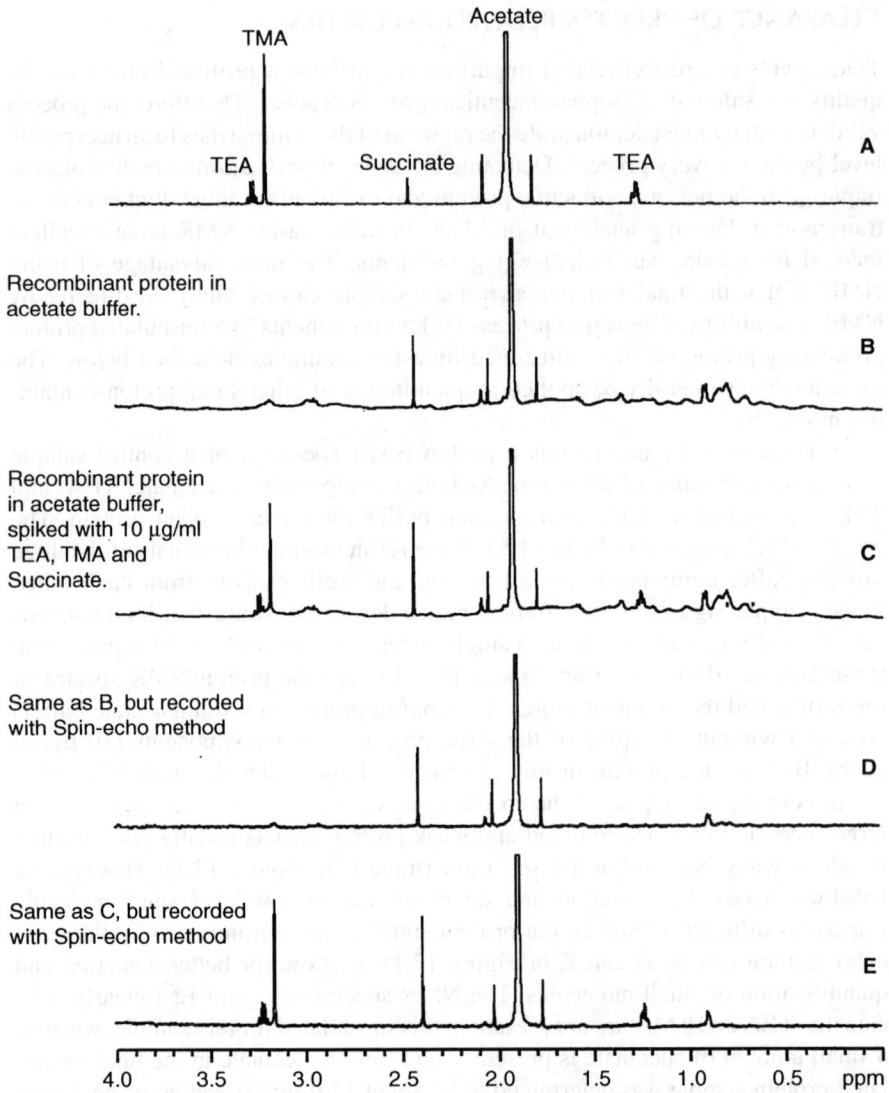

Figure 12.1 Clearance of small-molecule impurities from process buffers in a formulated protein product. Trace A: the NMR spectrum of a control sample containing a mixture of three components (succinate, tetraethylammonium, and tetramethylammonium) in the final formulation buffer (sodium acetate). These three components were used in the recovery process for a biopharmaceutical product. Traces B and D: the proton NMR spectra of the formulated protein product. No TEA or TMA were detected, but a small amount of succinate was observed in this sample. Traces C and E: the proton NMR spectra of a formulated protein product spiked with 10 µg/ml of succinate, TEA, and TMA. Traces D and E were recorded with CPMG spin-echo method to reduce the protein signals. The reduction of NMR signals from the protein allows for better observation of the small-molecule signals.

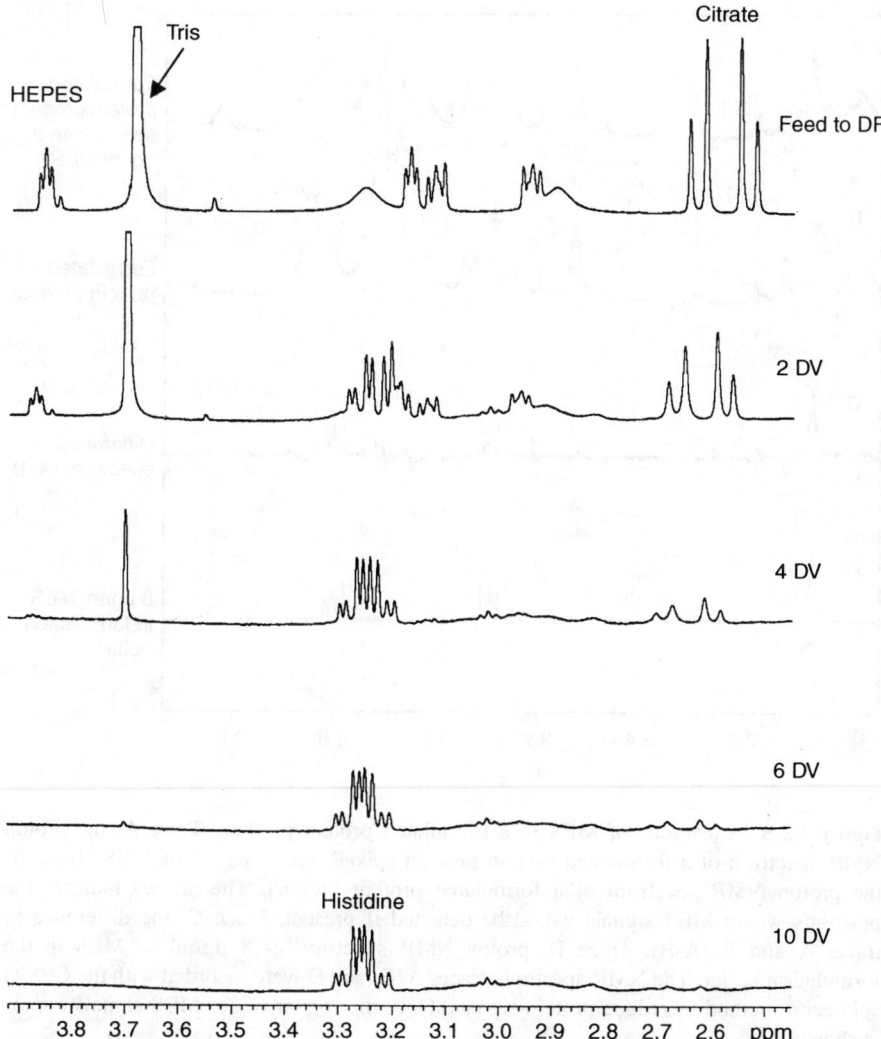

Figure 12.2 Using NMR to monitor the clearance of penultimate buffer components by diafiltration. The traces from top to bottom are the NMR spectrum of the feed to the dialfitration and the spectra of the samples taken after different diavolumes of buffer exchange have been completed. The penultimate buffer components (including HEPES, Tris, and citrate) are clearly removed after 10 diavolumes of buffer exchange.

that the NMR method can be used to verify and validate the performance of the diafiltration operation.

In some cases, the NMR signals of a small-molecule analyte might be broad and poorly resolved because of a chemical exchange process, for example,

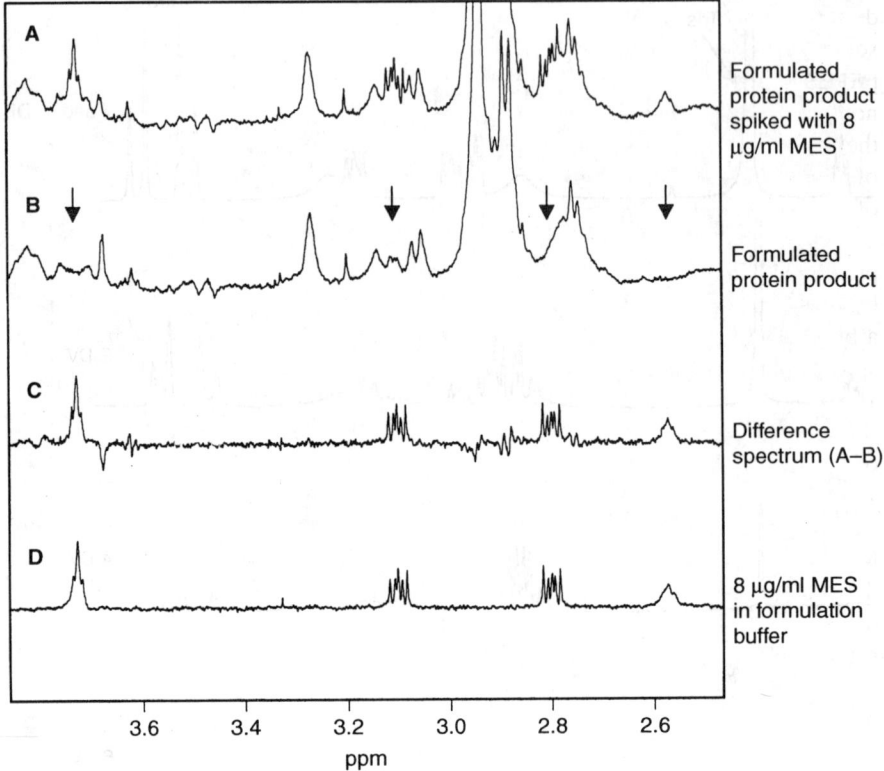

Figure 12.3 Clearance of MES in a formulated protein product. Trace A: the proton NMR spectrum of a formulated protein product spiked with 8 μg/ml of MES. Trace B: the proton NMR spectrum of a formulated protein product. The arrows indicate the positions where MES signals would be detected if present. Trace C: the difference of traces A and B (A-B). Trace D: proton NMR spectrum of 8 μg/ml of MES in the formulation buffer. The NMR spectra in traces A, B, and D were recorded with the CPMG spin-echo method to reduce protein signals. Only the region where MES signals appear is shown.

exchanging between two protonation states or two different conformational isomers. The spin-echo method will not work well in this situation. However, by carefully inspecting the spectrum one may still unambiguously determine if the analyte is present in the protein product. Figure 12.3 illustrates an example of detecting MES in a recombinant protein product by NMR. Trace D of Figure 12.3 is a proton NMR spectrum of MES (8 μg/ml) in the formulation buffer. At the pH of formulation buffer, the MES is in exchange between different protonation states. As a result, the NMR signals of MES are broader than those arising from other molecules with a similar size. These broad NMR signals make the detection of MES in the formulated protein sample (trace B) somewhat trouble-some. However, by carefully comparing the NMR spectrum of the formulated protein sample spiked with 8 μg/ml of MES (trace A) with the spectrum of the nonspiked sample (trace B), one concludes that the MES signals are absent in the spectrum of formulated protein sample (trace B). Subtracting the spectrum of nonspiked sample from the spectrum of the spiked sample leads to full recovery of the NMR signals of 8 μg/ml MES (trace C).

Another way to detect small molecules in the final formulated protein product without the interference from the protein signals is to remove the protein by ultrafiltration. Figure 12.4 compares a section of the proton NMR spectra of a biopharmaceutical protein product before (upper spectrum) and after (bottom spectrum) the protein was removed by ultrafiltering the sample with a Centricon-10 (Millipore Corp, Bedford, MA). Removing protein results in a flatter baseline (bottom spectrum). If small molecules are present in a protein sample, the removal of the protein may allow for unobstructed detection of the small molecules. In this case, a small amount of acetate (~ 1 μg/ml) is detected in the sample [bottom trace, Figure 12.4]. Figure 12.5 shows that spikes of 10 μg/ml of acetate and MES into the protein sample are fully recovered after the ultrafiltration to remove the protein. This example demonstrates that the interference of protein with the detection and quantitation of small-molecule impurities in a formulated protein product can be effectively eliminated by ultrafiltration.

Removing the protein offers an opportunity to detect not only process-related impurities, but also any small-molecule impurities. In fact, NMR is per-haps the easiest way to quickly determine if any significant amounts of small-molecule impurities are present in the final bulk of a protein product. Because sharp lines from small organic compounds can be readily detected at levels of 1 to 10 μg/ml and above, their absence in the NMR spectrum provides compelling evidence for the clearance of low-molecular weight impurities. Figure 12.6 dis-plays two sections of the proton NMR spectrum for a protein product recorded after the protein has been removed. After the removal of the protein, only the formulation buffer (25 m*M* citrate) and a small amount (~1 μg/ml) of residual Tris are observed. For this product, Tris is a component of the penultimate buffer used in the recovery process prior to the final UF/DF formulation step. The Tris signal has a signal-to-noise ratio of about 70, indicating that NMR sensitivity is more than enough to detect small-molecule impurities at 1 μg/ml. Moreover, the absence of any other significant proton NMR signals in Figure 12.6 provides convincing evidence that other small-molecule impurities are unlikely to be present at levels greater than 1 μg/ml.

The removal of protein by filtration is a convenient way to avoid the interference of protein in the NMR measurement. However, this method requires that the analyte is not physically associated with the protein; otherwise, ultrafil-tration will remove the analyte along with the protein. In addition, extractables from the devices employed to remove the protein may be introduced into the sample. Therefore, proper controls must be prepared and analyzed. Despite these

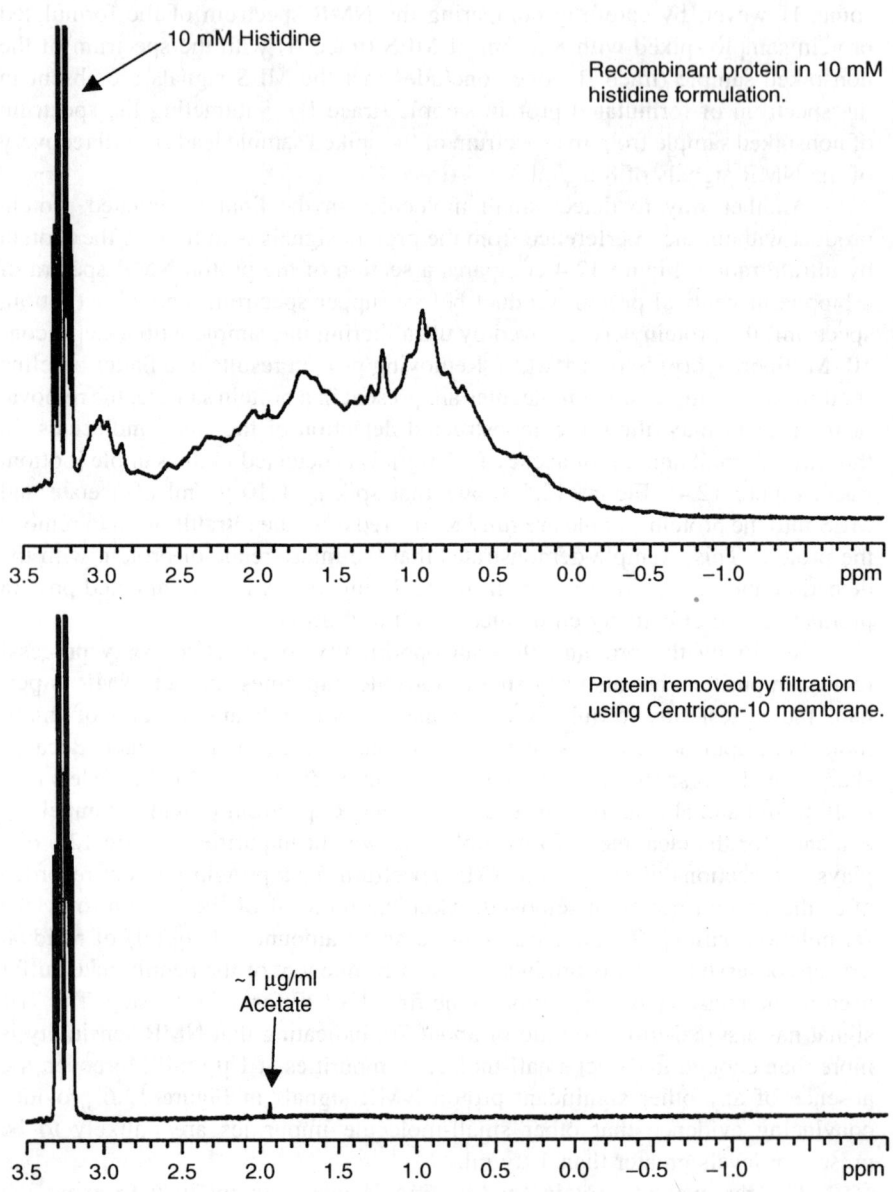

Figure 12.4   A section of the proton NMR spectra of a biopharmaceutical protein product before (upper spectrum) and after (bottom spectrum) the protein was removed by ultrafiltering the sample with Centricon-10.

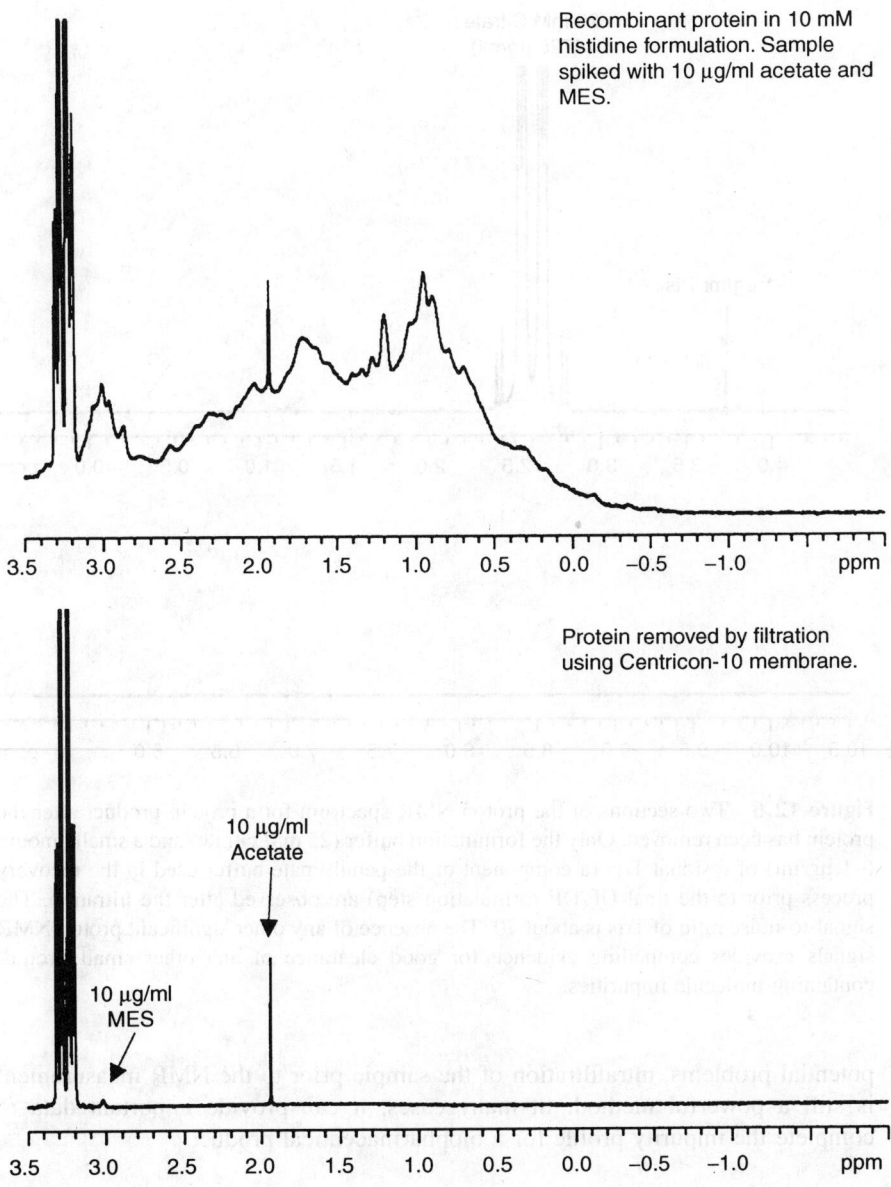

Recombinant protein in 10 mM histidine formulation. Sample spiked with 10 µg/ml acetate and MES.

Protein removed by filtration using Centricon-10 membrane.

10 µg/ml Acetate

10 µg/ml MES

Figure 12.5 A section of the proton NMR spectra of a biopharmaceutical protein product, spiked with10 µg/ml of acetate and MES, before (upper spectrum) and after (bottom spectrum) the protein was removed by filtering the sample with Centricon-10. The acetate and MES are recovered after the filtration to remove the protein.

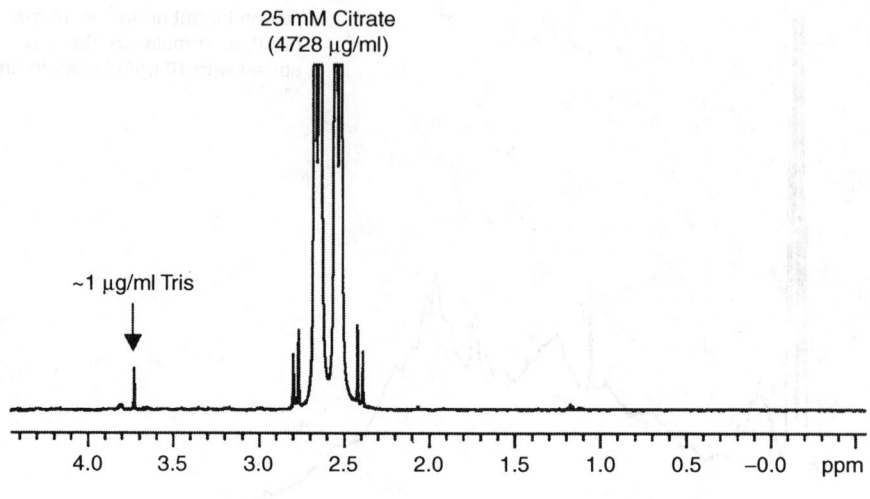

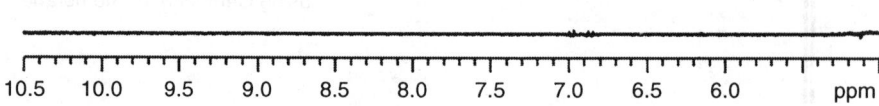

Figure 12.6 Two sections of the proton NMR spectrum for a protein product after the protein has been removed. Only the formulation buffer (25 m*M* citrate) and a small amount (~1 µg/ml) of residual Tris (a component of the penultimate buffer used in the recovery process prior to the final UF/DF formulation step) are observed after the filtration. The signal-to-noise ratio of Tris is about 70. The absence of any other significant proton NMR signals provides compelling evidence for good clearance of any other small proton-containing molecule impurities.

potential problems, ultrafiltration of the sample prior to the NMR measurement is still a powerful method. In many cases, it can provide important data to complete the impurity profile for a biopharmaceutical product.

QUANTITATION OF WATER-SOLUBLE POLYMERS

Quantitative NMR analysis may also be applied to water-soluble polymers or copolymers in some cases. Polymers usually have large molecular weights and are heterogeneous in size. However, the proton NMR signals of some polymers

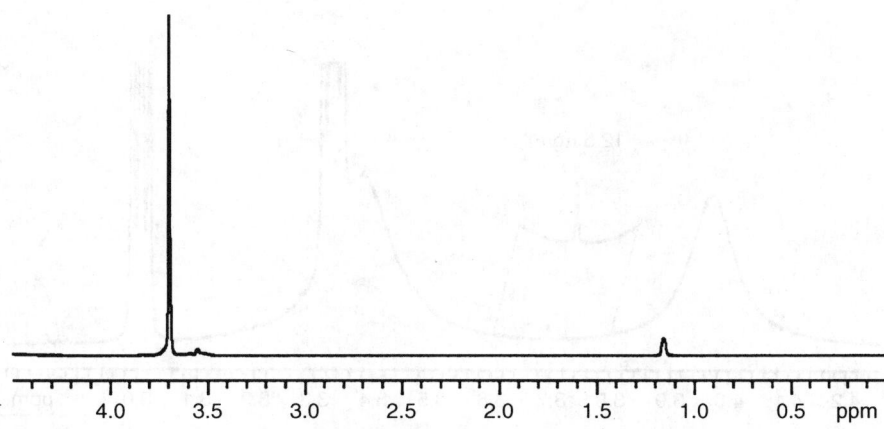

Figure 12.7 Proton NMR spectrum of a typical polymer used in mammalian cell culture processes: Pluronic F68.

are very simple and can be easily quantified by NMR. A good example is Pluronic F-68, an inhomogeneous polyethylene oxide/polypropylene oxide copolymer often used in cell culture media to reduce mechanical lysis of cells. The NMR spectrum of Pluronic F68 [Figure 12.7] contains relatively broad signals near 1.2 ppm (methyl protons) and between 3.5 and 3.9 ppm (methylene protons). The nonsymmetrical peak shape and the large line width of these signals are typical features of inhomogeneous polymers. The peak at about 3.7 ppm can be used to quantify Pluronic F-68 because it is the strongest signal. As demonstrated in Figure 12.8, a standard curve covering a range from 12.5 µg/ml to 50 µg/ml can easily be obtained for Pluronic F-68 in a buffer containing a significant amount of MOPS and acetic acid. Note that Pluronic F-68 may be detectable at a level lower than 12.5 µg/ml because the peak at 3.7 ppm still has a relative high signal-to-noise ratio (about 150) at 12.5 µg/ml. Normally, the purification process can clear the Pluronic F-68 to below this detection limit. The top spectrum in Figure 12.9 shows that the Pluronic F-68 signals are absent from the proton NMR spectra of in-process samples of a protein product. If Pluronic F-68 were not cleared, its NMR signal would have been detected, as clearly evidenced in the NMR spectrum of the in-process protein sample spiked with 12.5 µg/ml of Pluronic F-68 [Figure 12.9, bottom spectrum]. Because of the inhomogeneous nature of Pluronic F-68, this type of analysis cannot be easily done with other analytical techniques, such as HPLC or mass spectrometry. It is therefore important to include the NMR analysis in the process validation to demonstrate the clearance of Pluronic F-68. Quantitation of other polymers such as polysorbate 20, polyethylene glycol, etc., by NMR can also be achieved.

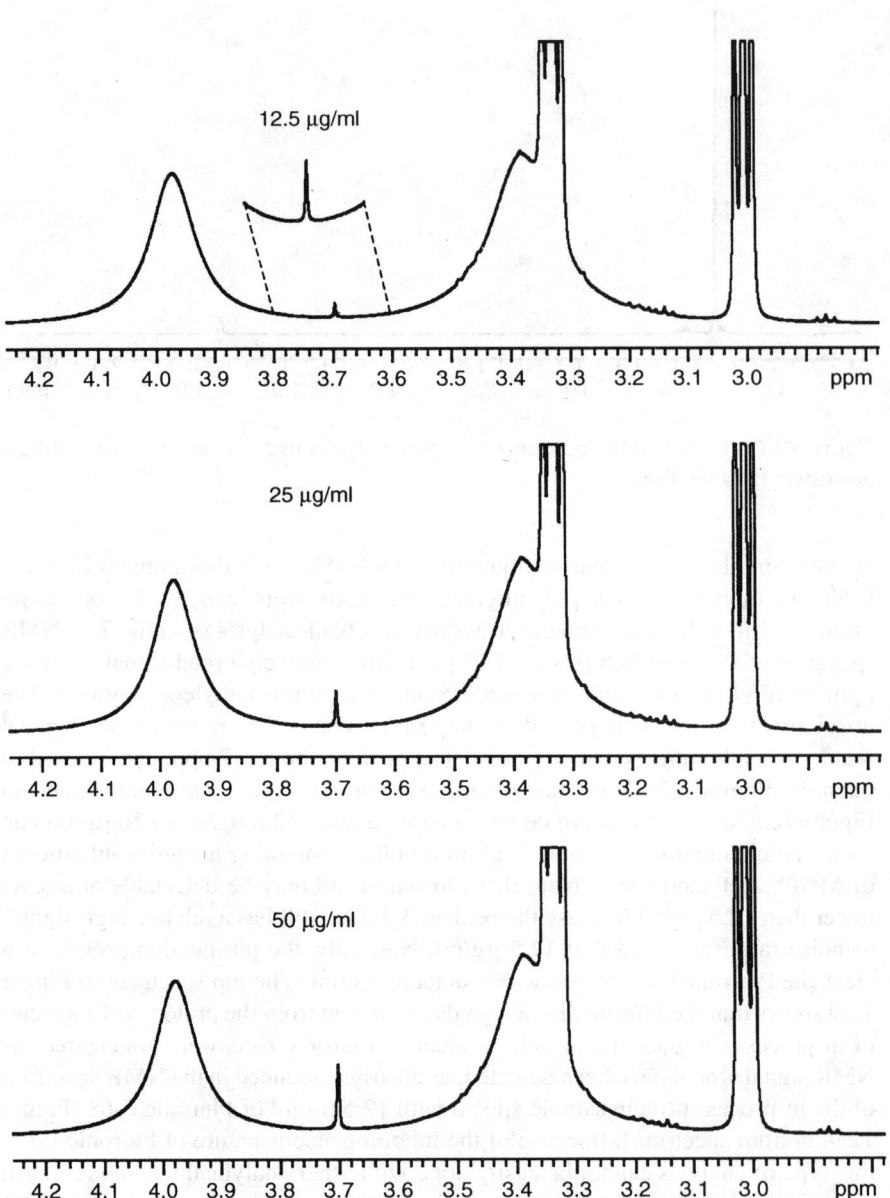

Figure 12.8 Proton NMR spectra of standards of Pluronic F68 prepared in a process buffer covering a concentration range from 12.5 µg/ml to 50 µg/ml.

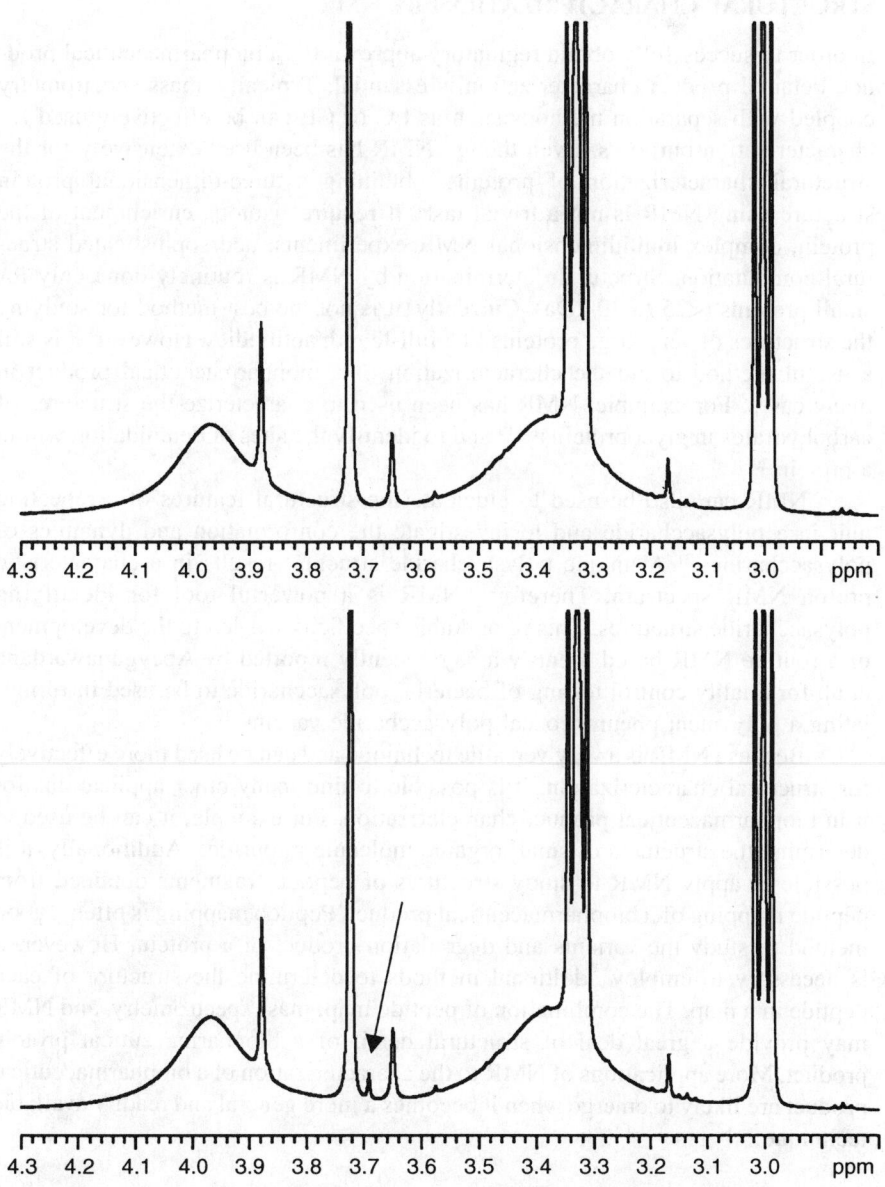

Figure 12.9 Proton NMR spectra for an in-process sample of protein product with (bottom spectrum) and without (upper spectrum) a spike of 10 μg/ml Pluronic F-68. The Pluronic F-68 signal is clearly absent in the upper spectrum. This example demonstrates that if the Pluronic F-68 level was greater than 12.5 μg/ml, it would have been observed.

STRUCTURAL CHARACTERIZATION BY NMR

In order to successfully obtain regulatory approval for a biopharmaceutical product, detailed product characterization is essential. Typically, mass spectrometry coupled with separation methods such as LC or CE can be effectively used for characterization purposes. Even though NMR has been used extensively for the structural characterization of proteins, obtaining a three-dimensional protein structure using NMR is not a trivial task. It requires isotope enrichment of the protein, complex multidimensional NMR experiments, and sophisticated structural computation. Structural determination by NMR is routinely done only for small proteins (<25 to 30 kDa). Currently, it is not the best method for studying the structures of very large proteins like full-length antibodies. However, it is still a useful method to aid the characterization of a biopharmaceutical product in many cases. For example, NMR has been used to characterize the structures of carbohydrates in glycoproteins[17–19] and to identify the sites of deamidation within a protein.[20]

NMR can also be used to elucidate the structural features of a repeating unit in a polysaccharide and to investigate the conformation and dynamics of polysaccharides.[21] A unique polysaccharide structure results in a characteristic proton NMR spectrum. Therefore, NMR is a powerful tool for identifying polysaccharide structures. This remarkable specificity has led to the development of a routine NMR-based identity assay, recently reported by Abeygunawardana et al. for quality control testing of bacterial polysaccharide to be used in formulating a polyvalent pneumococcal polysaccharide vaccine.[22]

Because NMR is a very versatile technique and can be used more effectively for structural characterization, it is possible to find many other applications for it in biopharmaceutical product characterization. For example, it can be used to determine the structures of small organic molecule impurities. Additionally, it is possible to apply NMR to study structures of peptide fragments obtained from peptide mapping of a biopharmaceutical product. Peptide mapping is often a good method to study the variants and degradation product of a protein. However, it is necessary to employ additional methods to determine the structure of each peptide in a map. The combination of peptide map, mass spectrometry, and NMR may provide a great deal of structural detail of a biopharmaceutical protein product. More applications of NMR to the characterization of a biopharmaceutical product are likely to emerge when it becomes a more general and readily available technique.

VALIDATION OF NMR ASSAY

Validation is critical for any assays to be used in pharmaceutical development. Validating an NMR assay is fairly straightforward. Some validation characteristics for an NMR assay are briefly discussed in this section.

Specificity: NMR spectroscopy is a specific analytical method by its nature. Each type of proton located in a different chemical or magnetic environment in the same or different molecule can be differentiated through its characteristic resonance frequency (Figure 12.1A). Thus, every proton-containing chemical has a unique set of proton NMR signals that can be readily identified. Although the NMR signals may be dependent on the solvent conditions (for example, temperature, pH, hydrophobicity, etc.), each molecule still can be uniquely identified in an NMR spectrum.

Linearity and range: Linearity is an inherent property of NMR spectroscopy. A standard curve can be easily obtained to cover a wide range of concentrations with a typical R^2 value of >0.99.

Precision and accuracy: Quantitative analysis by NMR is very precise with relative standard deviations for independent measurements usually much lower than 5%. The largest errors in NMR measurements are likely due to sample preparation, not the NMR method itself. If a good set of standards is available and all NMR measurements for the test and standard samples are performed under the same acquisition conditions, the quantitative results can be readily reproduced on different instruments operated by different analysts at different times. Therefore, good intermediate precision can also be achieved. An accurate quantitative NMR assay will require accurately prepared standards. The accuracy of an NMR assay can be assessed, for example, by measuring an independently prepared standard or an accurate reference sample with the assay. In many cases, a spike recovery experiment can also be used to demonstrate the accuracy of an NMR assay.

Detection and quantitative limit: Unless the NMR signals from the analyte are obscured by the NMR signals from buffer components, the detection limit of NMR measurement is typically 1 to 10 µg/ml for small organic molecules. The actual detection limit for an analyte may vary depending on the acquisition parameters and instrument settings. In particular, the more transients that are used for acquisition, the higher the sensitivity. Therefore, increasing the acquisition time can lower the detection limit. A signal-to-noise ratio of >5:1 for the signal of interest can be used to define the detection limit. A detection limit of 1 to 10 µg/ml for small organic molecules is probably sufficient for the detection of nontoxic impurities. Although this detection limit is not extraordinary, the ability of demonstrating that no small-molecule impurities are present in a protein product at a level above 1 to 10 µg/ml with a single NMR spectrum is, in fact, very valuable.

The quantitation limit is the lowest concentration of analyte in the standard curve with a signal-to-noise ratio of at least 10. Typically, the quantitation limit for most small organic molecules is 10 µg/ml. If necessary, the quantitation limit can be lowered, as long as the acquisition parameters are adjusted to yield sufficient sensitivity.

Robustness: Even if the NMR instrument is not properly calibrated (for example, the probe tuning and pulse length calibration are not optimized), as long as the acquisition parameters are kept constant during the assay, high precision and accuracy can still be achieved. In addition, the relative resonance frequencies (chemical shifts) of NMR signals do not depend on the experimental parameters; therefore, the spectrum of the same sample recorded using different acquisition parameters by different operators will contain the same peaks with the same chemical shifts and the same relative intensities. This significantly reduces the variations arising from instrumentation and acquisition parameters. Thus, the NMR assay is a very robust quantitative assay.

Suitability: The main requirement for the NMR measurement to be suitable for quantitating small organic molecules is that the NMR peaks of analyte must be resolved from any background signals (for example, from buffers or solvents). In some cases, sample conditions must be changed in order to resolve the peak of interest. For example, the Tris signal (a singlet at 3.7 ppm) is pH dependent. By changing the solution pH, one can move the Tris signal to a position where background signals do not interfere with the quantitation. As a result, the suitability must be evaluated individually for each analyte. However, the NMR assay normally can be qualified for the quantification of most small organic molecules without difficulties.

CONCLUDING REMARKS

NMR is a remarkably flexible technique that can be effectively used to address many analytical issues in the development of biopharmaceutical products. Although it is already more than 50 years old, NMR is still underutilized in the biopharmaceutical industry for solving process-related analytical problems. In this chapter, we have described many simple and useful NMR applications for biopharmaceutical process development and validation. In particular, quantitative NMR analysis is perhaps the most important application. It is suitable for quantitating small organic molecules with a detection limit of 1 to 10 µg/ml. In general, only simple one-dimensional NMR experiments are required for quantitative analysis. The other important application of NMR in biopharmaceutical development is the structural characterization of molecules that are product related (e.g., carbohydrates and peptide fragments) or process related (e.g., impurities and buffer components). However, structural studies typically require sophisticated multidimensional NMR experiments.

There has been significant advancement in the applications of NMR to the development of small-molecule pharmaceutical products. For example, advances in NMR automation (e.g., flow-injection analysis) and directly coupled methods (e.g., LC-MS-NMR analysis) have made analysis and characterization of small-molecule drugs much easier.[23–25] These improvements have helped chemists to develop and characterize small-molecule combinatorial libraries and to screen for active compounds.[4–6] It is likely some of these techniques can also be used in biopharmaceutical product development.

NMR offers many unique advantages that other methods cannot provide in spite of some limitations. Biopharmaceutical product development will certainly benefit from including NMR as an option for solving analytical problems. NMR instrumentation and methodology are constantly being improved. As better and more sensitive NMR techniques become available, the use of NMR as a standard analytical tool in biopharmaceutical process development and validation is expected to increase.

ACKNOWLEDGMENTS

The authors would like to thank Adeyma Arroyo and Chithkala Harinarayan for providing some samples used in our studies; Dian Feuerhelm for critical reading of the manuscript; Viswanatham Katta and Ken Skidmore for their comments on the manuscript; and Ken Skidmore for recording some small-molecule clearance data.

References

1. Montelione, G.T., Zheng, D., Huang, Y.J., Gunsalus, K.C., and Szyperski, T. (2000), *Nature Struct. Biol. 7,* Suppl: 982–985.
2. Clore, G.M. and Fronenborn, A.M. (1998), *Curr. Opin. Chem. Biol. 2,* 564–570.
3. Pellegrini, M. and Mierke, D.F. (1999), *Biopolymers 51,* 208–220.
4. Hajduk, P.J., Meadows, R.P., and Fesik, S.W. (1999), *Q. Rev. Biophys. 32,* 211–240.
5. Chen, A. and Shapiro, M.J. (1999), *Anal. Chem. 71,* 669A–675A.
6. Moore, J.M. (1999), *Curr. Opin. Biotechnol. 10,* 54–58.
7. Kao, Y.-.H., Bender, J., Hagewiesche, A, Wong, P, Huang, Y., and Vanderlaan, M. (2001), *PDA J. Pharm. Sci. Technol.* 55, 268–277.
8. Ernst, R.R., Bodenhausen, G., and Wokaun, A. (1987), *Principles of Nuclear Magnetic Resonance in One and Two Dimensions,* Oxford University Press, Oxford.
9. Braun, S., Kalinowski, H.-O., and Berger, S. (1998), *150 and More Basic NMR Experiments,* John Wiley & Sons, New York.
10. Croasmun, W.R. and Carlson, R.M. (Editors) (1998), *Two-Dimensional NMR Spectroscopy: Applications for Chemists and Biochemists,* John Wiley & Sons, New York.
11. Piotto, J., Saudek, V., and Sklenar, V. (1992), *J. Biomol. NMR 2,* 661–665.
12. Hwang, T.-L. and Shaka, A.J. (1995), *J. Magn. Reson. Ser. A 112,* 275–279.
13. Gueron, M., Plateau, P., and Decorps, M. (1991), *Prog. NMR Spectrosc. 23,* 135–209.
14. Silverstein, R.M., Bassler, G.C., and Morrill, T.C. (1991), *Spectrometric Identification of Organic Compounds,* 5th ed., John Wiley & Sons, New York.
15. Carr, H.Y. and Purcell, E.M. (1954), *Phys. Rev. 94,* 630.
16. Meiboom, S. and Gill, D. (1958), *Rev. Sci. Instrum. 29,* 688.
17. Vliegenthart, J.F., Dorland, L., and van Halbeek, H. (1983), *Adv. Carbohydr. Chem. Biochem. 41,* 209–374.
18. Spellman, M.W., Basa, L.J., Leonard, C.K., Chakel, J.A., O'Connor, J.V., Wilson, S., and van Halbeek, H. (1989), *J. Biol. Chem. 264,* 14100–14111.

19. Harris, R.J., van Halbeek, H., Glushka, J., Basa, L.J., Ling, V.T., Smith, K.J., and Spellman, M.W. (1993), *Biochemistry 32*, 6539–6547.
20. Chazin, W.J., Kordel, J., Thulin, E., Hofmann, T., Drakenberg, T., and Forsen, S. (1989), *Biochemistry 28*, 8646–8653.
21. Mulloy, B. (1996), *Mol. Biotechnol. 6*, 241–265.
22. Abeygunawardana, C., Williams, T.C., and Sumner, J.S. (2000), *Anal. Biochem. 279*, 226–240.
23. Keifer, P.A., Smallcombe, S.H., Williams, E.H., Salomon, K.E., Mendez, G., Belletire, J.L., and Moore, C.D. (2000), *J. Comb. Chem. 2*, 151–171.
24. Keifer, P.A. (2000), *Prog. Drug Res. 55*, 137–211.
25. Holt, R.M., Newman, M.J., Pullen, F.S., Richards, D.S., and Swanson, A.G. (1997), *J. Mass Spectrom. 32*, 64–70.

13

Microcalorimetric Approaches to Biopharmaceutical Development

Richard L. Remmele, Jr.

INTRODUCTION

A goal of biopharmaceutical development is to quickly move an investigative new drug into the clinic where it can be tested as a potential therapeutic to treat disease. To realize this goal, an enormous amount of effort must be carried out from conception in discovery to upstream, downstream, and formulation development stages. Much of this effort is guided by the generation of analytical data to help support optimization of yield, purification, and stability. Even with this strategy in place, there is room for better approaches to minimize any developmental bottlenecks. More rapid or "high-throughput" strategies and data-assimilation technologies are currently being used and evaluated at various stages of the process development assembly line. Within the past 20 years, mutagenic and recombinant technologies have shed light on structure–function relationships that have spawned from the protein engineering era. Given these strides, it is now possible to begin thinking about discovery biologics in terms of rational protein drug design.[1,2] As the science continues to advance, it would seem inevitable that one day accurate predictions about structure, function, and stability will greatly enhance therapeutic drug discovery, development, and approval. This also means that by simply understanding a disease, it could be possible to implement *de novo* design strategies to develop appropriate candidate biologics with the robust qualities of functionality and stability to cure or treat it. A necessary part of protein engineering would then involve building into the molecule the necessary conformational attributes that make it efficacious, nonimmunogenic, and stable. Moreover, it is important to acquire and apply knowledge that has its roots firmly planted in the laws of thermodynamics and kinetics. Thermodynamic and kinetic properties of biologics as they pertain to mode of action (i.e., ligand–receptor binding, antagonist), structure, and overall stability will therefore continue to play an integral role in biopharmaceutical development strategies.

Among the many biophysical techniques that may be considered, microcalorimetry offers some clear advantages. It can directly measure enthalpic change and is useful for defining important thermodynamic parameters (i.e., thermostability, heat capacity, free energies, entropies, binding affinities) that can aid the discovery, upstream, downstream, and formulation development operations. As more knowledge about the physicochemical characteristics of a broad array of protein therapeutics is amassed, it should be possible to bioengineer a protein therapeutic with the right properties that make it efficacious, robust to the stresses of the process, and stable in the desired dosage form. This is the end result of biopharmaceutical development. If successful in the prediction of a well-characterized product, this approach could greatly reduce bottlenecks in the manufacturing sector.

This chapter will explore the relationship of thermodynamic and kinetic data as it pertains to characterizing the stability of various protein systems in the liquid state. Finally, from the wealth of information generated over the past few decades, it should be possible to assess the practical use of microcalorimetry for predicting stability. This technique used in combination with several other bioanalytical methods can serve as a powerful tool in the measurement of thermodynamic and kinetic phenomena.[3–9] Attention will be given to limitations of the technique rendered from different applications as well as to areas where it is advantageous. Ultimately, the practical utility of this technique will rest with those familiar with the art.

FUNDAMENTAL PARAMETERS OF MICROCALORIMETRIC MEASUREMENTS

There are several parameters that can be measured or determined using microcalorimetry that enable one to study the energetics of biomolecular processes at the molecular and cellular levels. Heat or enthalpic effects are intimately linked to biochemical processes that involve a change of state that can include a transconformational event often leading to enhanced instabilities.[10–15] Systems that undergo state changes will either liberate (exothermic reactions) or absorb (endothermic reactions) heat. Because the amount of heat involved in such reactions is often quite small (on the order of microcalories), highly sensitive instruments are needed to quantitatively measure protein reactions.[16] These instruments operate on the basis of measuring the microcalories of heat absorbed or liberated by a given process (i.e., protein unfolding, association, binding) as a function of the scanning rate against a suitable reference that contains the identical solution environment without protein. The change of state accompanying the reaction, although sometimes characterized as "two-state" behavior, is really descriptive of an ensemble of microstates that predominantly populate two different states of the molecule.[17] This could take the form of a native and fully denatured state, or it could involve a transition from the native state to a partially folded intermediate state that depends on the conditions of the solution environment and the stresses involved (i.e., pH, temperature, chaotrope). This type of information can be important in describing the stability of a given biologic in a given environment (i.e., solution matrix). Other parameters that can be assessed using this technique include the free energy of unfolding, denaturation heat capacity, and entropy. Because all of these parameters are associated with changes of state described by an equilibrium process, they are all thermodynamic quantities.

The Melting Transition Temperature (T_m)

Consider a two-state system in which the protein exists predominantly in either its native or thermally unfolded state. The equilibrium describing such a system depends on temperature that in turn determines the population of the two states. At low temperatures the protein will largely exist in its native conformation. Although not always the case (i.e., α-lytic protease of bacterium *Lysobacter enzymogenes*), the native state is generally described as having relatively lower free energy than the unfolded or denatured state. A delicate balance exists between the forces that act to maintain the native form and those that would cause it to denature. Unfolding occurs as heat energy (enthalpy) loosens the protein structure by weakening the bonds and physical forces that maintain conformational integrity. The major destabilizing force is conformational entropy, and those that stabilize the native state are hydrogen bonding and the hydrophobic effect.[18,19] The thermally unfolded state is a denatured state. For sake of clarity, the term *denaturation*, used in this regard, will refer to "a process (or sequence of processes) in which the conformation of polypeptide chains within the molecule are changed from that typical of the native protein to a more disordered arrangement."[20] Hence, partially folded or intermediate states can be considered denatured states. During a typical heating experiment within the microcalorimeter, the denaturation transition is observed as an endothermic quantity arising from significant alterations in structure associated with unfolding. This process is identified as melting of the protein, and has an associated temperature, the T_m or melting temperature. This parameter is taken as the temperature at peak-maximum, where approximately half of the protein is denatured.[21,22] It is important as an integral parameter of the modified Gibbs–Helmholtz equation (defined in the following text) and also because it indicates thermostability.

An important consequence of thermostability is the attainable level of stability indicated by the T_m. For example, different solution environments can cause the T_m to shift.[23–27] A shift to higher temperatures by adding different excipients and/or changing the pH of the solution can be extremely helpful in defining solution properties that are optimal for stabilization. The more easily unfolded or structurally perturbed (described by lowering T_m) from the native conformation, the more vulnerable the protein potentially becomes to instability. Protein unfolding can often augment a multiplicity of instabilities that include aggregation,[10–12,28,29] deamidation,[15] oxidation,[13] and proteolysis.[30] In these cases it is important to recognize that the prominence of the T_m to accurately describe stability is related to whether or not the degradation depends on unfolding. Therefore, the strategy in formulation development is always to impose a solution environment that prefers the native state in order to minimize the deleterious consequences associated with the instability of the product. By shifting the equilibrium in the direction of the native state, resulting in higher T_m, thermostability can be a useful parameter to define the solution variables that make it more difficult for the protein to unfold.[31,32]

An illustration of the unfolding for a multidomain protein is presented in Figure 13.1. It is the unfolding of IL-1 receptor (type I) that shows two predominant unfolding features (inlay) with T_m at 53.3°C and 66.7°C. After baseline correction, the T_m changes slightly to become 53.1° and 66.3°C, thus resulting in a small adjustment in the T_m of a few tenths of a degree. The inlay in Figure 13.1 also shows the reproducibility of two completely independent scans that are virtually superimposable. This is testimonial to the kind of precision attainable using this technique, given that it portrays complete alignment of the T_m in both scans.

Reversible vs. Nonreversible Systems

An inherent property of thermodynamic systems is an equilibrium between at least two states. In many cases the unfolding of a protein can be more complex than two states. Such an equilibrium can be verified by determining the reversibility of the process defining the two states. Using microcalorimetry, this may be evaluated simply by examining the enthalpy of two subsequent heating scans (given by the equation: % reversibility = $[\Delta H_2/\Delta H_1] \times 100$, where the subscript numbers designate the first and second scan results). Both the T_m and the enthalpies of unfolding should afford a significant measure of reversibility. The thermal reversibility of a number of protein systems has been described in the literature.[9,33–36] The thermodynamic significance of the calorimetric parameters requires evidence of thermal reversibility. Despite this need, can useful information about the energetics of a process be obtained from systems that exhibit nonreversible behavior?

To answer this question, information obtained from studies of irreversible systems needs to be examined. Irreversible protein processes may occur as a result of intermolecular interactions (i.e., aggregation, chemical modification, intermolecular cross-linking). Although an attempt is generally made to search for conditions that provide maximal reversibility, perhaps by altering the solution conditions (i.e., pH, salt content, lowering the protein concentration) that minimize contact and electrostatic interactions, many systems can still exhibit little or no reversibility. This would be the case for the core protein obtained by limited

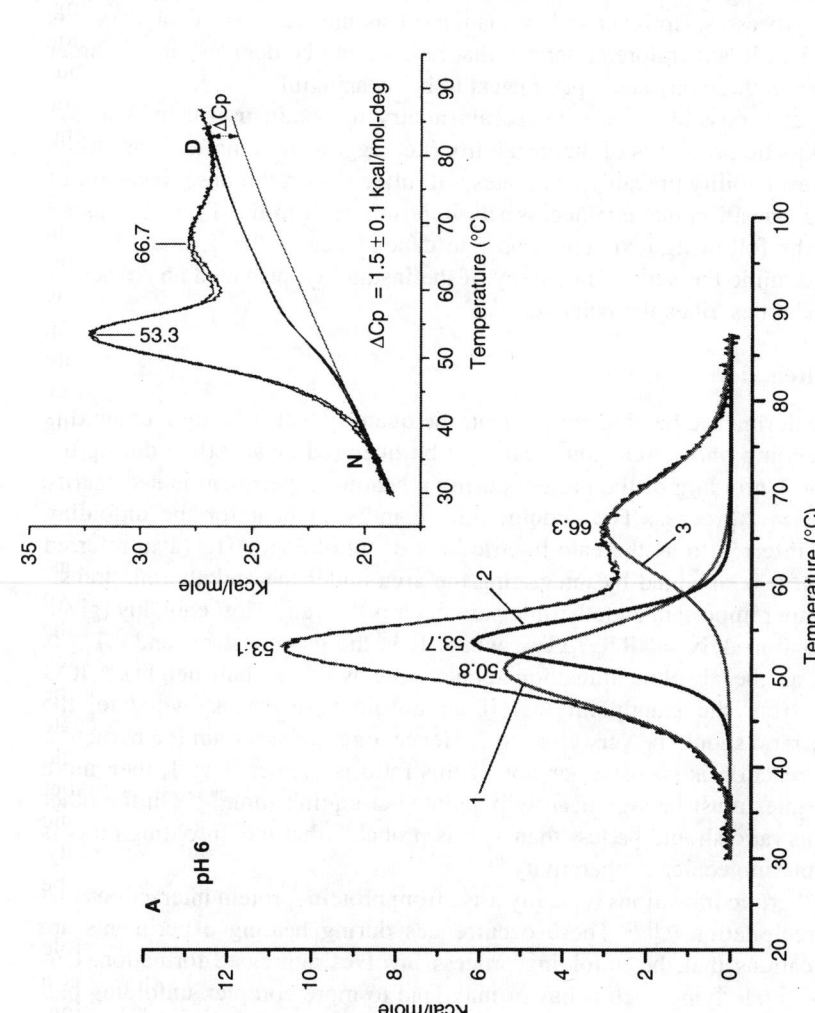

Figure 13.1 Microcalorimetry scans displaying T_m values for interleukin-1 receptor (IL-1R type I). The inlay displays the unfolding of IL-1R (I) showing the ΔCp measured as the baseline difference between the native (N) and denatured (D) states for two independent scans. Thermal unfolding of IL-1R (I) is composed of three cooperative unfolding transitions, labeled 1, 2, and 3.

proteolysis from the *lac* repressor of *E. coli*. It is tetrameric at ambient temperatures and irreversibly denatures during differential scanning calorimetry (DSC) heating experiments, yet it yields estimates of ΔHv (van't Hoff enthalpy) using three different methods (e.g., curve fitting of the endotherm, calculation of the van't Hoff enthalpy, slope of a van't Hoff plot) that were all found to be within reasonable agreement.[22] These results suggest that it is possible to apply equilibrium thermodynamics to irreversible processes. However, this evaluation still assumes that a reversible process is in operation. It is therefore preferable that reversibility be demonstrated in order for equilibrium thermodynamic parameters to be meaningful.

It is also possible to use microcalorimetry to obtain useful information about the kinetic processes of the instability (i.e., aggregation, proteolysis) when thermal irreversibility prevails. Scan rates will often distort the onset behavior of the melting transition that can necessarily impose a shift in the T_m, as discussed further in the following text. The scan rate dependence of the T_m may then be used to determine the activation energy of the instability, provided an Arrhenius kinetic model describes the behavior.

Heats of Reaction

During a calorimetric heating experiment, the quantity that is being measured is heat. Depending on the reaction, heat may be liberated or absorbed during the experiment. Unfolding of the protein during a heating experiment is necessarily an endothermic process. The endothermic quantity of heat for the unfolding process is referred to as the calorimetric heat of unfolding, ΔH_{cal} (also referred to as ΔH_m). It is obtained by integrating the area under the endothermic transition.[37] Another important endothermic parameter is the van't Hoff enthalpy (given by the equation $\Delta Hv = 4RT_m^2/\Delta T_{1/2}$, where R is the gas constant, and $\Delta T_{1/2}$ is expressed as the absolute transition temperature width at half-height).[38] It is important from the standpoint that if an unfolding event is two-state, the $\Delta H_{cal}/\Delta Hv$ ratio should be very close to 1. Hence, one can ascertain if a particular unfolding reaction is two-state or not. If this ratio is greater than 1, then more than two states must be significantly populated at equilibrium.[38,39] On the other hand, if this ratio should be less than 1, it is probable that the unfolding process involves intermolecular cooperativity.[40]

Exothermic transitions typically arise from protein–protein interactions that lead to precipitation.[28,41,42] These occurrences during heating experiments are good indications that the unfolding process involves aggregate formation. Circumstances underlying such behavior may lead to more complex unfolding patterns than can be described by a tri-state model.[43] For example, when aggregation is a competing reaction that depends upon the equilibrium that defines the unfolded state, there is a tendency for the reaction to conform to some aspect of nonreversibility. When such conditions exist, the irreversible step (or steps) are kinetically controlled and thus are not directly applicable to the treatment of equilibrium thermodynamics.[37,44]

Denaturation Heat Capacity

The heat capacity of thermal unfolding or denaturation (designated here as ΔCp) is positive because the denatured protein typically has a larger heat capacity than the native state. This parameter is believed to originate from the exposure of hydrophobic residues that are buried in the native state and subsequently become solvent-exposed in the unfolded state.[45] It can be measured in three ways. It can be determined as the heat capacity difference between the pre- and posttransitional baselines associated with the unfolding enthalpic envelope.[21] A representation of this behavior is described in the inlay of Figure 13.1. Note that the low (pretransition) temperature baseline (below the onset or incipient melting of the protein) does not coincide with the same at the posttransition temperatures. In fact, it is the difference in heat between these two regions (the pretransition associated with the native and the posttransition associated with the denatured states) that determines the heat capacity change between the two states. Alternatively, the ΔCp can be obtained from the slope of a linear plot of ΔH_m as a function of T_m, where T_m may be altered by changing the solution pH environment.[26,29] This latter method implies that the ΔCp is a constant parameter and is independent of temperature.[37] However, this issue will be discussed further in the following text because it has remained the subject of some debate. Finally, with the acquisition of greater sensitivity in microcalorimeter instrumentation, a new method has emerged that determines the heat capacity change from the slope of a linear plot of Cp as a function of the mass of protein in the calorimeter cell.[46] The denaturation heat capacity can vary in magnitude from 0.59 kcal/mol-°K (for Ovomucoid III) to 7.51 kcal/mol-°K (for phosphoglycerate kinase).[47]

Modified Gibbs–Helmholtz Equation

Protein stability has generally been defined in terms of the free energy change between the native and the unfolded states (ΔGu). This parameter of unfolding can be used to decipher stability of the native state across a wide temperature range.[48] The relation that incorporates all of the terms described above is the modified Gibbs–Helmholtz equation and is a function of temperature.

$$\Delta Gu(T) = \Delta H_m (1 - T/T_m) - \Delta Cp \left[(T_m - T) + T\ln (T/T_m) \right] \qquad (1)$$

Knowing the T_m (°K), ΔH_m (kcal/mol), ΔCp (kcal/mol-°K) permits calculation of the free energy change (kcal/mol) for the unfolding reaction at any temperature. This equation assumes a two-state mechanism in which the predominant forms are represented by the folded (native) and unfolded (denatured) states. Free energies of protein unfolding typically range from as little as 5 kcal/mol to 25 kcal/mol (a small energy threshold to overcome). It follows that proteins are essentially vulnerable to unfolding. However, the degree and extent of unfolding may not involve complete conversion to a random coil state. For example, it is known that denatured proteins retain a significant portion of native structure suggesting that, in some cases, partial unfolding is all that is required to form aggregated products.[49-51] Equation 1 describes a parabolic function. At the temperature where $T = T_m$, $\Delta Gu = 0$. Given the fact that the function describes a parabola, there must not only be a T_m where $\Delta Gu = 0$ at elevated temperatures, but the equation predicts that the same is true at low temperatures (subzero in most cases) as well. The temperature where $\Delta Gu = 0$ at low temperatures is defined as cold denaturation.[21] Questions can arise about the validity of this equation if ΔCp is not constant over the temperature range considered. Evidence gathered would suggest that it is essentially independent of temperature within a finite range extending from approximately 20 to 80°C.[21] When considering temperatures beyond this range, in some cases it might be possible to check the accuracy of ΔCp over a broader temperature range by considering the accuracy of Equation 1 to predict cold denaturation.

Entropy

During a thermal heating experiment, the entropy (ΔS) of the protein logically should increase. If ΔGu can be determined by Equation 1, it follows then that ΔS may be determined from the expression $\Delta Gu = \Delta H_m - T\Delta S$. As the thermal energy of the system increases, the tendency of the protein to approach a state that is more structurally disordered is likewise expected. In fact, it has been shown for metmyoglobin (approximately two-state) that the enthalpy and entropy changes associated with Equation 1 increase with the temperature.[52] Knowing that ΔGu is essentially the work required to disrupt the structure of a folded protein, the temperature at which maximal stability is achieved (described by Equation 1) should coincide with the condition where the native and denatured states do not differ in their entropy values.[48] Entropy from the formation of a complex (i.e., protein–protein interactions) relative to the free molecule in solution is largely driven by hydration effects that involve polar and nonpolar groups. The nonpolar or hydrophobic groups will tend to order water molecules into networks of clathrate-like structure.[19] The apparent cost of this structure is a reduction in the entropy of the solvent. When such groups are removed from the aqueous environment as a result of protein binding (i.e., aggregation, receptor–ligand), there is a significant reduction of hydrophobic surface accessible to the water environment. This will lead to a large overall entropy change that is often positive.[53] This favorable change can be offset by a reduction in side-chain mobility as a consequence of binding that translates into an unfavorable contribution to the entropy gained by the solvent.

LIQUID FORMULATION DEVELOPMENT STRATEGIES

There are two primary questions that a formulator will often ask regarding a biopharmaceutical. First and foremost, what excipients (inactive ingredients like chemical stabilizers, buffers, and toxicifiers) and solution conditions (i.e., pH, temperature) offer the greatest stabilization? Second, how long will the chosen candidate formulations remain stable? The first question addresses rank-order assessment of formulation composition based upon thermodynamic and chemical stability. If there were a way to predict the outcome of massive screening efforts that were designed to determine the optimal formulation candidates, this would prove to be a valuable step toward reducing time and cost that would otherwise be spent serendipitously trying to reach that goal. The second question deals with the kinetic processes associated with the degradation of the biologic. Typical projections for liquid biopharmaceuticals to remain stable usually target a 2-year duration. Microcalorimetry can be used to aid in the screening of excipients that stabilize the native state. This technique is well suited as already established (e.g., sensitivity to solution environment, protein-unfolding properties) to assess how different solution environments perturb protein unfolding. The primary objective in most cases is to maintain the native structure or conformation of the molecule in the desired dosage form.

pH Profiles

The first step in ascertaining the appropriate solution environment to store and stabilize a given biologic is to identify the pH conditions where it remains conformationally stable. This can be approached by examining the behavior of the T_m as a function of pH. This has been a successful strategy for the development of formulations for M-CSF and CD40L.[14,32] In these two particular cases, the T_m optimum correlated with the pH conditions in which aggregation was minimized. Figure 13.2 illustrates the pH profiles of several protein biologics manufactured at Immunex Corporation. Interestingly, preformulation conditions generally align with regions where T_m attains its highest levels. In the case of IL-1R (type II), optimal pH corresponded with the region of the highest T_m and low susceptibility to breakdown or aggregation instabilities. Once the optimal pH region is determined, appropriate buffers should then be chosen for the preparation of candidate formulations.

Here it is important to point out that buffers should be considered that exhibit little change with temperature because such dependence is proportional to the enthalpy of ionization of the buffering agent.[52] Alternatively, in order to minimize ionization effects and environmental changes during a given heating experiment, polybuffer systems have been used for pH profiling studies.[32] The polybuffer system used in the pH profiles shown in Figure 13.2 consisted of 50 mM sodium citrate, 20 mM NaCl, 20 mM phosphate, 17 mM L-arginine, 40 mM L-glycine, 25 mM HEPPS, and 20 mM HEPES. When all the ingredients are mixed together, the pH of the mixture is nearly neutral. One can simply add acid or base to adjust the pH from neutrality to the desired condition. Use of this polybuffer system was intended to normalize the influence of the buffering components and minimize the enthalpy of ionization. All of the buffering agents chosen minimize the pH dependence on temperature and permit adequate buffer capacity across a broad pH range extending from pH 3 to 11.

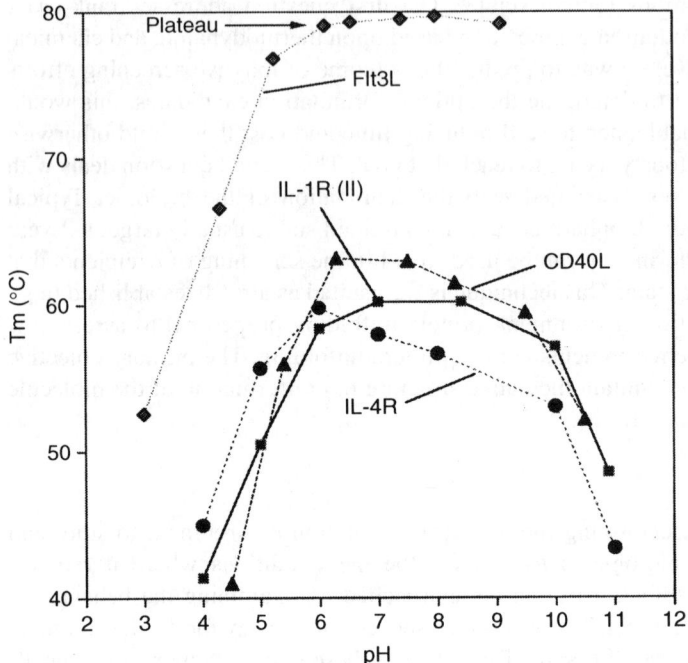

Figure 13.2 The pH profiles of several protein therapeutics. The data illustrate the behavior of T_m as a consequence of change in pH conditions. At the point where T_m essentially achieves a maximum across a broad pH range, a plateau is observed (as indicated for Flt3L). Note that these studies were conducted in the same polybuffer previously described.[54]

Thermal Reversibility

The second step toward defining stable conditions for the biologic is to evaluate (if possible) the thermal reversibility within the region of optimal pH. This was a successful approach in the development of a stable formulation for Flt3L.[54] In this case the solution conditions that best characterized minimal aggregation were found to occur where the thermal reversibility was greatest within the region of the T_m plateau (Figure 13.2). Even in the case of systems that are not thermally reversible (e.g., CD40L), pH profiling can still be advantageous in identifying the best conditions where stabilization is optimally achieved.[14] This is clearly illustrated in Figure 13.3, where the pH profile of the T_m behavior mirrors the tendency of the CD40L molecule to aggregate. The optimal pH region of the T_m profile matches the optimal pH region obtained from accelerated stability studies conducted at 37°C over 7 d.

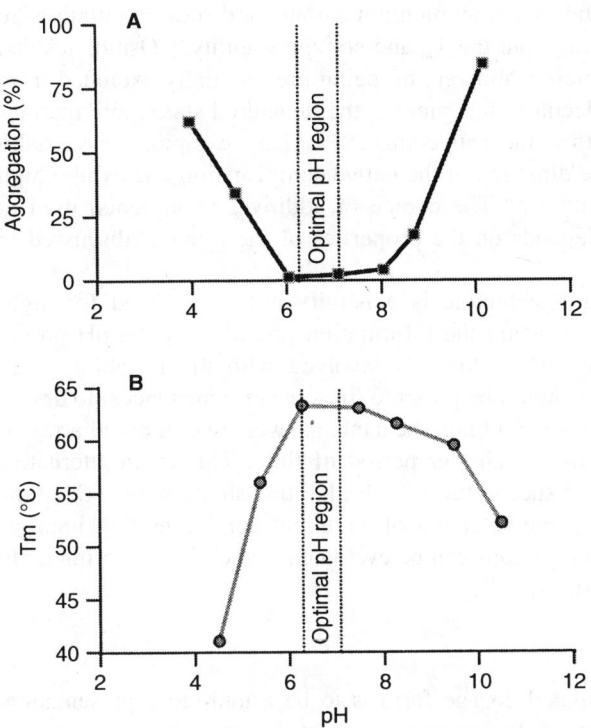

Figure 13.3 Stability behavior of CD40 ligand (coiled-coil CD40L trimer), showing the aggregation response obtained from accelerated studies (7 d at 37°C) as a percentage determined from size exclusion chromatography (A), and the T_m response (measured by microcalorimetry) as a function of pH (B). The bracketed vertical dashed lines represent the optimal pH region in which aggregation was minimized, corresponding to the same region in which T_m was maximized. Optimal pH conditions may be attained between pH 6 and 7.5. (Permission to use the figure granted by BioPharm.)

Excipient Screening

Once optimal pH conditions are determined and thermal reversibility is evaluated, manipulation of the solution conditions by screening stabilizing excipients (i.e., salt, stabilizers, other inactive ingredients) that further augment T_m and/or reversibility can be valuable. This was true for IL-1R (type I), where among 23 different excipient conditions, NaCl was shown to increase T_m to its highest attainable levels.[31] Additionally, the role of excipients on thermal reversibility is of some benefit for the stabilization of recombinant human megakaryocyte growth and development factor.[55] In another study concerning the stabilization of recombinant human keratinocyte growth factor where a variety of osmolytes and salts were screened, NaCl, sodium phosphate, ammonium sulfate, and sodium citrate were highly effective in increasing both the T_m and storage stability.[56] Osmolytes and certain salts behave as protein stabilizers by being preferentially excluded from the vicinity of protein molecules, thus making the denatured state more thermo-dynamically unfavorable than the native state.[56–58] These excipients essentially drive the equilibrium in the direction of the native conformation and exhibit high melting transition temperatures.[56] The choice of additives (excipients) used to stabilize a biologic also depends on the properties of the protein (discussed in the following text).

The microcalorimetric technique is generally not well suited for high-throughput testing. However, using the information provided by the pH profile and a basic understanding of the kinetics involved with the instability (i.e., aggregation, breakdown), it should be possible (in some circumstances) to design appropriate accelerated studies to obtain the same answers over a broad array of screening conditions and over a shorter period of time. This is an alternative approach that has been used successfully to abridge this shortcoming. Once the accelerated studies reveal some selection of excipient conditions that improve stability, a check on these conditions can be evaluated in the microcalorimeter to confirm the effects on thermostability.

Use of Preservatives

At this point, if the formulated dosage form is to be a multidose presentation, preservatives will be necessary. Preservatives protect the product from microbial contamination during reentrance into the vial. A choice of several preservatives might be considered, such as phenol, benzyl alcohol, and metacresol. All can have a destabilizing effect on the protein as shown for IL-1R (type I).[31] Protein unfolding as a function of preservative concentration generally shifts the T_m to lower temperature accompanied by band broadening.[14] This behavior is presented in Figure 13.4, showing the effects of benzyl alcohol concentration on the T_m of a protein. While broadening of the transition envelope with decreasing T_m has been regarded as an indication of a large positive ΔCp for protein unfolding, penetration of the preservative into folded regions of the protein, facilitating an overall loosening of the folded structure, could also be a contributing factor.

The unfolding behavior of multidomain proteins in the presence of preser-vatives has also been evaluated. For example, IL-1R (type I) has three domains that correspond to three unfolding transitions as measured by microcalorimetry and depicted in Figure 13.1.[31] All three transitions exhibit some shift to lower T_m in the presence of the three preservatives tested (i.e., 0.065% phenol, 0.1% metacresol, and 0.9% benzyl alcohol), in comparison to a control containing no preservative. Such a destabilization could have consequences for the shelf-life of the product. Another example of the impact of preservatives on a multidomain protein is illustrated in Figure 13.5. The protein is fused to a single IgG_1 Fc. The

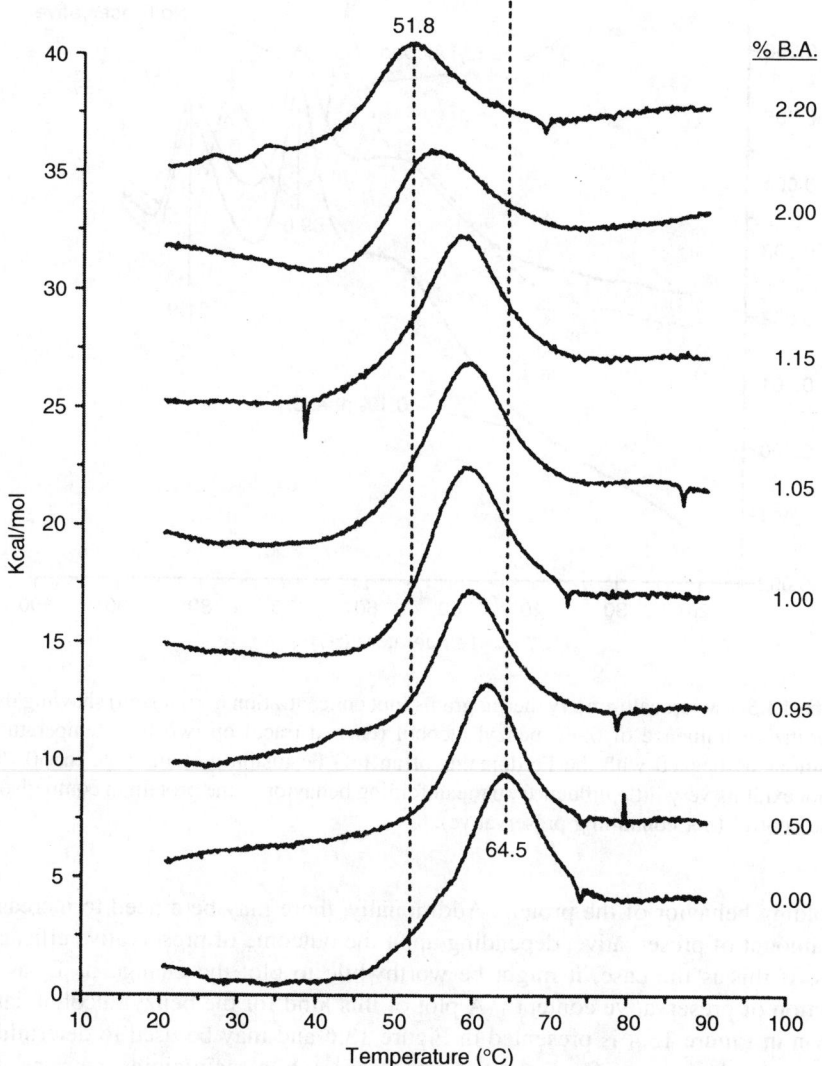

Figure 13.4 Perturbation of the melting transition temperature of a single domain protein as a function of benzyl alcohol content (described as % B.A.). The liquid protein formulation consists of 10 mM Tris, 4% mannitol, and 1% sucrose, pH 7.4. (Permission to use the figure granted by BioPharm.)

data show that 0.9% benzyl alcohol (dashed scan) destabilizes the two upper transitions by shifting them several degrees to a lower temperature. In contrast to the effect of the benzyl alcohol, 0.1% phenol had very little impact on the

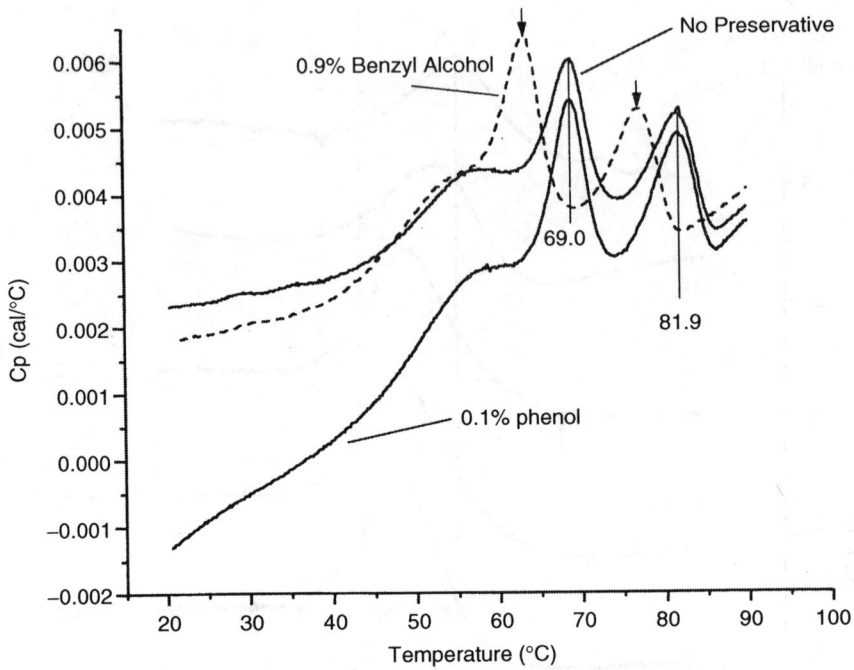

Figure 13.5 Microcalorimetry thermograms (not concentration normalized) showing the destabilizing influence of 0.9% benzyl alcohol (dashed trace) on two high-temperature transitions associated with the Fc domains of an IgG_1Fc fusion protein. Note that 0.1% phenol exhibits very little influence on the unfolding behavior of the protein in comparison to the control (not containing preservative).

unfolding behavior of the protein. Additionally, there may be a need to increase the amount of preservative, depending upon the outcome of preservative efficacy tests. If this is the case, it might be worthwhile to plot the change in T_m as a function of preservative content.[14] A plot of this kind for the benzyl alcohol data shown in Figure 13.4 is presented in Figure 13.6 and may be used to determine the type and amount of preservative necessary, while maintaining appropriate stability. The slope of such a plot indicates the destabilizing force a given preservative can exert (dT_m/dPv, where Pv represents the concentration of the preservative as a percent or as a mole ratio). If, for example, another preservative is evaluated that produces a smaller negative slope (described by the dashed line in Figure 13.6, denoted A) than that described by benzyl alcohol, it would exert less of a destabilizing effect on the protein, and a higher concentration could be used to achieve the desired preservative efficacy. The converse would be true for the situation described by the dashed line denoted B in Figure 13.6.

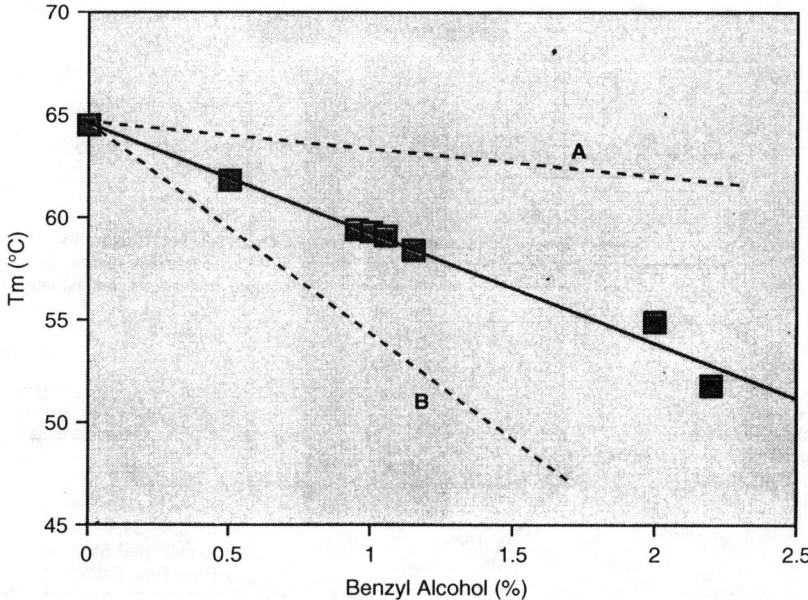

Figure 13.6 The relationship between T_m (°C) and benzyl alcohol content (obtained from the data in Figure 13.4) expressed as a percentage (w/w). Note that the relationship can be fitted to a line having a slope that describes the destabilizing force of the preservative. The dashed line designated as "A" illustrates a preservative force that is less destabilizing in contrast to the dashed line designated as "B" that is more destabilizing than benzyl alcohol.

It is also important to point out that stabilization against preservative effects may be augmented with appropriate excipient ingredients. This was the case for recombinant human interferon-γ (rhIFN-γ), where the interaction with benzyl alcohol was formulation dependent.[59] This study concluded that a stable preservative multidose liquid formulation could be achieved by minimizing the preservative concentration and using an acetate buffer at pH 5 (thermally stable condition). These results at least support the strategy described above, given that optimal pH conditions coupled with excipient screening (e.g., buffer, ionic strength) and benzyl alcohol concentration permitted a stable formulation to be derived.

Accelerated Stability

Finally, the last step in the preformulation development process is to test the candidate formulations that exhibited the greatest overall stability, using microcalorimetry. The true test of the procedure summarized in Figure 13.7 is to

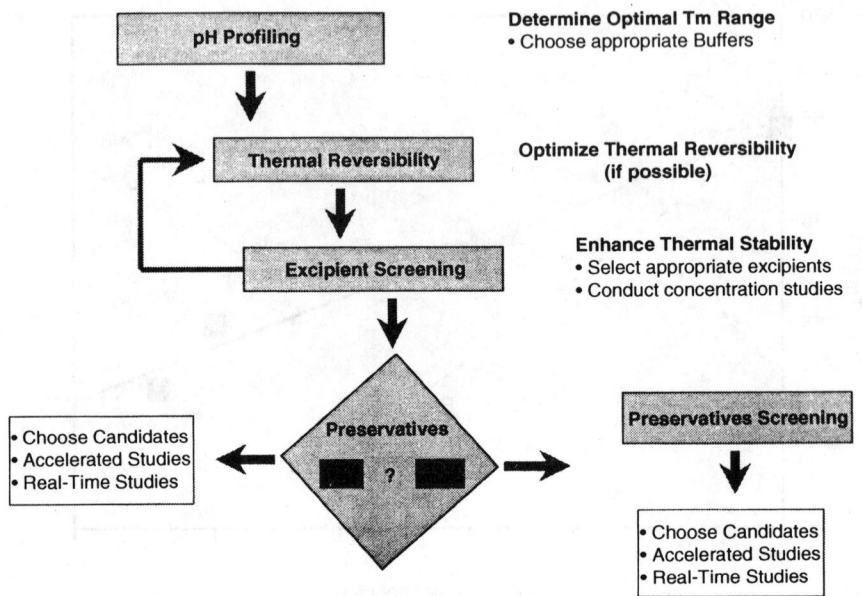

Figure 13.7 A formulation strategy using microcalorimetry aimed at deriving stable liquid candidates.

evaluate the predictive outcome by carrying out accelerated stability studies. In the case of IL-1R (type I), using the T_m as a measure of the influence of preservatives to destabilize the three transitions in this molecule, the predicted stability would have followed control (no preservatives) > 0.065% phenol > 0.1% m-cresol > 0.9% benzyl alcohol.[31] Accelerated studies were conducted at the physiological temperature of 37°C for a period of 60 d and yielded the results shown in Table 13.1. Clearly, the quantities of aggregation observed by size exclusion chromatography correlate nicely with the predicted ranking of the T_m in each preservative case. This analysis of the microcalorimetric data justifies using the technique to correctly rank different formulations in terms of their predicted impact on aggregation. The evidence gathered supports the principles presented herein. However, so as not to misconstrue the importance of microcalorimetry, this technique, in combination with several other bioanalytical methods, can provide a powerful alliance in testing the scheme presented. For example, HPLC, SDS-PAGE, CD, and mass detection methods (i.e., AUC, light scattering, mass spectrometry) in conjunction with microcalorimetry can all participate in describing how the conformational state impacts the stability of a given biologic.

It has already been mentioned how high T_m's correlate with a lower susceptibility to aggregation and the preferred solution conditions in which M-CSF, IL-1R (types I and II), Flt3L, and CD40L are stabilized. Additionally, G-CSF

Table 13.1 Microcalorimetric Melting Temperatures of IL-1R (type I) and SEC *In Vitro* Stability Data at 37°C

| | Microcalorimetry | | | SEC | | | |
| | | | | 7 d | | 60 d | |
	T_m1 (C)	T_m2 (C)	T_m3 (C)	Agg (%)	Native (%)	Agg (%)	Native (%)
Control	50.8	53.7	66.3	0.66	98.93	1.50	97.54
0.065% phenol	50.3	53.4	66.5	1.02	98.62	3.07	96.02
0.1% m-cresol	48.4	51.9	65.8	1.37	98.25	5.10	93.92
0.9% benzyl alcohol	45.2	48.5	63.6	2.93	96.92	16.46	83.09

Note: Preservatives added to the same citrate formulation (20 m*M* sodium citrate, pH 6, and 100 m*M* NaCl).

Source: Data from Remmele, R.L., Jr., N.S. Nightlinger, S. Srinivasan, and W.R. Gombotz. 1998. Interleukin-1 receptor (IL-1R) liquid formulation development using differential scanning calorimetry. *Pharm Res* 15: 200–208.

(Neupogen), a four-helix bundle cytokine, is formulated at pH 4 but has been shown to maintain both thermal stability and tertiary structure at pH 2.[60] In fact, the secondary structure of this molecule was shown to remain highly helical at pH 4 (T_m approximately 62°C) and 2 (T_m approximately 63°C) as compared to pH 7 (T_m approximately 55°C) where a less conformationally stable form was observed. In the same study, FTIR and CD data corroborated the tendency of the protein to unfold as measured by the loss of helical structure in the order pH 7 > pH 4 > pH 2. Moreover, after determining optimal pH conditions of thermostability, several studies have shown that excipient screening at such conditions can successfully predict the rank of formulation cocktails that offer the most favorable stability.[14,23,31,56]

A number of examples have now been described where T_m has been shown to be a good parameter to indicate and rank appropriate stability. However, it is only one parameter in Equation 1 that is equated to conformational stability as described by ΔGu. It is important to remember that the current usage of T_m to confer stabilization is predicated upon the notion that high T_m's represent a greater resistance of the protein to unfold and that this behavior holds at lower temperatures where the protein is to be stored. Equation 1 permits a better understanding of the relationship of stability as a function of temperature that may not necessarily coincide with the paradigm so far outlined. In some cases, it may not be possible to accurately obtain the thermodynamic parameters that make the application of Equation 1 valid (e.g., non-two-state behavior). Furthermore, where parameters are measurable, maximal conditions of conformational stability may indicate more favorable free energies at low temperatures in the presence of some excipients that tend to lower the T_m. A listing of the thermodynamic parameters pertaining to several protein systems[34,38,39,61,62] that can be plugged into the modified Gibbs–Helmholtz equation are shown in Table 13.2. The thermodynamic behaviors of a few of these systems are plotted in Figure 13.8. Maximal conformational stability is achieved when the free energy difference between the native and denatured states is the greatest and positive. The compact native state is assumed to depict a state of lower energy than the denatured state.[63] Therefore, an enhancement of stability as a consequence of increasing the difference between these states by optimizing the solution conditions that make this possible is the premise for the microcalorimetric approach described in the preceding text. The temperature where this is true is called the temperature of maximum stability, or T_{ms}.[47] It is the point where the parabolic maximum occurs as described by the behavior shown in Figure 13.9.

Consider what happens to pepsinogen near its optimal T_m of 339.4°K (66.2°C), when 20% ethanol is added to the solution (Figure 13.9). The temperature of maximum stability occurs near 300°K (26.9°C). When 20% ethanol is added, the T_m is lowered to 329.0°K (55.9°C) as expected, suggesting some destabilization of the protein between the two solution conditions. However, the temperature of maximum stability occurs at a lower temperature near 273°K

Table 13.2 Thermodynamic Parameters of Different Protein Systems

Protein	Conditions	T_m (°K)	ΔH_m (kcal/mol)	ΔCp (kcal/mol-°K)	Ref.
α-Chymotryp-sinogen (bovine)	pH 2.3	316.2	78	3.8	38
	pH 2.6	322.2	102	3.8	38
	pH 2.8	324.2	110	3.8	38
	pH 3.4	331.2	130	3.8	38
	pH 4.0	334.2	140	3.8	38
	pH 5.0	335.2	148	3.8	38
α-Chymotrypsin	pH 6.0	333.3	166	3.013	38
FGF-1 (human)	20 mM N-(2-acetamido)-iminodiacetic acid, 0.1 M NaCl, pH 6.6, with added 0.6 M GuHCl	315.7	73.04	0.608	34
	0.7 M GuHCl	312.6	61.50	1.554	34
	0.8 M GuHCl	310.1	56.02	1.995	34
	0.9 M GuHCl	307.0	47.08	2.360	34
	1.0 M GuHCl	304.8	41.09	2.193	34
	1.1 M GuHCl	299.9	29.42	2.428	34
Papain	pH 5	358.9	215.5	3.307	39
Pepsin	5 mM Na-phosphate and pH 6.5 (2 pks)	322.2	135	4.15	60
	pH 6.5 + 20% ethanol	312.0	227	2.77	60
	pH 5.0 + 20% ethanol	325.2	215	2.77	60
	pH 4.0 + 20% ethanol	332.7	233	2.77	60
	pH 3.0 + 20% ethanol	331.7	238	2.77	60
	pH 2.2 + 20% ethanol	329.2	212	2.77	60
	pH 2.0 + 20% ethanol	329.2	214	2.77	60
Pepsinogen	5 mM Na-phosphate and pH 6.0	339.4	254	5.8	61
	pH 6.4	336.0	248	5.8	61
	pH 7.2	329.3	195	5.8	61
	pH 7.7	328.2	173	5.8	61
	pH 8.0	324.3	182	5.8	61
	pH 5.9 + 20% ethanol	329.0	245	4.2	61
	pH 6.4 + 20% ethanol	325.3	248	4.2	61
	pH 6.8 + 20% ethanol	320.4	221	4.2	61
	pH 7.3 + 20% ethanol	313.0	192	4.2	61
	pH 8.0 + 20% ethanol	308.5	169	4.2	61
	pH 8.2 + 20% ethanol	307.0	170	4.2	61

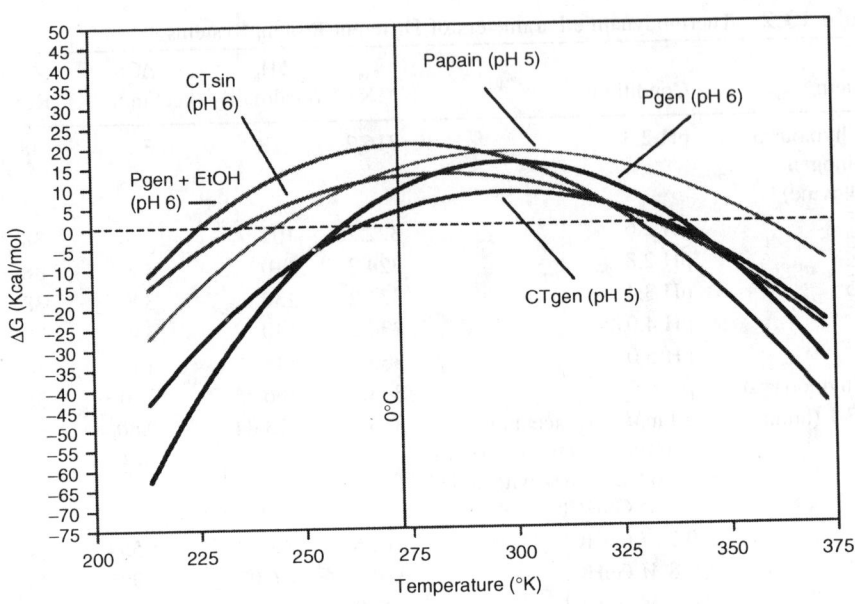

Figure 13.8 Temperature response trace (at optimal T_m solution conditions) of the free energy of unfolding (ΔGu), calculated from Equation 1. Pgen = pepsinogen, Ctsin = α-chymotrypsin, Ctgen = α-chymostrypsinogen, and papain. Note the pH conditions are listed in parentheses and the thermodynamic parameters used in Equation 1 are from those listed in Table 13.2.

(0°C) and indicates a greater ΔGu. Considering T_m alone, one might be inclined to think that the stability of this protein would be lower at conditions near 2 to 8°C. In fact, the cold denaturation temperature is also lowered, making this protein more stable in the presence of 20% ethanol at low temperature. This is something not readily apparent from T_m. It suggests that other excipients may produce similar behavior. Although not evaluated in this light, this could explain why the presence of polyethylene glycol (PEG) is somewhat more stabilizing at lower temperatures even though it exhibits a destabilizing shift in the T_m.[25] It should be pointed out that in both cases there was an observed change in the heat capacity term. In the case of pepsinogen with 20% ethanol, the ΔCp was lowered from 5.8 to 4.2 kcal/mol°K. A similar result was observed in the presence of PEG in terms of concentrations and molecular weight of the PEG added to various test protein systems analyzed.[25]

An explanation for these observances in the cases of ethanol and PEG may arise from hydrophobic interactions (i.e., methylene groups, methyl) with the unfolded state of the protein at elevated temperatures. This idea is supported by studies of the interaction of several alkylureas (methyl-, *N,N*dimethyl-, ethyl-, and butylureas) with the thermal unfolding of ribonuclease A, where it was shown

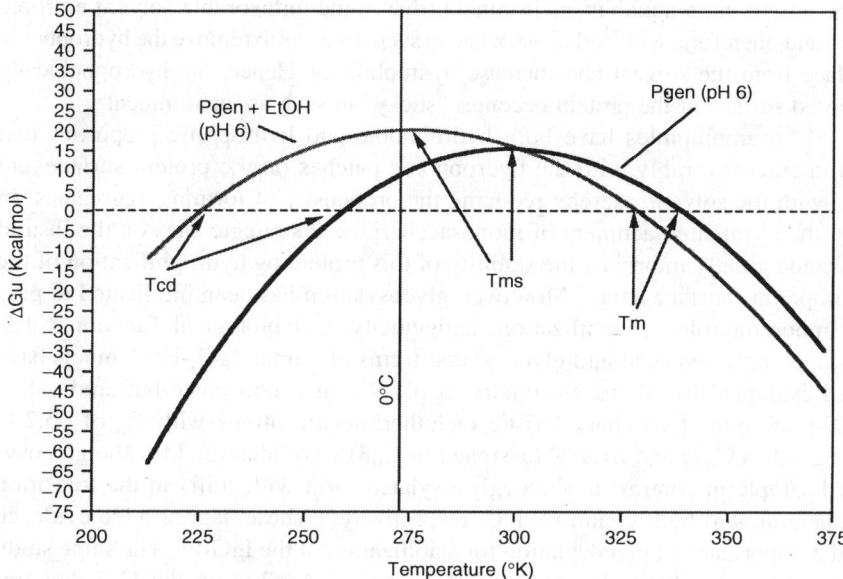

Figure 13.9 Temperature response trace (at optimal T_m solution conditions) of ΔGu, for pepsinogen at pH 6, with and without 20% ethanol (EtOH). Labels shown are identified as Tcd = cold denaturation temperature, T_{ms} = temperature of maximum stability, and T_m = melting transition temperature.

that the thermal stability of the protein decreased with increasing concentration and size of the hydrophobic group substituted on the urea molecule.[64] Another class of excipients commonly used in biopharmaceuticals are surfactants. They can afford significant stabilization at low temperatures during agitation where air–water interface reactions can lead to unfolding. However, they commonly result in lower T_m's that may stem from nonspecific interactions with solvent-exposed hydrophobic regions of the denatured protein at elevated temperatures.[65]

Protein Properties

Physical properties of the protein structure should be considered in designing strategies to achieve stable formulations because they can often yield clues about which solution environment would be appropriate for stabilization. For example, the insulin molecule is known to self-associate via a nonspecific hydrophobic mechanism.[66] Stabilizers tested include phenol derivatives, nonionic and ionic surfactants, polypropylene glycol, glycerol, and carbohydrates. The choice of using stabilizers that are amphiphilic in nature to minimize interactions where protein hydrophobic surfaces instigate the instability is founded upon the hydrophobic effect.[19] It has already been mentioned that hydrophobic surfaces prefer to be buried in an aqueous environment (due to the unfavorable solvent entropic cost) and, therefore, will tend to associate in such a way as to remove the hydrophobic surface from the solvent and increase hydrophilicity. Hence, the hydrophobically exposed surface of the protein becomes "sticky" in such an environment.

The amphiphiles have both hydrophobic and hydrophilic properties that can interact favorably with the hydrophobic patches on the protein surface and also with the solvent, thereby reducing the propensity of forming aggregates. In fact, the covalent attachment of monosaccharides at strategic sites on the insulin molecule greatly improved the stability of this protein by hydrophilization of the hydrophobic surface area.[67] Moreover, glycosylation has been implicated to play an important role in stabilization, antigenicity, and biological function.[68] For example, aglycosylated and glycosylated forms of human IgG_1-Fc domains have been evaluated by microcalorimetry at pH 7.4 in a phosphate-buffered saline (PBS) solution. Two characteristic endothermic transitions with T_m of 65.2°C (assigned to C_H2) and 81.9°C (assigned to C_H3) were identified for the glycosylated sample in contrast to the deglycosylated form with shifts in the transition temperatures to 60.9°C and 68.1°C, respectively.[68] These data provide evidence of the importance of glycosylation for stabilization of the IgG-Fc. The same study also emphasized how glycosylation (localized at Asp297 on the C_H2 domain) affects structure required for FcγRI, FcγII, FcγRIII, and C1 binding and activation. It should be noted that the second transition of the Fc assigned to the C_H3 domain actually constitutes the melting of two domains (C_H3/C_H3) as a single cooperative block.[69] Furthermore, the C_H2 transition is accompanied by disruption of interdomain interactions (*cis* between C_H2 and C_H3 domains) that contribute to the high enthalpy of the first transition (57 kcal/mol) in contrast to the C_H3 transition (36 kcal/mol).[68,69] The afforded thermal stability of the glycan in this case bears a similar resemblance to the insulin case given that the IgG_1-Fc is predominantly hydrophobic. Salt in this case should benefit stabilization of the hydrophobic core.

Other studies of the role glycans play concerning protein stabilization suggest that deglycosylated forms exhibit lower thermostability as judged by a decrease in T_m and ΔH_m although not substantially affecting the conformation as indicated by far-UV CD.[70] In these latter studies, the thermal stability seemed to depend on the amount of glycosylation present (i.e., maximum stabilization was observed for the most heavily glycosylated protein, irrespective of the types, N-linked or O-linked, or patterns, mono- or multibranched, of the covalently attached carbohydrate chains). Additionally, it was shown that thermal reversibility of deglycosylated proteins was significantly diminished and that they were prone to aggregate in comparison to their glycosylated counterparts. Hence, glycosylation of recombinant proteins using appropriate expression vectors (i.e., CHO, yeast) can potentially afford improved stability (as described by microcalorimetric data) and robustness to withstand the stresses associated with the biopharmaceutical development train.

Understanding protein unfolding as it pertains to interactions with contact surfaces is also important to achieving a stable dosage form. Adsorption onto container or device surfaces (i.e., syringes, vials, catheters) can lead to a loss of drug potency. In very dilute dosage forms, this can result in substantial differences between the expected concentration and that delivered to the patient. Microcalorimetry has been used successfully to quantify losses in secondary structure of globular proteins upon adsorption to solids.[71] For example, the T_m, ΔH_{P-D}, ΔH_{N-D}, and ΔCp_{P-D} were all determined (where the subscripts P = adsorbed perturbed state, N = native, D = denatured). The magnitude of the ΔH_{P-D} relative to the ΔH_{N-D} was used as an unambiguous gauge of the extent of secondary structure loss in proteins as a result of adsorption. The high T_m was indicative of a highly stable adsorbed-state structure. Examination of lysozyme adsorption to polystyrene latex and hematite revealed that it adsorbed irreversibly, yet the extent of the unfolded structure was less for hematite than for polystyrene latex. This comparison indicated that lysozyme retained most of its native structure when adsorbed to a more hydrophilic surface. Studies of this kind make it possible for one to not only obtain a better understanding of the thermodynamics of protein adsorption, but also allow one to delineate between solid surfaces where the protein is less susceptible to adsorb.

Most of what has been presented has dealt with the propensity of the unfolded state to form aggregates. However, deamidation reactions can also be affected by the conformational flexibility of the protein in the vicinity of Asn residues.[15,72] Protein sequences possessing Asn-Gly are particularly labile as a result of decreased steric hindrance associated with the lack of a side chain on the Gly residue that allows free rotation of the asparaginyl side chain to permit nucleophilic attack from the peptide bond nitrogen forming a cyclic imide intermediate. This intermediate is particularly unstable toward hydrolysis, leading to the formation of a mixture of α- and β-aspartyl residues, which may affect the efficacy of the drug. Deamidation studies at pH 7 of polyanion-stabilized (heparin) acidic fibroblast growth factor (aFGF) have revealed that, among the three Asn sites located near the N-terminus (Asn3-Leu4, Asn8-Tyr9, and Asn19-Gly20), only the Asn8-Tyr9 deamidated.[72] The unexpected absence of deamidation at the notoriously labile Asn19-Gly20 suggested that there was no simple correlation between local amino acid sequence, conformational flexibility, and deamidation potential. Yet an explanation offered for this anomalous result was that the Asn19-Gly20 site was involved in heparin binding, thus permitting greater conformational rigidity and affording some protection against deamidation. It should be pointed out that heparin had been shown to stabilize aFGF against thermal, pH, and proteolytic degradation.[72] This example illustrates how binding a particular substrate to a protein can sometimes increase stabilization of chemically labile residues.

Folding and Stability

Structural factors are important regarding rational design approaches that lead to predicting stable protein folds. Can anything be learned about protein stability from different structural elements, amino acids, and packing of the native folds common for most protein systems? The answer to this important question is yes! Efforts to understand the tendencies of amino acids in peptides to contribute to ordered and disordered structure have been and continue to be examined. For example, scales based upon probabilities associated with a given amino acid's contribution to structure in a peptide segment have been used to classify amino acids from order-promoting to disorder-promoting in the following manner: Trp > Phe > Ile > Tyr > Val > Leu > Cys > Met > Ala > Asn > Thr > Arg > His > Gly > Asp > Ser > Gln > Glu > Lys > Pro.[73] Using such an index, one would expect tryptophan to have a high probability as an order-promoting amino acid in contrast to proline that would be disorder promoting. Deciphering the role of each amino acid as it pertains to fold recognition requires an understanding of the thermodynamic tendency of each amino acid to stabilize native-state ensembles. It has been observed that amino acid types partition unequally into high, medium, and low thermodynamic stability environments and suggest that calculated thermodynamic stability profiles have the potential to encode sequence information.[74] In this context, 44 nonhomologous proteins were analyzed using the COREX algorithm[75] where energy differences between each partially unfolded microstate and fully folded reference state were determined by the energetic contributions of all amino acids comprising the folding units that unfolded in each microstate, plus the energetic contributions associated with exposing additional (complementary) surface area on the protein. This can be described by the equation $\Delta ASA_{total} = (ASA_{unf} + ASA_{comp}) - (ASA_{native})$. The results suggested that aromatic amino acids Phe, Trp, and Tyr are mostly found in high-stability environments, whereas Gly and Pro were overwhelmingly found in low-stability environments. This was in contrast to amino acid residues such as Ala, Met, and Ser that showed distributions not significantly different from the randomized data. Additional studies using computational methods to test different combinations of secondary structure fragments of 240 nonhomologous proteins have shown that an overwhelming majority can be successfully recognized by the energies of interaction between residues of secondary structure.[76] The results of this study found that β-structures contributed more significantly to fold recognition than α-helices or loops. Additionally, it has been shown from studies of unfolding thermodynamics of the all β-sheet structure of interleukin-1β that the Gibbs free energy of a hydrogen bond in the β-sheet structure is greater than in α-helices.[77]

In a different study, a 29-residue dicyclic helical peptide, shown to be two-state, thermally reversible, and to unfold in a cooperative manner within the temperature range extending from 10 to 100°C, was intentionally designed with side-chain to side-chain covalent links at each terminal to stabilize the helix structure.[78] Complete thermodynamic characterization of helix unfolding could then be examined. Unfolding was found to proceed with a small positive heat capacity increment, consistent with the solvation of nonpolar (hydrophobic) groups upon unfolding. The conclusion was that hydrogen bonds were not the only factors responsible for the formation of the α-helix, but that hydrophobic interactions also contributed to its stabilization. At 30°C, the calorimetric enthalpy and entropy values were estimated to be 18.85 (±1.45) kcal/mol and 58 (±5.8) cal/mol-°K, respectively. With regard to protein systems, human plasma apolipoprotein A-2 (apoA-2) also unfolds reversibly and has an enthalpy of unfolding of 17 (±2) kcal/mol for helical structure.[79] Thermal unfolding of $\gamma\gamma$ and $\beta\beta$ homodimeric coiled-coils of tropomyosin exhibit multistate unfolding properties with overall unfolding enthalpies near 300 kcal/mol.[80] The electrostatic stabilization effects of coiled-coils have been investigated using a sequence shuffling strategy without changing the overall content of amino acids in the peptides, and have shown that in solutions of low ionic strength, ionic pairs contribute significantly to the stability of the coiled-coil conformation.[81]

Structure stabilization of protein systems has also focused on design parameters of the protein's core. In one such study, the hydrophobic core of a four-helix bundle protein, Rop, was altered mutagenically to study packing patterns and various side-chain shapes and sizes as it related to properties of stabilization and destabilization.[82] It was discovered that overpacking the core with larger side chains caused a loss of native-like fold, decreasing the associated thermal stability of the molecule. This study defined the role of tight residue packing and burial of hydrophobic surface area in the construction of compact native-like proteins. Solvent influences on hydration properties of protein secondary structures have also been evaluated, indicating that the strength of water binding to carbonyl groups is lower for β-sheet proteins than in α-helical proteins, owing to the differences in the geometry of the water carbonyl group interactions.[83]

What structural properties are necessary to make a protein thermally stable? Much of what is currently known comes from examination of structural properties of hyperthermophilic and thermophilic proteins. At high-temperature conditions where many proteins from mesophilic organisms would naturally denature, hyperthermophilic proteins thrive and exhibit higher T_m. Clues to answer the question may be obtained by comparing the structural fingerprint features of mesophilic proteins to those of the hyperthermophilic class. For example, the structure of the hyperthermophilic tungstopterin enzyme, aldehyde ferredoxin oxidoreductase, has a relatively small solvent-exposed surface area and a relatively large number of both ion pairs and buried atoms.[84] The ion pairs constitute salt bridges between acidic amino acids (i.e., aspartic and glutamic acids) and the basic amino acids (i.e., arginine, histidine, and lysine).

In another study of *Thermus aquaticus* D-glyceraldehyde-3-phosphate dehydrogenase, among the various structural factors evaluated (i.e., salt bridges, hydrogen bonds, buried surface area, packing density, surface-to-volume ratio, stabilization of α-helices and β-turns), a strong correlation between thermostability and the number of hydrogen bonds between charged side chains and neutral partners was discovered.[85] Charge-neutral hydrogen bonds were believed to provide electrostatic stabilization without the heavy desolvation penalty of salt bridges. What is interesting concerning ΔGu of mesophilic and thermophilic proteins is the fact that the free energy for thermophiles is not much larger than that of the mesophiles, essentially equivalent to a few hydrogen bonds or ion pairs. Oligomerization of the protein has also been cited as a means for *Thermus thermophilus* pyrophosphatase stability in which it is tightly packed, having a rigid structure, and comprising hydrogen bond and ionic interactions that form an interlocking network which covers all of the oligomeric surfaces.[86] This in turn translates into an increase in buried surface area upon oligomerization by about 16%.

Recently, investigation of how proteins are packed has led to the idea that although they have average packing densities as high as crystalline solids, they look more like liquids or glasses by their free volume distributions.[87] Moreover, the distributions are broad and indicate that the interiors may better be described as randomly packed spheres where larger proteins are packed more loosely than smaller proteins. Citrate synthase from hyperthermophilic archeon *Pyrococcus furiosus* (an organism that optimally grows at 100°C) was compared to mesophilic and thermophilic versions of the same protein. The most significant feature was an increased compactness of the enzyme, a more intimate association of the subunits (homodimer), an increase in intersubunit ion pairs, and a reduction of thermolabile residues.[88] Compactness was achieved by shortening the number of loops, increasing the number of buried atoms in the core, and optimizing the packing of side chains in the interior, resulting in a reduction of cavities. A common thread emerges concerning each of these investigations, namely, thermal stability is related to a native compact structure that involves an enhancement of hydrogen bonding, packing density, and salt bridges that are important to the protein fold.

Finally, assimilating information about the structural properties of a given biologic, combined with microcalorimetric data about stabilizing solution environments and the results of accelerated studies to characterize major instabilities (i.e., aggregation, deamidation, oxidation, proteolysis, hydrolysis), it should be possible to select respectable candidate solutions to evaluate in real time to determine achievable longevity. The approaches outlined in this section are suggested as a guide in the preformulation and formulation development design. They have been shown to work well in some cases presented. Currently, there is no acceptable way of predicting the real-time shelf-life apart from carrying out real-time studies. However, there have been attempts to characterize degradation products using activation energies obtained from reactions that follow Arrhenius behavior that have met some measure of success.[13,89,90] This subject will be discussed in more detail in a following section.

ISOTHERMAL TITRATION CALORIMETRY (ITC)

The study of protein function as it applies to covalent and noncovalent molecular changes associated with specific binding is of fundamental interest to those involved with the discovery of therapeutic proteins. Discovering soluble receptors that bind the cytokines that elicit inflammatory immune responses is one of many approaches to treating physical disorders (i.e., rheumatoid arthritis, asthma). In such cases where intermolecular recognition is to be characterized, the ITC technique can offer some advantages. This technique is capable of thermodynamic characterization of the binding affinities of new drugs that act as receptors, antagonists, or agonists of biological function. It has been acclaimed to rival most alternative methods used to acquire information of this kind (i.e., enzyme-linked immunosorbent assays, absorption and fluorescence spectroscopy, magnetic resonance, chromatography, etc.). Although radiolabeling techniques and plasmon resonance methods (e.g., BIAcore) are capable of obtaining data at lower concentrations than microcalorimetry, they are not necessarily superior techniques for measuring binding constants. Among the true "in-solution" methods that do not require protein modification (i.e., ITC and analytical ultracentrifugation), only ITC can provide a complete thermodynamic characterization of the system under study.[91] This is an important advantage when comparing results to BIAcore techniques in that binding reactions that exhibit entropy/enthalpy compensation often yield free energies of binding that exhibit little change.[53] An understanding of whether entropy or enthalpy drives the reaction can be very important yet not obvious from the free-energy parameter obtained from the on/off rate equilibrium constants typically obtained from BIAcore measurements. In enthalpy-driven reactions, the enthalpy is large, and the entropy component of the reaction is unfavorable (negative). The inverse is generally true for entropy-driven reactions. In situations in which a reaction is enthalpy driven, a loss of conformational mobility may result upon binding. Alternatively, reactions that are entropy driven may suggest that hydrophobic contacts remove hydrophobic surface area from the water environment, thereby increasing entropy of the solvent upon binding. As for binding constant comparisons between the BIAcore and ITC, there have been studies that would indicate that the two methods yield similar results.[92]

Protein–Ligand Interactions

Just like conventional scanning microcalorimetry, the liberation and absorption of heat associated with the chemical reaction are measured in an ITC experiment. However, in this case, the heat is due to a reaction involving binding between the protein and a ligand. The ligand may be either another protein, nucleic acid, lipid, or a small molecule. It is the only method that is capable of determining ΔG_b, ΔH_b, and ΔS_b of binding (subscript b denotes binding), the binding constant (Ka) and stoichiometry (N), in a single experiment.[91] Additionally, background interferences commonly observed in optical absorption or fluorescence methods (where solution components may also have absorption, fluorescence, or scattering properties) are normally very low in ITC experiments, which permits the study of heterogenous mixtures.

In principle, the experiment is carried out at constant temperature by titrating one binding partner into a solution containing the other in the sample cell of the calorimeter. After a small increment of titrant is added to the sample cell, the heat liberated or adsorbed in the sample cell is quantitatively measured with respect to a reference cell filled with the identical solution without protein. The change in heat is the electrical power required to maintain a constant temperature between the sample and reference cells. Upon continued titration, the enthalpic response decreases until saturation is achieved. The contents of the cell are stirred continuously during the experiment by the spinning motion of the syringe-paddle from which the titrant is dispensed to achieve rapid equilibrium between the reagents.

An example of such a titration is illustrated in Figure 13.10, where 0.061 mM RNase A in the sample compartment was titrated with 2.13 mM 2′CMP (titrant). In the upper panel, the heat signal shown by the downward directional peaks are exotherms of the reaction mixture. In a typical experiment, the total heat of each injection of titrant is computed as the area of the downward peak. When the total heat is plotted against the molar ratio of ligand (2′CMP) added to the sample (RNase A) as depicted in the lower panel of the figure, one obtains the complete binding isotherm. This experiment was performed using a MicroCal VP-ITC system with accompanying software containing the algorithms to compute the stoichiometry of binding, the binding constant, and enthalpy of binding. The free energy of binding may then be determined by the equation $\Delta G_b = -$ RT ln Ka, where R is the gas constant and T is the absolute temperature. The entropy can be indirectly determined using the Gibbs–Helmholtz expression ($\Delta G_b = \Delta H_b - T\Delta S_b$) once ΔG_b is calculated from Ka. In this case, $\Delta G_b = -8.07$ kcal/mol. Saturation occurs when the heat signal becomes small, approximating the heat of dilution of the titrant into the sample. This is the case in the example presented when the mole ratio exceeds 2.5 (Figure 13.10). This technique can be a powerful tool used in conjunction with scanning microcalorimetry for characterizing a given protein biologic.[53]

One limitation of this technique is that the largest binding constants that may be reliably measured are about 10^9 M^{-1} for typical protein–ligand interactions.[93] Options to get around this issue involve experiments conducted at different pH or temperature conditions in which the binding constant may be evaluated with greater ease. Although this approach can be useful in acquiring information about linked protonation effects, the observed binding enthalpy and heat capacity change can undergo extreme deviations from their intrinsic values depending upon pH and buffer conditions.[94] There could be concerns regarding the value at physiological conditions where the protein biologic is expected to be efficacious. Alternatively, competitive binding by titration displacement schemes may be applied to estimate tight binding affinities that are greater than 10^9 M^{-1}.[95] However, even with this approach there is a possibility of error and added complexity associated with the reaction. Recently, competitive binding by displacement titration has been improved using cubic binding equations that may allow for greater accuracy for very tight binding reactions.[96] Such an approach requires independent experiments with a competitive binding inhibitor that is a few orders of

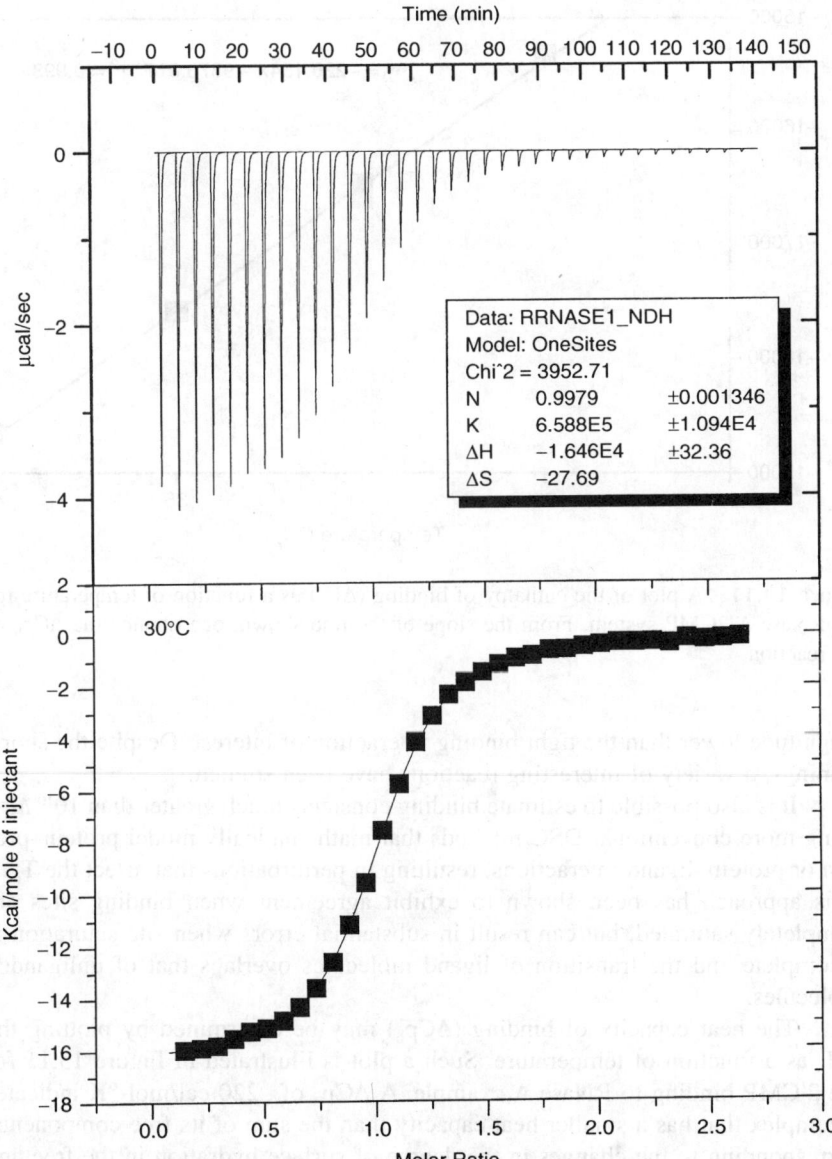

Time (min)

Data: RRNASE1_NDH
Model: OneSites
Chi^2 = 3952.71
N   0.9979   ±0.001346
K   6.588E5   ±1.094E4
ΔH   −1.646E4   ±32.36
ΔS   −27.69

30°C

Figure 13.10 Calorimetric titration response showing the exothermic raw (downward-projecting peaks, upper panel) heats of the binding reaction over a series of injections titrating 0.061 m*M* RNase A (sample) with 2.13 m*M* 2CMP at 30°C. Bottom panel shows the binding isotherm obtained by plotting the areas under the peaks in the upper panel against the molar ratio of titrant added. The thermodynamic parameters were estimated (shown in the inlay of the upper panel) from a fit of the binding isotherm.

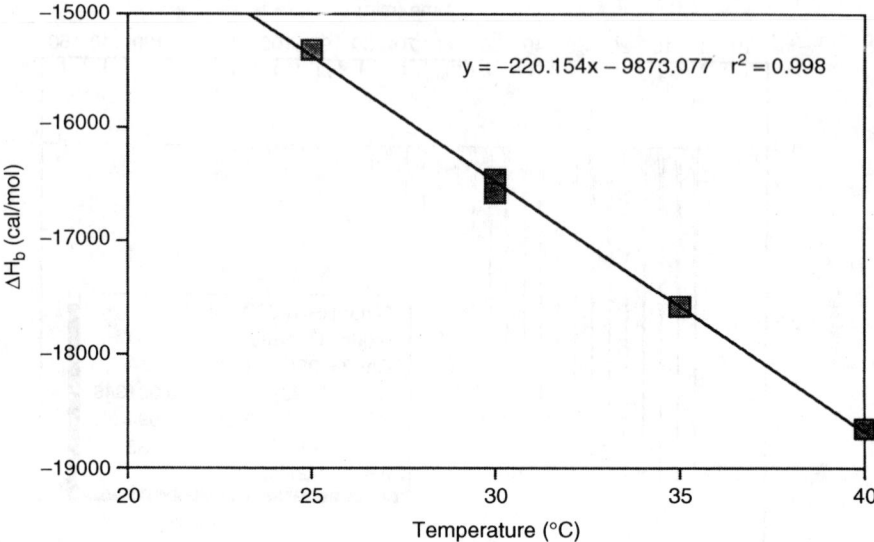

Figure 13.11   A plot of the enthalpy of binding (ΔH_b) as a function of temperature for the RNase A/2CMP system. From the slope of the line shown, one obtains the ΔCp_b of the reaction.

magnitude lower than the tight binding interaction of interest. Despite the shortcomings, a variety of interesting reactions have been studied.

It is also possible to estimate binding constants much greater than 10^{10} M^{-1}, using more conventional DSC methods that mathematically model protein–protein or protein–ligand interactions, resulting in perturbations that affect the T_m.[97] This approach has been shown to exhibit agreement when binding sites are completely saturated, but can result in substantial errors when site saturation is incomplete and the transition of ligand molecules overlaps that of unliganded molecules.

The heat capacity of binding (ΔCp_b) may be determined by plotting the ΔH_b as a function of temperature. Such a plot is illustrated in Figure 13.11 for the 2'CMP binding to RNase A example. A ΔCp_b of –220 cal/mol-°K indicates a complex that has a smaller heat capacity than the sum of its free components, corresponding to the changes in the degree of surface hydration in the free and complexed forms.[53] Results from liquid hydrocarbon transfer and solid dissolution experiments support the idea that polar and nonpolar interactions differ in the sign of the change in heat capacity.[98] Hence, mutation studies that weaken hydrophobic contributions of the complex form of the protein would be expected to make ΔCp_b less negative. Conversely, decreasing polar interactions would necessarily make ΔCp_b more negative. For example, the ΔCp_b values for a triple alanine mutant (–927 cal/mol-°K) of the human growth hormone (hGH)–receptor

.nterface was shown to be significantly more negative than the ΔCp_b for the wild-type interaction (−767 cal/mol-°K).[99] Such negative ΔCp_b values are consistent with the proposal that the hydrophobic effect is the primary contributor to the free energy of binding at the protein complex interface.

In the development of a biopharmaceutical, it is often desirable to determine binding affinities of a target protein–ligand system. Despite the shortcomings associated with binding affinities within a narrow range ($\leq 10^9$ M^{-1} and $\geq 10^3$ M^{-1}), many systems can be directly evaluated by this technique. The remaining discussion will focus on some of the work that has been performed looking at protein–ligand interactions using ITC, where the ligand may be a receptor, an antigen, an antibody, a carbohydrate, or a lipid.

Receptors

This technology is particularly well suited for the characterization and study of protein design factors as they affect stability and binding to different receptors. For example, receptor dimerization has been considered a primary signaling event during binding of a growth factor to its receptor at the cell surface. Analysis of binding epidermal growth factor (EGF) to the extracellular domain of its receptor (sEGFR) has been performed using ITC.[100] The data indicated the stoichiometry of binding to be one EGF monomer bound to one sEGFR monomer followed by dimerization of two monomeric (1:1) EGF–sEGFR complexes. Rational mutagenic approaches involving the modification of protein binding sites to acquire a biophysical understanding of the protein engineered changes that modulate binding affinities in solution phases have been performed on a number of systems.[99–104] For example, the homopentameric B subunit of vertoxin 1 (VT1) binds to the glycosphingolipid receptor globotriaosylceramide (Gb₃). Mutants with alanine substitutions near the cleft between adjacent subunits were studied to understand the effects of binding affinity.[102] Substitution of alanine with phenylalanine (Phe-30) resulted in a fourfold reduction in binding affinity for Gb₃. Examination of the binding of VT1 subunit B with trisaccharide in solution by ITC showed that the Phe-30 mutant was markedly impaired in comparison to the wild type and too weak to calculate the binding constant. To ensure that the mutant did not undergo a significant change in structure, it was crystallized and evaluated by x-ray diffraction. The data showed that the structure was in fact identical to the wild type except for the Ala30Phe substitution. This evidence suggested that the aromatic ring of Phe-30 greatly affected the binding to the Galα1–4Galβ1–4Glc trisaccharide portion of Gb3.[101] In another mutagenesis study of the contact side chains in the hGH–receptor interface, a triple mutant comprised of Phe25Ala, Tyr42Ala, and Gln46Ala was evaluated by ITC.[99] Binding kinetics of the triple-alanine mutant showed that neither association nor dissociation rates were significantly affected, and only slight local disorder was seen in the crystal structure. However, ITC showed large and compensating changes in the enthalpy and entropy of binding. The triple mutant bound with a favorable enthalpy of $\Delta H_b = -12.2$ kcal/mol and a corresponding less favorable entropy of $\Delta S_b = -2.3$ cal/mol-°K in comparison to the wild type of $\Delta H_b = -9.4$ kcal/mol and $\Delta S_b = 7.7$ cal/mol-°K. Dissection of the triple-alanine mutant into single Phe25Ala and double Tyr42Ala/Gln46Ala mutants showed that the more favorable enthalpy was derived from the removal of the Phe-25 side chain on helix-1 of the hormone. These results demonstrate that multiple-alanine mutations at contact residues may not affect binding kinetics, affinity, or global structure, yet they produced local structural changes that cause large compensating effects on the entropy of binding.

In another study, the thermodynamics of the binding of cyclic adenosine monophosphate (cAMP) and its nonfunctional analog, cyclic guanosine monophosphate (cGMP), to cyclic AMP receptor protein (CRP) and its Thr127Leu mutant were investigated by ITC at 24°C and 39°C.[102] The binding of the first cAMP molecule to CRP was found to be exothermic. However, successive binding of cAMP to CRP was endothermic and cooperative, indicating that the binding of the first cAMP molecule increased the affinity for the second one by more than an order of magnitude at 24°C. In the same study, it was shown that the overall binding of cGMP to CRP was exothermic and noncooperative. The point mutation of Thr127Leu switched off the cooperativity between cAMP-ligated binding sites without affecting the binding constant of cAMP and changed the specificity of the protein so that transcription was activated only upon cGMP binding. All of the binding reactions to CRP and the mutant were found to be entropically driven at 24°C.

The principal loci for binding interactions between aspartate and serine receptors of *E. coli* and methytransferase has been investigated and found to involve the last five amino acids of the receptor.[104] Truncation experiments performed on the C-terminal fragments of an aspartate receptor involving either the last 297, 88, or 38 amino acids gave comparable values of binding by ITC (N = 1, $\Delta H_b \sim 13$ kcal/mol, and Ka $\sim 4 \times 10^5$ M^{-1}). Truncating either 16 or 36 amino acids from the C-terminus eliminated detectable interactions. In the same study, a pentapeptide consisting of Asn-Trp-Glu-Thr-Phe (corresponding to the last five amino acids of the receptor and conserved in *E. coli* serine and aspartate receptors and also in the *Salmonella typhimurium* aspartate receptor) was found to have complete binding activity of the full-length receptor and the C-terminal fragments. In an *in vitro* methylation assay, this peptide was able to completely block receptor methylation.

The ITC technique was used to show that aFGF forms a 1:1 complex with the soluble extracellular domain of the FGF receptor (sFGFR).[103] At 25°C, the Ka of this reaction was shown to be approximately 1.89×10^6 M^{-1} ($\Delta H_b = -5.3$ kcal/mol). This binding reaction exhibits a favorable binding entropy ($\Delta S_b = 10.93$ cal/mol-°K), perhaps indicating that hydrophobic surface area is removed from the solvent phase upon binding. As mentioned earlier, the labile Asn19–Gly20 linkage of this molecule does not deamidate appreciably when liganded to a stabilizing polyanion (i.e., heparin, sulfated β-cyclodextrin, sucrose octasulfate).

The stoichiometry for the binding reactions of 4.8 kD heparin was found to be about 4:1 (aFGF:heparin); of 16 kD heparin, 11:1 (aFGF:heparin); and for sucrose octasulfate, 1:1 (aFGF:sucrose octasulfate). Binding affinities for the same reactions were approximately 2.17×10^6 M^{-1} (ΔH_b = −4.8 kcal/mol), 1.98×10^6 M^{-1} (ΔH_b = −5.1 kcal/mol) and 0.294×10^6 M^{-1} (ΔH_b = −5.8 kcal/mol), respectively. Hence, the binding affinities for both heparin treatments were essentially the same as that for the sFGFR with comparable enthalpies of binding. However, sucrose octasulfate, showing comparable enthalpy of binding, exhibited a weaker binding affinity with a less favorable ΔS_b (5.56 cal/mol-°K), suggesting that less hydrophobic surface area was buried in the complex. This study indicates the potential use of ITC to evaluate excipients that bind a given drug, realizing that in this particular case, stabilization against deamidation was minimized by adding an appropriate substrate that could bind to aFGF.

Antibodies

Considerable progress has been made regarding the design of antibodies that target a given disease.[98] A fundamental aspect involves an improved understanding of the forces surrounding the high affinity and specificity of antibody–antigen interactions. To this end, thermodynamic characterization is essential for improving protein design and deriving structural and functional information about binding. The role of hydrogen bonding and solvent structure as it pertains to the removal of specific hydrogen bonds in an antigen–antibody interface of three Fv (antibody fragment consisting of the variable domains of the heavy and light chains) mutants complexed with lysozyme has been examined using ITC.[105] The thermodynamic parameters for a series of mutants indicate a Ka = 2.6×10^7 M^{-1} for a VL-Tyr50Ser mutant (where VL represents the variable light chain portion of the antibody), Ka = 7.0×10^7 M^{-1} for a VH-Tyr32Ala mutant (where VH represents the variable heavy chain portion of the antibody), and Ka = 4.0×10^6 M^{-1} for the VH-Tyr101Phe mutant. All mutants yielded binding constants that were less than the WT complex (Ka = 2.7×10^8 M^{-1}). In each mutant case, entropy compensation was attributed to the affinity losses. This study provided evidence that the complex was considerably tolerant, both structurally and thermodynamically, to the removal of antibody side chains that form hydrogen bonds with the antigen because the absence of these hydrogen bond sites resulted in minimal shifts in the positions of the remaining protein atoms. Furthermore, modified Fv fragments of antilysozyme monoclonal antibodies have been characterized using ITC.[105–109] Both enthalpy- and entropy-driven binding reactions have been observed. X-ray diffraction data at 2.5-Å resolution indicated that conformational changes of several D1.3 (antilysozyme monoclonal antibody) contacting residues located in the complimentary determining regions might explain the entropy-driven binding observed.[106] In another study, thermodynamic characterization of variable domains linked covalently with a flexible linker yielded constructs with reduced activity in comparison to the Fv of the monoclonal antibody (HyHEL10) that was attributed to a loss of entropy upon antigen binding.[107] Mutagenic experiments of the antigen hen egg white lysozyme (HEL) and its binding affinity to HyHEL10 confirmed experimentally that structural perturbations involving the rotation of Trp-62 of HEL contributed to the gain in enthalpic energy upon binding to the antibody, even though it was proposed not to be in direct contact.[108] Tyrosine fragments have been mutated on the HyHEL10 antibody to show that the role of Tyr residues contributed to formation of hydrogen bonds through the hydroxyl group, permitting more favorable interactions through the aromatic ring and less entropic loss upon binding.[109]

   Purification steps involving columns packed with protein A resin are often used to bind and elute proteins possessing IgG Fc domains (i.e., antibody fusion proteins, antibodies). Protein A has five immunoglobulin binding domains, identified as domains A–E. Thermodynamic characterization of the Fc binding and localization of the Fv-binding site to domains of protein A were evaluated by ITC.[110] Analysis of the binding isotherms obtained for the titration of hu4D5 antibody Fab fragment (containing the antigen-binding site) with intact full-length protein A indicated that three to four of the five domains can simultaneously bind to Fab with an approximate binding affinity of 5.5×10^5 M^{-1}. Both D and E domains can functionally bind hu4D5 Fab fragments. Thermodynamic parameters for the titration of the E domain with hu4D5 Fab were found to be N = 1, $K_a = 2.0 \times 10^5$ M^{-1}, ΔG_b = –7.228 kcal/mol, ΔH_b = –7.1 kcal/mol, ΔS_b = –0.4298 cal/mol-°K.

Protein/Carbohydrate

Developing a better understanding of how to manipulate the cellular machinery to engineer the desired carbohydrate composition of cellular proteins is currently being considered.[111] This process utilizes a substrate-based approach in which novel cellular properties such as complex carbohydrates may be engineered by using analogs of small-molecule metabolites instead of relying on the manipulation of enzymes that process such compounds. Because it has already been mentioned that carbohydrates can structurally play a role concerning stability and recognition, such an approach could be quite valuable in the development of more robust biopharmaceuticals.

   Chemokines selectively recruit and activate a variety of cells during inflammation. Chemokines bind glucosaminoglycans (GAGs) on human umbilical vein endothelial cells with affinities in the μmolar range (10^6 M^{-1}): RANTES > MCP-1 > IL-8 > MIP-1α.[112] Isothermal titration calorimetry was used to confirm that this binding depended upon the length of GAG fragment and optimally required both N- and O-sulfation.[112] Suffice to say that GAG–protein interactions regulate myriad physiologic and pathologic processes. Yet our structural understanding of such binding reactions has been found lacking. Heparin is an example of a GAG that interacts with proteins, yet the general structural requirements for protein or peptide–GAG interaction have not been well characterized. Study of the nature of electrostatic interactions between sulfate and carboxylate groups of GAG and basic amino acid residues in protein or peptides (i.e., arginine, lysine) has been carried out using ITC.[113] The results of blocked- and unblocked-charge experiments suggested that arginine-containing peptides bound more tightly to GAGs than the analogous lysine species owing to a high affinity of the guanidino cation to the sulfate anion of GAGs. This interaction was thought to originate from the combination of stronger hydrogen-bonding and exothermic response related to the electrostatic association between the arginine side chain and the glycan.

Close examination of the spacing of arginine and lysine residues as it pertained to heparin-binding sites revealed that binding affinity varied directly with arginine enrichment in a set of peptide sequences tested by ITC.[114] Given the high affinity of peptides of this kind, long stretches of basic amino acids are uncommon in heparin-binding proteins. Protein-binding sites commonly contain single isolated basic amino acids separated by one nonbasic amino acid. Comparisons between heparin (highly sulfated) and heparan (with fewer and greatly spaced sulfate groups) showed that both interact with complimentary peptide side chains that most appropriately reflect the same spacing among basic amino acids.[114] Hence, in contrast to heparin, heparan interacted most tightly with peptides with more widely spaced cationic residues. As a side note regarding the importance of appropriately positioned cationic charge to augment the binding interaction between protein and heparin, it has also been demonstrated that the removal of such charged side chains by replacement with alanine has deleterious consequences on heparin binding.[115] Collectively, this knowledge may be used to modulate the affinity of protein binding to heparin and to design effective peptide inhibitors to act as antagonists to frustrate protein–GAG interactions in some situations.

Serum-type and liver-type mannose-binding proteins (MBP) found in higher animals are composed of a carbohydrate-recognition domain (CRD) and a collagenous domain. These proteins can bind *N*-acetylglucosamine and other related sugars. The substrate preference of CRDs has been investigated and found to show little discrimination among monosaccharide specificities of the CRDs of the two different MBPs.[116] However, there are notable differences for the affinities of natural glycoproteins and mannose-containing cluster glycosides. Synthetic cluster ligands with two terminal GlcNAc moieties have affinities equal to monovalent GlcNac ligands of both CRDs, whereas a series of structurally similar mannose terminal divalent ligands displayed about 20-fold enhanced affinity for the liver-type CRD exclusively. An explanation for this observation is that liver MBP-CRDs have two sugar-binding sites per subunit, one of which only binds mannose, and the other both mannose and *N*-acetylglucosamine. In contrast, the serum MBP-CRD has only one site of the latter type. These findings were confirmed by ITC measurements.

The thermodynamics of the maltose-binding protein of *E. coli* (MalE) have been studied using both DSC and ITC.[117] This protein is a periplasmic component of the transport system for malto-oligosaccharides and is used widely as a carrier protein for the production of recombinant fusion proteins. The MalE protein was shown to exhibit a ΔHcal/ΔHv ratio of 1.3 to 1.5, suggesting that the endotherm of unfolding comprises two strongly interacting thermodynamic domains. Binding of maltose resulted in an increase of the T_m by 8 to 15°C, depending on pH. The binding of maltose to MalE is characterized by very low enthalpy changes (on the order of 1 kcal/mol). Thermal melting of MalE was found to be accompanied by an exceptionally large change in heat capacity, consistent with a large amount of nonpolar surface area, approximately 0.72 Å^2/g protein that becomes accessible to the solvent in the unfolded state. It is worth noting that the high ΔCp determines a very steep ΔG vs. temperature profile for this protein and predicts that cold denaturation should occur above freezing temperatures. In fact, evidence for this was provided by changes in fluorescence intensity upon cooling the protein, showing a sigmoidal cooperative transition with a midpoint near 5°C at low pH conditions.[117] Furthermore, analysis of several fusion proteins containing MalE have illustrated the feasibility of assessing the folding integrity of recombinant products prior to separating them from the MalE carrier protein.

Protein and Lipid

Typical ranges for the binding affinities of proteins with lipids or fatty acids range from 10^3 to 10^6 M^{-1}. This is well within the range of direct detection using ITC. The binding of neuropeptide substance P (SP) agonist to lipid membranes and to neurokinin-1 (NK-1) receptor is one example.[118] The rationale of the study was to correlate the physical-chemical properties of three different SP analogs with lipid-induced conformation and membrane-binding affinity, and with receptor binding and functional activity. Hence, one analog was made more hydrophilic at the C-terminus with an (Arg9)SP or more hydrophobic in the (Nle9)SP form. The third analog was made with a reduced charge at the N-terminal address (AcPro2, Arg9)SP. In solution, all three analogs exhibited random coil conformations as confirmed, using circular dichroism. Addition of SDS micelles and negatively charged vesicles induced partially α-helical structures for (AcPro2, Arg9)SP and (Arg9)SP, but both α-helix and β-sheet structures were manifested for (Nle9)SP. The thermodynamic parameters of lipid binding were determined with monolayer expansion measurements and high-sensitivity titration calorimetry. The apparent binding constants for membranes containing 100% POPG (1-palmitoyl-2-oleoyl-sn-glycero-3-phosphoglycerol) were of the order of 10^3 to 10^5 M^{-1}. Binding was found to involve electrostatic attraction of the cationic peptides to the negatively charged membrane surface. Measurement of binding affinities to the NK-1 receptor and of the *in vitro* activities showed that all three peptides behaved as agonists. The fact that even the highly charged (Arg9)SP had agonistic activity provided evidence that the binding epitope at the receptor was in a hydrophilic environment. The results indicated that a membrane-mediated receptor mechanism was unlikely, but rather provided insight to suggest that the agonist approached the receptor-binding site from the aqueous phase.

Magainins are another example of positively charged amphiphatic peptides that permeabilize cell membranes and display antimicrobial activity. Magainins are known to bind to negatively and neutral charged membranes.[119] The binding properties have been characterized using ITC and are believed to occur by a surface partition equilibrium mechanism. Binding was found to proceed via large exothermic reaction enthalpies ranging from −15 to −18 kcal/mol (at 30°C). Determination of the ΔCp_b by plotting the ΔH_b as a function of temperature (i.e., in the same way illustrated in Figure 13.11) yielded a large positive value of 130 cal/mol-°K. The corresponding ΔG_b was found to be between −6.4 and −8.6 kcal/mol, resulting in a large, less favorable negative ΔS_b. Hence, the binding of magainin to small unilamellar vesicles is an enthalpy-driven reaction. Circular dichroism experiments provided further evidence that the membrane-bound fraction of magainin was approximately 80% helical at 8°C, decreasing to about 60% at 45°C. Knowing that the random coil-to-helix transition in aqueous solution is known to be exothermic, the same process occurring at the membrane surface may account for up to 65% of the measured enthalpy.[119] Hence, membrane-facilitated helix formation is followed by insertion of the nonpolar amino side chains into the lipid bilayer. In the same study it was shown that cholesterol drastically reduced the extent of magainin binding and exhibited enthalpy–entropy compensation.

The effects of pH, ionic strength, and temperature on the thermodynamics of medium-chain acyl-CoA dehydrogenase (MCAD)–octenoyl-CoA interaction have been studied.[120] The binding reaction is characterized by a stoichiometry of 0.89 mol of octenoyl-CoA/mole MCAD subunit. Thermodynamic parameters at 25°C include ΔG_b = −8.75 kcal/mol, ΔH_b = −10.3 kcal/mol, and ΔS_b = −5.3 cal/mol-°K, indicating that the formation of the complex is enthalpy-driven in a solution containing 50 mM phosphate buffer, pH 7.6 and ionic strength of 175 mM. The study showed that although ionic strength did not significantly influence the complex reaction, the pH of the buffer media had a pronounced effect. The ΔCp_b of this complex reaction was found to be −0.37 kcal/mol-°K, indicating that the binding of octenoyl-CoA to MCAD is dominated by hydrophobic forces. However, for some systems like cytochrome c interactions with negatively charged dioleoylphosphatidylglycerol (DOPG), it has been shown that the binding constant and enthalpy of association decrease with increasing ionic strength, where no binding is detected above 0.5 M NaCl.[121] Hence, in some cases, salt can be a powerful modulator of binding activity when electrostatics define the binding mechanism.

It is interesting that modifying solution conditions by adding different concentrations of ethanol can produce a biphasic effect on melting transition temperatures of lipid-like systems (e.g., acyl chains of hydrocarbons). For example, low concentrations of ethanol reduce the T_m of phosphatidylcholine bilayers, whereas higher concentrations increase the T_m of the same system.[122] This effect has been shown to depend upon acyl chain length and can be explained by the fact that interdigitated phases are formed in which there is an energy gain because van der Waals energy is augmented in more closely packed phases of this type. Interdigitated phases are favored with high ethanol concentrations in which the hydrocarbon chains are more highly ordered, resulting in narrow phase transitions.

Lipid systems are of special interest to those working in the area of drug delivery. Liposomes are currently being considered as vehicles for target delivery of biologics and peptides.[123,124] In order to facilitate rational approaches in the development of novel delivery systems of this type, the properties of the liposome and the active drug need to be taken into account to better understand stabilization. Microcalorimetry has been useful in the determination of such interactions between different proteins and their interactions with various lipid systems (i.e., neutral lipid, dipalmitoylphophatidylcholine, DPPC; negatively charged lipid, dipalmitoylphosphatidylglycerol, DPPG). According to Papahajopoulos and coworkers,[125] proteins can be classified into three categories, based upon their thermotropic behavior with a given phospholipid. For example in the first category, proteins that are predominantly hydrophilic (i.e., polylysine and ribonuclease A) have been thought to absorb onto the bilayer surface by simple electrostatic interactions, exerting a stronger effect on the phase transitions of charged rather than zwitterionic phospholipids. The second category is ascribed to proteins that are partially hydrophilic and hydrophobic, able to associate on the surface via electrostatic forces, but that also may become embedded into the bilayer (i.e., cytochrome c, myelin basic protein [A-1]). Finally, the third category is characterized by proteins that can penetrate into the core of anionic or zwitterionic lipid bilayers characterized by strong hydrophobic interactions (i.e., gramicidin A and proteolipid apoprotein [N-2]). More recent investigation of this scheme of classification has revealed that hydrophilic proteins do not bind to liposomes exclusively at the surface by electrostatic interactions, and some degree of penetration is observed in most cases.[124] Moreover, superoxide dismutase was found to unexpectedly bind DPPG and, as a result, was thought to protect lipid membranes against oxygen-mediated injury.

Recombinant human granulocyte colony-stimulating factor (rhG-CSF) interacts with liposomes composed of the anionic phospholipid DOPG, and this interaction enhances the stability of the protein.[126] Recombinant hG-CSF inserts into bilayers of anionic, but not zwitterionic phospholipids, making this system conform to category 2 of the Papahajopoulos scheme. Isothermal titration analysis of this protein with dimyristoylphosphatidylglycerol (DMPG) at 25°C indicated that the binding was saturable, involving 10 lipids/rhG-CSF.[127] This titration experiment is endothermic, having a $Ka = 4.3 \times 10^5 \, M^{-1}$, $\Delta H_b = 5.42$ kcal/mol, $\Delta G_b = -7.69$ kcal/mol, and a $\Delta S_b = 44$ cal/mol-°K. The large favorable ΔS_b indicates that this reaction is entropically driven. In the same study it was shown that the stabilization of rhG-CSF by anionic phospholipids required a chain length that was $\geq C_{10}$. Moreover, the stabilization of other growth factors that were structurally similar to rhG-CSF (i.e., recombinant porcine somatotropin, recombinant human interleukin 4, recombinant human interleukin 2, recombinant human granulocyte-macrophage CSF) were stabilized by the anionic phospholipid DMPG. This study showed that a group of structurally ˙similar proteins can interact preferentially with anionic phospholipids and that the complexation of the growth factors with vesicles composed of anionic phospholipids improves the stability of the proteins under conditions where they normally denature.

The nature of lipid structure variation and its influence on the stability of integral membrane proteins has been investigated by microcalorimetry. For example, in the structural stability of erythrocyte anion transporter, band 3 has been studied in different lipid environments.[125] This protein system falls into the third category defined by Papahajopoulos and coworkers. The data revealed that the stability of the 55-kDa membrane-spanning domain of band 3 was exquisitely sensitive to the acyl chain length of its phospholipid environment, increasing almost linearly from a T_m of 47°C in DPPC (C14:1) to 66°C in dinervonylphosphatidylcholine (C24:1). It was shown that although band 3 was native in all reconstituted lipid systems, the transport protein's stability was found to be much greater in zwitterionic lipids (i.e., phosphatidylethanolamine, phosphatidylcholine). In a subsequent study of this system, fatty acids that most effectively stabilized band 3 also yielded the highest affinity for the transporter protein.[128]

PRESSURE PERTURBATION CALORIMETRY (PPC)

The PPC method is a relatively new technique that is based upon the measurement of volumetric properties of proteins in dilute solutions. This technique, developed by MicroCal, LLC (Northampton, MA), measures the heat change (ΔQ) that results from a change in pressure (ΔP) above a solution containing dissolved protein. The total fill volume of the sample (containing the protein analyte) and reference (identical buffer solution) cells have identical volumes of 0.5 ml. The experimental design involves access of both cells to a common pressure chamber (containing a pressure sensor that transmits data to the computer). Pressure in the chamber alternates between P1 and P2 (Figure 13.12), operating at selected pressures ranging from 0 to 5 atm. The calorimetric baseline is first allowed to equilibrate before a pressure P1 is applied. The excess pressure is then changed to P2, causing heat to be absorbed in both cells. Because the solutions in the sample and reference cells are identical except for the small amount of dissolved protein in the sample cell, counterbalanced by the corresponding volume of buffer in the reference cell, differential heats are quite small. The resulting peaks of compression and decompression are of identical size but of opposite sign. Integration of each peak provides the heat change between the reference and sample cells. Experiments can be carried out via computer control to automatically obtain data at numerous temperatures. It can be shown that ΔQ is a function of the coefficient of thermal expansion, α.[129]

What is obtained by performing such experiments across an appropriate temperature range is the precise measurement of the coefficient of thermal expansion for protein partial specific volumes that may be detected at concentrations

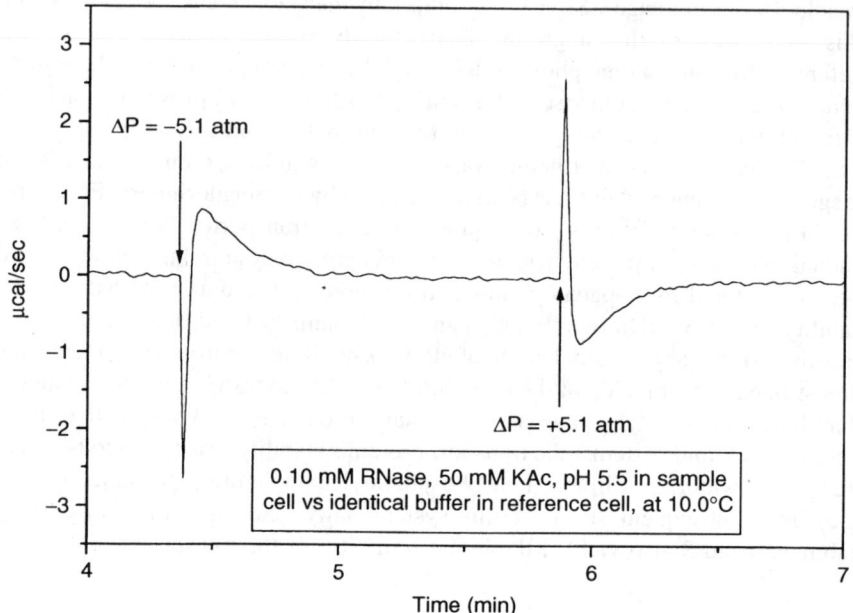

Figure 13.12 Pressure perturbation calorimetry (PPC) experiment showing alternating pressure pulses of 5.1 and +5.1 atm as a function of time. The reaction mixture comprises 0.10 m*M* RNase A, 50 m*M* potassium acetate (pH 5.5) in the sample cell with identical buffer in the reference cell. The experiment is carried out at 10°C. (Permission to use the figure granted by MicroCal, LLC.)

as low as 0.25%. The partial specific volume of a protein in solution is equal to the sum of the intrinsic volume of the protein plus the change in the volume of solvent resulting from its interaction with the accessible protein side chains. Solutes that favor an increase in water structure are expected to have a small positive (or even negative) α at low temperature with a correspondingly large positive coefficient as the temperature is raised.[130] The converse is true for solutes that decrease water structure (i.e., hydrophilic and salt compounds). The nature of the solvation contribution to α can therefore be assessed for a given class of amino acids using this technique. For example, the aliphatic side chains, alanine, valine, and isoleucine, are similar in that their values are negative at low temperature, with a strong positive slope and negative curvature as temperature increases (Figure 13.13). Because water tends to be arranged in an orderly fashion around apolar surfaces (as already mentioned, this is accommodated by a decrease in entropy), such side chains may be considered "structure makers." In contrast, aromatic side chains like tryptophan and phenylalanine are positive and nearly independent of temperature. By comparison, polar side chains such as asparagine,

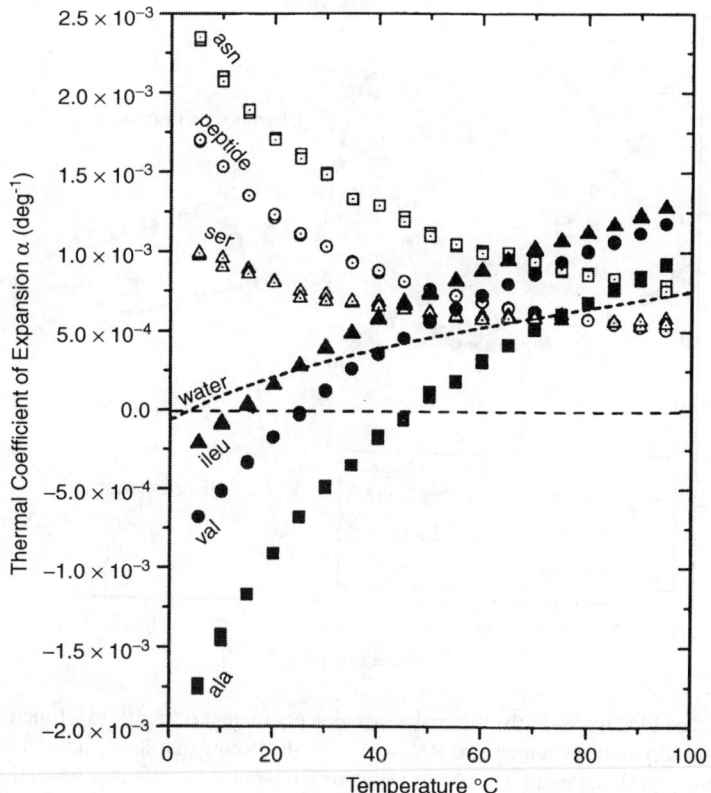

Figure 13.13 A plot showing the behavior of the thermal coefficient of expansion α (deg⁻¹) for different amino acids, peptide, and water as a function of temperature measured by PPC. The dashed curve displays the estimated progress baseline for the pre- and posttransition region. (Permission to use the figure granted by MicroCal, LLC.)

serine, and other highly polar peptide groups exhibit highly positive values at low temperature with a characteristic negative slope and positive curvature with increasing temperature (Figure 13.13). Such behavior is consistent with the "structure-breaking" properties of solvated water in these cases, given the tendency for water to adopt a more disordered structure (entropy is favored). This is also the case with other hydrophilic side chains and various electrolytes.

Globular proteins typically have their hydrophobic side chains buried to remove them from the polar water solvent. Hence, a proportionately greater hydrophilic surface is in contact with the solvent at a given instant. Proteins like ribonuclease A and chymotrypsinogen exhibit decreasing α values with increasing temperature that is consistent with "structure breakers" (Figure 13.14). Generally,

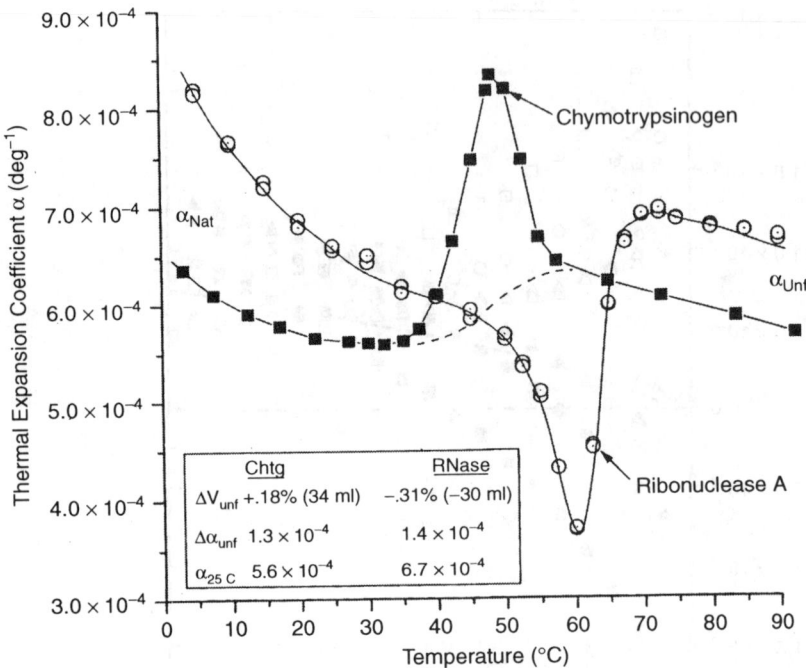

Figure 13.14 The PPC traces of the thermal expansion coefficient α (deg^{-1}) as a function of temperature for chymotrypsinogen and RNase A. The data show that both native (low-temperature region) proteins exhibit a strong negative temperature coefficient as well as a large positive curvature. (Permission to use the figure granted by MicroCal, LLC.)

when the protein unfolds at high temperature, the solvent-accessible surface area (ASA) will increase and so will α (as in the case of chymotrypsinogen). Computational studies have been performed on the mutant stability of T4 lysozyme, and they correlate with the surface thermal expansion of the protein.[131] In contrast to the positive change in the volume of unfolding (ΔV_{unf}, obtained by integrating over the temperature range where the transition occurs) concerning chymotrypsinogen, ribonuclease A exhibits a negative change in unfolding volume (Figure 13.14). The correlation of such data may one day lead to the ability to predict the stability outcome of a given engineered biologic based upon its thermal expansion properties within a given formulation environment. Alternatively, it might be possible to use such information to modify specific amino acid sites that confer greater stability in a given biologic. The combination of PPC data with improved computational approaches could facilitate a more detailed understanding about the effects of thermal behavior and the prediction of protein stability in different solution environments.

CHALLENGES

Thermodynamic–Kinetic Relationships

The study of reaction rates or kinetics of a particular denaturation process of a protein therapeutic can provide valuable information about the mechanism, i.e., the sequence of steps that occur in the transformation of the protein to chemically or conformationally denatured products. The kinetics tell something about the manner in which the rate is influenced by such factors as concentration, temperature, excipients, and the nature of the solvent as it pertains to properties of protein stability. The principal application of this information in the biopharmaceutical setting is to predict how long a given biologic will remain adequately stable.

Applying kinetics to denaturation studies must take into account the thermodynamic properties associated with the energetics of the reaction process. Thermodynamic properties are ascribed to the equilibrium between the native and the denatured states. For example, M-2 glycoprotein does not bind thyroxine in its native state, but when heat denatured, acquires thyroxine-binding properties with an associated ΔH_b of 1.5 kcal/mol.[132] Likewise, the reaction rate will be affected by the proportion and extent of thermally unfolded forms that may participate in the binding reaction. The concept that protein conformational changes are related to the energetics and mechanism of reactions was suggested by Rufus Lumry and Henry Eyring in 1954.[133] They reasoned that because the native state could be perturbed by any large change in solvent composition, temperature, or pressure producing alterations attributed to conformational isomers that made up a denatured ensemble (of relatively higher energy states), complex kinetics would ensue. Therefore, a reaction originating from an altered or more unfolded conformation would in fact emanate not just from one major species, but from many. Reaction paths of lower free energy would be available for selected conformations with emphasis on the participation of one or a few forms. The overall reaction, however, would be the sum of the processes that originated from the native state (or a number of closely related states) and end at another group of closely related states. This situation could be taken into account by summing the rate of reactant loss, $-dRi/dt$, over all reaction paths i and thus over all activated complexes of concentration Ci^*. They had noted that Arrhenius plots of protein "transconformation" and enzymatic reactions were frequently nonlinear, as expected when several simultaneous processes produced the same product. They surmised that this resulted from different activation energies or a changing distribution of species with temperature, and concluded that the kinetic products formed from heat-denatured protein must depend on the speed of the initial reaction, no matter whether the intermediate product formed in equilibrium or as a steady-state intermediate.

Arrhenius plots permit the determination of activation energies (Ea) associated with a particular pathway of degradation that allows one to estimate reaction rates as a function of temperature. Such information, if demonstrated to accurately model the reaction, can make it possible to predict longevity of biopharmaceutical products stored at real-time conditions. To be useful, it is important to factor in the consequences of protein unfolding as it affects *Ea*. In fact, there is evidence of change in reaction rates at different temperatures and different amounts of unfolding. An example is described by the unusual temperature dependence of ovalbumin at the first stage of urea denaturation.[134,135] At low temperature, the rate constant decreases sharply as temperature increases with an apparent *Ea* of about −28 kcal/mol at 0°C. Near 20°C, the rate constant becomes independent of temperature so that *Ea* is zero. Finally, at higher temperatures the rate constant increases with temperature and *Ea* approaches 50 kcal/mol. In the absence of urea the *Ea* is 130 kcal/mol. In another example, IL-1R (type II) aggregation has been recently shown to exhibit two different activation energies across a temperature regime spanning from 30 to 75°C. What was interesting about the discontinuity in the two activation energies was that it happened to occur in the vicinity of the T_m for this molecule.[136] This latter result provided compelling evidence that the aggregation pathway depended on the unfolding of the IL-1R (II) molecule.

Microcalorimetry can be used to determine activation energies of denaturation reactions. In general, scan rates of 60°C/h are considered to be adequate for simple, reversible unfolding transitions in which the melting transition should be unaffected by different scan rates.[137] However, in some cases, kinetically controlled irreversible processes can occur (i.e., aggregation, or chemical degradation at elevated temperatures) that can affect the onset slope of the endotherm describing the reaction, causing the T_m to shift in a scan rate-dependent manner. In such situations, a change in the apparent T_m behavior (because the true melting temperature in this instance is not precisely known) as a function of different scan rates using a modified form of the Arrhenius equation can permit determination of the *Ea* of the reaction.[138] A plot of ln (v/T_m^2) as a function of $1/T$ results in a straight line with slope of Ea/R (where v is the scan rate in units of °K/min). The validity of this approach to determining activation energies has been demonstrated for the thermal denaturation of thermolysin[138] and carboxypeptidase B.[139] In both cases this treatment was applied to systems that were shown to be completely irreversible. An example of such a plot for the case of thermolysin is illustrated in Figure 13.15. A tri-state model was used to describe the experimental behavior, shown by the scheme below.

$$N \underset{k_{-1}}{\overset{k_1}{\rightleftharpoons}} D \overset{k_2}{\longrightarrow} D_I$$

Here, in association with the dynamics of the native (N) to denatured (unfolded, D) state, there are associated rate constants for the forward and reverse reactions. In some cases of thermal denaturation, the forward reaction rate denoted by k_1 is much greater than the reverse reaction (k_{-1}) associated with refolding and $k_2 \gg k_1$.[138–140] When this is true (i.e., completely irreversible process), the scheme above may be reduced to the following:

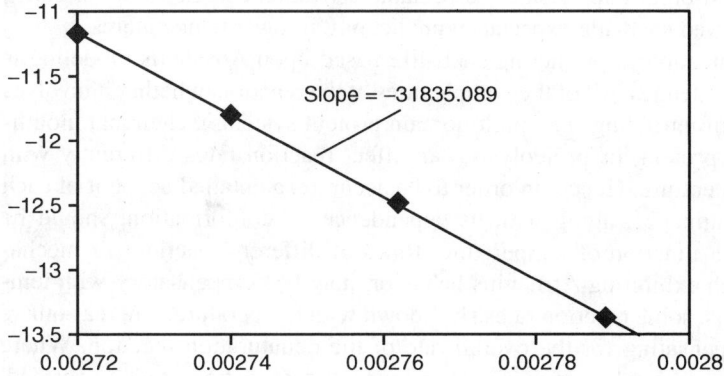

Figure 13.15 A plot of ln (v/T_m^2) vs. $1/T_m$ for the thermal denaturation of thermolysin using DSC scan rates of 1.9, 1.0, 0.5, and 0.2 °K/min. The plot assumes Arrhenius behavior holds. The slope of the plot is equal to $-Ea/R$. Using R = 1.9872 cal/deg-mol, one can calculate the activation energy to be ~ 63.3 kcal/mol. (Data plotted were obtained from Reference 138.)

$$N \xrightarrow{k'} D_I$$

In this approach, the rate of the reaction is kinetically controlled and k' is the rate constant for the formation of the irreversibly denatured product, D_I.

It should be noted that the unfolding kinetics can sometimes involve quite complex unfolding schemes of different substates in equilibrium with the native state. Staphylococcal nuclease is an example of such behavior, known to unfold with three different substates that exhibit an equilibrium that does not appear to shift with temperature.[49] Irreversible aggregation processes of proteins have been known to involve first- or second-order reactions.[132,141] The mechanism of recombinant human interferon-γ aggregation is an example where thermodynamic and kinetic aspects of the reaction provided a powerful tool for understanding the pathway of instability and permitted a rationale for screening excipients that inhibited the process.[141]

The stability of a biological is a key aspect of quality, and regulatory authorities recognize that useful-life claims need to be realistic and demonstrable. A shelf-life is tentatively assigned on the basis of accelerated stability data. When predicting a shelf-life from accelerated storage conditions, a reaction order is determined with an associated rate constant and activation energy. This information may then be used to predict the rate of decomposition at low temperatures where the biopharmaceutical is to be stored in real time.[142] Arrhenius kinetics have been applied to other forms of instability with some success toward predicting shelf-life, e.g., the chemical decomposition of human epidermal growth factor (hEGF) by oxidation, deamidation, and succinimide processes.[89] Arrhenius plots of the apparent first-order rates yielded respectable activation energies that correlated well with observed shelf-life expectancy predictions at lower temperatures.

The application to predicting shelf-life based upon Arrhenius modeling is fraught with challenges. All of the complexities of the reaction, whether it involves intermediates of unfolding (i.e., multidomain protein systems), chemical modifications of the protein, or proteolysis, can affect reaction rates differently with respect to temperature. Hence, in order to be accurate, a detailed account of each destabilizing influence, along with its dependence on conformation, should be considered as a function of temperature. Rates of different reactions or mechanisms, although exhibiting Arrhenius behavior, may be compensatory with temperature. That is, some reaction rates slow down with temperature, whereas others increase, compensating for the overall rate of the denaturation reaction. Where conformation plays a role in the amount and type of degradation products formed, attention should be given to ascertaining the degree of unfolding acquired at low-temperature conditions. Sometimes very little unfolding is required to propagate an aggregation product.[49–51,141,143] In the case of proteins composed of several cooperative unfolding domains, it may be necessary to characterize the influence of each domain as it contributes to the overall reaction process.

There are four underlying assumptions regarding Arrhenius kinetics. First, that the concentration dependence (assignable to the order of the reaction) associated with the kinetics of the reaction can be described in terms of a single mechanism of instability (i.e., aggregation, hydrolysis, proteolysis, deamidation) without influence of competing reactions. Second, that the predominant mechanisms at elevated temperatures are the same at lower temperatures. Third, reaction by-products have little influence on the reaction rate. Fourth, accurate control of the temperature is achieved. Regarding this point, the determination of Ea using microcalorimetry has an advantage over conventional reaction studies with incubators, or temperature baths, because of the greater insulation and lower temperature flux associated with such instruments. Finally, it is assumed that solution conditions remain constant for the duration of the experiment (i.e., pH, ionic strength) and correctly represent kinetic behavior at accelerated conditions. The best way to understand the role of the underlying assumptions is to carry out kinetic studies at several accelerated temperature conditions (being mindful of the T_m) and compare the rates to those stored at real-time conditions (using Ea to estimate performance). It is anticipated that microcalorimetry will play an integral role in the accumulation of such data, providing knowledge of the predominant mechanisms involved and adding significance to the Arrhenius models used for making shelf-life predictions acceptable. Implied in this endeavor is the development of good stability-indicating assays that can support and quantitatively characterize protein reactions of instability.

High-Throughput Methodologies

Another area of development that could greatly improve the ability to rapidly derive stable liquid formulation candidates is the development of high-throughput microcalorimetry instruments. As already mentioned, this has been a limitation that does not lend itself to screening multiple excipients in an efficient manner. The availability of such instruments will make it possible to not only determine which excipients to include in a given formulation, but can also be used to determine optimization of excipient concentrations as they affect stability. This challenge has been undertaken by MicroCal, LLC, where a high-throughput capillary-DSC instrument has been developed with the capability of performing up to 50 scans during 24 h of unattended operation. Automated instruments of this kind are now readily available.

References

1. Jiang, X., J. Kowalski, and J.W. Kelly. 2001. Increasing protein stability using a rational approach combining sequence homology and structural alignment: stabilizing the WW domain. *Protein Sci* 10: 1454–1465.
2. van Gunsteren, W.F. and A.E. Mark. 1992. Prediction of the activity and stability effects of site-directed mutagenesis on a protein core. *J Mol Biol* 227: 389–395.
3. Bartolucci, S., A. Guagliardi, E. Pedone, D. De Pascale, R. Cannio, L. Camardella, M. Rossi, G. Nicastro, C. de Chiara, P. Facci, G. Mascetti, and C. Nicolini. 1997. Thioredoxin from *Bacillus acidocaldarius*: characterization, high-level expression in *Escherichia coli* and molecular modelling. *Biochem J* 328: 277–285.
4. Bombardier, H., P. Wong, and C. Gicquaud. 1997. Effects of nucleotides on the denaturation of F actin: a differential scanning calorimetry and FTIR spectroscopy study. *Biochem Biophys Res Commun* 236: 798–803.
5. Boye, J.I., A.A. Ismail, and I. Alli. 1996. Effects of physicochemical factors on the secondary structure of beta-lactoglobulin. *J Dairy Res* 63: 97–109.
6. Brass, O., J.M. Letoffe, A. Bakkali, J.C. Bureau, C. Corot, and P. Claudy. 1995. Involvement of protein solvation in the interaction between a contrast medium (iopamidol) and fibrinogen or lysozyme. *Biophys Chem* 54: 83–94.
7. Waldner, J.C., S.J. Lahr, M.H. Edgell, and G.J. Pielak. 1999. Nonideality and protein thermal denaturation. *Biopolymers* 49: 471–479.
8. Welfle, K., R. Misselwitz, G. Hausdorf, W. Hohne, and H. Welfle. 1999. Conformation, pH-induced conformational changes, and thermal unfolding of anti-p24 (HIV-1) monoclonal antibody CB4-1 and its Fab and Fc fragments. *Biochim Biophys Acta* 1431: 120–131.
9. Zaiss, K. and R. Jaenicke. 1999. Thermodynamic study of phosphoglycerate kinase from *Thermotoga maritima* and its isolated domains: reversible thermal unfolding monitored by differential scanning calorimetry and circular dichroism spectroscopy. *Biochemistry* 38: 4633–4639.
10. Chen, B.L., T. Arakawa, C.F. Morris, W.C. Kenney, C.M. Wells, and C.G. Pitt. 1994. Aggregation pathway of recombinant human keratinocyte growth factor and its stabilization. *Pharm Res* 11: 1581–1587.
11. De Young, L.R., K.A. Dill, and A.L. Fink. 1993. Aggregation and denaturation of apomyogiobin in aqueous urea solutions. *Biochemistry* 32: 3877–3886.
12. Mitraki, A., J.M. Betton, M. Desmadril, and J.M. Yon. 1987. Quasi-irreversibility in the unfolding-refolding transition of phosphoglycerate kinase induced by guanidine hydrochloride. *Eur J Biochem* 163: 29–34.

13. Powell, M.F. 1994. In *Formulation and Delivery of Proteins and Peptides*. J.L. Cleland and R. Langer, editors. American Chemical Society, Washington, D.C., 100–117.

14. Remmele, R.L., Jr. and W.R. Gombotz. 2000. Differential scanning calorimetry: a practical tool for elucidating stability of liquid biopharmaceuticals. *Biopharm* 13: 36–46.

15. Wearne, S.J. and T.E. Creighton. 1989. Effect of protein conformation on rate of deamidation: ribonuclease A. *Proteins* 5: 8–12.

16. Plotnikov, V.V., J.M. Brandts, L.N. Lin, and J.F. Brandts. 1997. A new ultrasensitive scanning calorimeter. *Anal Biochem* 250: 237–244.

17. Brandts, J.F. 1967. Heat effects on proteins and enzymes. In *Thermobiology*. A.H. Rose, editor. Academic Press, New York, 25–75.

18. Pace, C.N., B.A. Shirley, M. McNutt, and K. Gajiwala. 1996. Forces contributing to the conformational stability of proteins. *Faseb J* 10: 75–83.

19. Tanford, C. 1973. In *The Hydrophobic Effect: Formation of Micelles and Biological Membranes*. John Wiley & Sons, New York.

20. Kauzmann, W. 1959. Some factors in the interpretation of protein denaturation. *Adv Protein Chem* 14: 1–64.

21. Privalov, P.L. and S.J. Gill. 1988. Stability of protein structure and hydrophobic interaction. *Adv Protein Chem* 39: 191–234.

22. Sturtevant, J.M. 1987. Biochemical applications of differential scanning calorimetry. *Ann Rev Phys Chem* 38: 463–488.

23. Chan, H.K., K.L. Au-Yeung, and I. Gonda. 1996. Effects of additives on heat denaturation of rhDNase in solutions. *Pharm Res* 13: 756–761.

24. Chang, B.S., C.S. Randall, and Y.S. Lee. 1993. Stabilization of lyophilized porcine pancreatic elastase. *Pharm Res* 10: 1478–1483.

25. Lee, L.L. and J.C. Lee. 1987. Thermal stability of proteins in the presence of poly(ethylene glycols). *Biochemistry* 26: 7813–7819.

26. Liggins, J.R., F. Sherman, A.J. Mathews, and B.T. Nall. 1994. Differential scanning calorimetric study of the thermal unfolding transitions of yeast iso-1 and iso-2 cytochromes c and three composite isozymes. *Biochemistry* 33: 9209–9219.

27. Maneri, L.R. and P.S. Low. 1988. Structural stability of the erythrocyte anion transporter, band 3, in different lipid environments. A differential scanning calorimetric study. *J Biol Chem* 263: 16170–16178.

28. Davio, S.R., K.M. Kienle, and B.E. Collins. 1995. Interdomain interactions in the chimeric protein toxin sCD4(178)-PE40: a differential scanning calorimetry (DSC) study. *Pharm Res* 12: 642–648.

29. Martinez, J.C., M. el Harrous, V.V. Filimonov, P.L. Mateo, and A.R. Fersht. 1994. A calorimetric study of the thermal stability of barnase and its interaction with 3′GMP. *Biochemistry* 33: 3919–3926.

30. North, M.J. 1993. Prevention of unwanted proteolysis. In *Proteolytic Enzymes: a Practical Approach*. R.J. Beynon and J.S. Bond, editors. IRL Press, Oxford,

31. Remmele, R.L., Jr., N.S. Nightlinger, S. Srinivasan, and W.R. Gombotz. 1998. Interleukin-1 receptor (IL-1R) liquid formulation development using differential scanning calorimetry. *Pharm Res* 15: 200–208.

32. Schrier, J.A., R.A. Kenley, R. Williams, R.J. Corcoran, Y. Kim, R.P. Northey, Jr., D. D'Augusta, and M. Huberty. 1993. Degradation pathways for recombinant human macrophage colony-stimulating factor in aqueous solution. *Pharm Res* 10: 933–944.

33. Beldarrain, A., J.L. Lopez-Lacomba, G. Furrazola, D. Barberia, and M. Cortijo. 1999. Thermal denaturation of human gamma-interferon. A calorimetric and spectroscopic study. *Biochemistry* 38: 7865–7873.

34. Blaber, S.I., J.F. Culajay, A. Khurana, and M. Blaber. 1999. Reversible thermal denaturation of human FGF-1 induced by low concentrations of guanidine hydrochloride. *Biophys J* 77: 470–477.

35. Christensen, T., B. Svensson, and B.W. Sigurskjold. 1999. Thermodynamics of reversible and irreversible unfolding and domain interactions of glucoamylase from *Aspergillus niger* studied by differential scanning and isothermal titration calorimetry. *Biochemistry* 38: 6300–6310.

36. Medved, L.V., M. Migliorini, I. Mikhailenko, L.G. Barrientos, M. Llinas, and D.K. Strickland. 1999. Domain organization of the 39-kDa receptor-associated protein. *J Biol Chem* 274: 717–727.

37. Sturtevant, J.M. 1980. Differential scanning calorimetry: processes involving proteins. In *Bioenergetics and Thermodynamics: Model Systems*. A. Braibanti, editor. John Wiley & Sons, New York, 391–396.

38. Privalov, P.L., N.N. Khechinashvili, and B.P. Atanasov. 1971. Thermodynamic analysis of thermal transitions in globular proteins. I. Calorimetric study of chymotrypsinogen, ribonuclease and myoglobin. *Biopolymers* 10: 1865–1890.

39. Privalov, P.L. 1979. Stability of proteins: small globular proteins. *Adv Protein Chem* 33: 167–241.

40. Hu, C.Q. and J.M. Sturtevant. 1987. Thermodynamic study of yeast phosphoglycerate kinase. *Biochemistry* 26: 178–182.

41. Morris, A.E., R.L. Remmele, Jr., R. Klinke, B.M. Macduff, W.C. Fanslow, and R.J. Armitage. 1999. Incorporation of an isoleucine zipper motif enhances the biological activity of soluble CD40L (CD154). *J Biol Chem* 274: 418–423.

42. Steadman, B.L., P.A. Trautman, E.Q. Lawson, M.J. Raymond, D.A. Mood, J.A. Thomson, and C.R. Middaugh. 1989. A differential scanning calorimetric study of the bovine lens crystallins. *Biochemistry* 28: 9653–9658.

43. Lepock, J.R., K.P. Ritchie, M.C. Kolios, A.M. Rodahl, K.A. Heinz, and J. Kruuv. 1992. Influence of transition rates and scan rate on kinetic simulations of differential scanning calorimetry profiles of reversible and irreversible protein denaturation. *Biochemistry* 31: 12706–12712.

44. Vogl, T., C. Jatzke, H.J. Hinz, J. Benz, and R. Huber. 1997. Thermodynamic stability of annexin V E17G: equilibrium parameters from an irreversible unfolding reaction. *Biochemistry* 36: 1657–1668.

45. Livingstone, J.R., R.S. Spolar, and M.T. Record, Jr. 1991. Contribution to the thermodynamics of protein folding from the reduction in water-accessible nonpolar surface area. *Biochemistry* 30: 4237–4244.

46. Kholodenko, V. and E. Freire. 1999. A simple method to measure the absolute heat capacity of proteins. *Anal Biochem* 270: 336–338.

47. Ganesh, C., N. Eswar, S. Srivastava, C. Ramakrishnan, and R. Varadarajan. 1999. Prediction of the maximal stability temperature of monomeric globular proteins solely from amino acid sequence. *FEBS Lett* 454: 31–36.

48. Privalov, P.L. and N.N. Khechinashvili. 1974. A thermodynamic approach to the problem of stabilization of globular protein structure: a calorimetric study. *J Mol Biol* 86: 665–684.

49. Chen, H.M., J.L. You, V.S. Markin, and T.Y. Tsong. 1991. Kinetic analysis of the acid and the alkaline unfolded states of staphylococcal nuclease. *J Mol Biol* 220: 771–778.

50. Shortle, D. 1996. The denatured state (the other half of the folding equation) and its role in protein stability. *Faseb J* 10: 27–34.

51. Uversky, V.N., A.S. Karnoup, D.J. Segel, S. Seshadri, S. Doniach, and A.L. Fink. 1998. Anion-induced folding of Staphylococcal nuclease: characterization of multiple equilibrium partially folded intermediates. *J Mol Biol* 278: 879–894.

52. Privalov, P.L. and S.A. Potekhin. 1986. Scanning microcalorimetry in studying temperature-induced changes in proteins. *Methods Enzymol* 131: 4–51.

53. Jelesarov, I. and H.R. Bosshard. 1999. Isothermal titration calorimetry and differential scanning calorimetry as complementary tools to investigate the energetics of biomolecular recognition. *J Mol Recognit* 12: 3–18.

54. Remmele, R.L., Jr., S.D. Bhat, D.H. Phan, and W.R. Gombotz. 1999. Minimization of recombinant human Flt3 ligand aggregation at the T_m plateau: a matter of thermal reversibility. *Biochemistry* 38: 5241–5247.

55. Narhi, L.O., J.S. Philo, B. Sun, B.S. Chang, and T. Arakawa. 1999. Reversibility of heat-induced denaturation of the recombinant human megakaryocyte growth and development factor. *Pharm Res* 16: 799–807.

56. Chen, B.L. and T. Arakawa. 1996. Stabilization of recombinant human keratinocyte growth factor by osmolytes and salts. *J Pharm Sci* 85: 419–426.

57. Lee, J.C. and S.N. Timasheff. 1981. The stabilization of proteins by sucrose. *J Biol Chem* 256: 7193–7201.

58. Sola-Penna, M., A. Ferreira-Pereira, A.P. Lemos, and J.R. Meyer-Fernandes. 1997. Carbohydrate protection of enzyme structure and function against guanidinium chloride treatment depends on the nature of carbohydrate and enzyme. *Eur J Biochem* 248: 24–29.

59. Lam, X.M., T.W. Patapoff, and T.H. Nguyen. 1997. The effect of benzyl alcohol on recombinant human interferon-gamma. *Pharm Res* 14: 725–729.

60. Kolvenbach, C.G., L.O. Narhi, J.S. Philo, T. Li, M. Zhang, and T. Arakawa. 1997. Granulocyte-colony stimulating factor maintains a thermally stable, compact, partially folded structure at pH2. *J Pept Res* 50: 310–318.

61. Makarov, A.A., I.I. Protasevich, E.G. Frank, I.B. Grishina, I.A. Bolotina, and N.G. Esipova. 1991. The number of cooperative regions (energetic domains) in a pepsin molecule depends on the pH of the medium. *Biochim Biophys Acta* 1078: 283–288.

62. Makarov, A.A., I.I. Protasevich, N.P. Bazhulina, and N.G. Esipova. 1995. Heat denaturation of pepsinogen in a water-ethanol mixture. *FEBS Lett* 357: 58–61.

63. Lee, J.C. 2000. Biopharmaceutical formulation. *Curr Opin Biotechnol* 11: 81–84.

64. Poklar, N., N. Petrovcic, M. Oblak, and G. Vesnaver. 1999. Thermodynamic stability of ribonuclease A in alkylurea solutions and preferential solvation changes accompanying its thermal denaturation: a calorimetric and spectroscopic study. *Protein Sci* 8: 832–840.

65. Bam, N.B., J.L. Cleland, J. Yang, M.C. Manning, J.F. Carpenter, R.F. Kelley, and T.W. Randolph. 1998. Tween protects recombinant human growth hormone against agitation-induced damage via hydrophobic interactions. *J Pharm Sci* 87: 1554–1559.

66. Baudys, M., T. Uchio, D. Mix, D. Wilson, and S.W. Kim. 1995. Physical stabilization of insulin by glycosylation. *J Pharm Sci* 84: 28–33.

67. Yan, B., W. Zhang, J. Ding, and P. Gao. 1999. Sequence pattern for the occurrence of N-glycosylation in proteins. *J Protein Chem* 18: 511–521.
68. Ghirlando, R., J. Lund, M. Goodall, and R. Jefferis. 1999. Glycosylation of human IgG-Fc: influences on structure revealed by differential scanning micro-calorimetry. *Immunol Lett* 68: 47–52.
69. Tischenko, V.M., V.M. Abramov, and V.P. Zav'yalov. 1998. Investigation of the cooperative structure of Fc fragments from myeloma immunoglobulin G. *Biochemistry* 37: 5576–5581.
70. Wang, C., M. Eufemi, C. Turano, and A. Giartosio. 1996. Influence of the carbohydrate moiety on the stability of glycoproteins. *Biochemistry* 35: 7299–7307.
71. Norde, W. and C.A. Haynes. 1995. Reversibility and the mechanism of protein adsorption. In *Proteins at Interfaces II: Fundamentals and Applications.* T.A. Horbett and J.L. Brash, editors. American Chemical Society, Washington, D.C., 26–40.
72. Volkin, D.B., A.M. Verticelli, M.W. Bruner, K.E. Marfia, P.K. Tsai, M.K. Sardana, and C.R. Middaugh. 1995. Deamidation of polyanion-stabilized acidic fibroblast growth factor. *J Pharm Sci* 84: 7–11.
73. Williams, R.M., C.J. Brown, T. Soule, J.A. Foster, and A.K. Dunker. 2001. A scale of amino acid propensities for ordered and disordered structure. *Protein Sci* 10: 171–172.
74. Wrabl, J.O., S.A. Larson, and V.J. Hilser. 2001. Thermodynamic propensities of amino acids in the native state ensemble: implications for fold recognition. *Protein Sci* 10: 1032–1045.
75. Hilser, V.J. and E. Freire. 1996. Structure-based calculation of the equilibrium folding pathway of proteins. Correlation with hydrogen exchange protection factors. *J Mol Biol* 262: 756–772.
76. Reva, B. and S. Topiol. 2000. Recognition of protein structure: determining the relative energetic contributions of β-strands, α-helices and loops. *Pac Symp Biocomput* 5: 165–175.
77. Makhatadze, G.I., G.M. Clore, A.M. Gronenborn, and P.L. Privalov. 1994. Thermodynamics of unfolding of the all beta-sheet protein interleukin-1 beta. *Biochemistry* 33: 9327–9332.
78. Taylor, J.W., N.J. Greenfield, B. Wu, and P.L. Privalov. 1999. A calorimetric study of the folding-unfolding of an alpha-helix with covalently closed N- and C-terminal loops. *J Mol Biol* 291: 965–976.
79. Gursky, O. and D. Atkinson. 1996. High- and low-temperature unfolding of human high-density apolipoprotein A-2. *Protein Sci* 5: 1874–1882.
80. O'Brien, R., J.M. Sturtevant, J. Wrabl, M.E. Holtzer, and A. Holtzer. 1996. A scanning calorimetric study of unfolding equilibria in homodimeric chicken gizzard tropomyosins. *Biophys J* 70: 2403–2407.
81. Yu, Y., O.D. Monera, R.S. Hodges, and P.L. Privalov. 1996. Investigation of electrostatic interactions in two-stranded coiled-coils through residue shuffling. *Biophys Chem* 59: 299–314.
82. Munson, M., S. Balasubramanian, K.G. Fleming, A.D. Nagi, R. O'Brien, J.M. Sturtevant, and L. Regan. 1996. What makes a protein a protein? Hydrophobic core designs that specify stability and structural properties. *Protein Sci* 5: 1584–1593.
83. Barlow, D.J. and P.L. Poole. 1987. The hydration of protein secondary structures. *FEBS Lett* 213: 423–427.

84. Chan, M.K., S. Mukund, A. Kletzin, M.W. Adams, and D.C. Rees. 1995. Structure of a hyperthermophilic tungstopterin enzyme, aldehyde ferredoxin oxidoreductase. *Science* 267: 1463–1469.

85. Tanner, J.J., R.M. Hecht, and K.L. Krause. 1996. Determinants of enzyme thermostability observed in the molecular structure of *Thermus aquaticus* D-glyceraldehyde-3-phosphate dehydrogenase at 25 Angstroms Resolution. *Biochemistry* 35: 2597–2609.

86. Salminen, T., A. Teplyakov, J. Kankare, B.S. Cooperman, R. Lahti, and A. Goldman. 1996. An unusual route to thermostability disclosed by the comparison of *Thermus thermophilus* and *Escherichia coli* inorganic pyrophosphatases. *Protein Sci* 5: 1014–1025.

87. Liang, J. and K.A. Dill. 2001. Are proteins well-packed? *Biophys J* 81: 751–766.

88. Russell, R.J., J.M. Ferguson, D.W. Hough, M.J. Danson, and G.L. Taylor. 1997. The crystal structure of citrate synthase from the hyperthermophilic archaeon *Pyrococcus furiosus* at 1.9 Å resolution. *Biochemistry* 36: 9983–9994.

89. Senderoff, R.I., S.C. Wootton, A.M. Boctor, T.M. Chen, A.B. Giordani, T.N. Julian, and G.W. Radebaugh. 1994. Aqueous stability of human epidermal growth factor 1-48. *Pharm Res* 11: 1712–1720.

90. Senderoff, R.I., K.M. Kontor, J.K. Heffernan, H.J. Clarke, L.K. Garrison, L. Kreilgaard, G.W. Lasser, and G.B. Rosenberg. 1996. Aqueous stability of recombinant human thrombopoietin as a function of processing schemes. *J Pharm Sci* 85: 749–752.

91. Livingstone, J.R. 1996. Antibody characterization by isothermal titration calorimetry. *Nature* 384: 491–492.

92. Morton, T.A., D.B. Bennett, E.R. Appelbaum, D.M. Cusimano, K.O. Johanson, R.E. Matico, P.R. Young, M. Doyle, and I.M. Chaiken. 1994. Analysis of the interaction between human interleukin-5 and the soluble domain of its receptor using a surface plasmon resonance biosensor. *J Mol Recognit* 7: 47–55.

93. Doyle, M.L., G. Louie, P.R. Dal Monte, and T.D. Sokoloski. 1995. Tight binding affinities determined from thermodynamic linkage to protons by titration calorimetry. *Methods Enzymol* 259: 183–194.

94. Baker, B.M. and K.P. Murphy. 1996. Evaluation of linked protonation effects in protein binding reactions using isothermal titration calorimetry. *Biophys J* 71: 2049–2055.

95. Sigurskjold, B.W., C.R. Berland, and B. Svensson. 1994. Thermodynamics of inhibitor binding to the catalytic site of glucoamylase from *Aspergillus niger* determined by displacement titration calorimetry. *Biochemistry* 33: 10191–10199.

96. Sigurskjold, B.W. 2001. ITC analysis involving cubic binding equations: strong binding and self-associating ligands. Biocalorimetry 2001 Conference, Philadelphia Meeting (e-mail for a copy of algorithm: bwsigurskjold@aki.ku.dk).

97. Brandts, J.F. and L.N. Lin. 1990. Study of strong to ultratight protein interactions using differential scanning calorimetry. *Biochemistry* 29: 6927–6940.

98. Kelley, R.F., M.P. O'Connell, P. Carter, L. Presta, C. Eigenbrot, M. Covarruabias, B. Snedecor, R. Speckart, G. Blank, D. Vetterlein, and C. Kotts. 1993. Characterization of humanized anti-p185[HER2] antibody fab fragments produced in *Escherichia coli*. *ACS Symp Ser* 526: 218–239.

99. Pearce, K.H., Jr., M.H. Ultsch, R.F. Kelley, A.M. de Vos, and J.A. Wells. 1996. Structural and mutational analysis of affinity-inert contact residues at the growth hormone-receptor interface. *Biochemistry* 35: 10300–10307.

100. Lemmon, M.A., Z. Bu, J.E. Ladbury, M. Zhou, D. Pinchasi, I. Lax, D.M. Engelman, and J. Schlessinger. 1997. Two EGF molecules contribute additively to stabilization of the EGFR dimer. *Embo J* 16: 281–294.

101. Clark, C., D. Bast, A.M. Sharp, P.M. St Hilaire, R. Agha, P.E. Stein, E.J. Toone, R.J. Read, and J.L. Brunton. 1996. Phenylalanine 30 plays an important role in receptor binding of verotoxin-1. *Mol Microbiol* 19: 891–899.

102. Gorshkova, I., J.L. Moore, K.H. McKenney, and F.P. Schwarz. 1995. Thermodynamics of cyclic nucleotide binding to the cAMP receptor protein and its T127L mutant. *J Biol Chem* 270: 21679–21683.

103. Spivak-Kroizman, T., M.A. Lemmon, I. Dikic, J.E. Ladbury, D. Pinchasi, J. Huang, M. Jaye, G. Crumley, J. Schlessinger, and I. Lax. 1994. Heparin-induced oligomerization of FGF molecules is responsible for FGF receptor dimerization, activation, and cell proliferation. *Cell* 79: 1015–1024.

104. Wu, J., J. Li, G. Li, D.G. Long, and R.M. Weis. 1996. The receptor binding site for the methyltransferase of bacterial chemotaxis is distinct from the sites of methylation. *Biochemistry* 35: 4984–4993.

105. Fields, B.A., F.A. Goldbaum, W. Dall'Acqua, E.L. Malchiodi, A. Cauerhff, F.P. Schwarz, X. Ysern, R.J. Poljak, and R.A. Mariuzza. 1996. Hydrogen bonding and solvent structure in an antigen-antibody interface. Crystal structures and thermodynamic characterization of three Fv mutants complexed with lysozyme. *Biochemistry* 35: 15494–15503.

106. Tello, D., E. Eisenstein, F.P. Schwarz, F.A. Goldbaum, B.A. Fields, R.A. Mariuzza, and R.J. Poljak. 1994. Structural and physicochemical analysis of the reaction between the anti-lysozyme antibody D1.3 and the anti-idiotopic antibodies E225 and E5.2. *J Mol Recognit* 7: 57–62.

107. Tsumoto, K., Y. Nakaoki, Y. Ueda, K. Ogasahara, K. Yutani, K. Watanabe, and I. Kumagai. 1994. Effect of the order of antibody variable regions on the expression of the single-chain HyHEL10 Fv fragment in *E. coli* and the thermodynamic analysis of its antigen-binding properties. *Biochem Biophys Res Commun* 201: 546–551.

108. Tsumoto, K., Y. Ueda, K. Maenaka, K. Watanabe, K. Ogasahara, K. Yutani, and I. Kumagai. 1994. Contribution to antibody–antigen interaction of structurally perturbed antigenic residues upon antibody binding. *J Biol Chem* 269: 28777–28782.

109. Tsumoto, K., K. Ogasahara, Y. Ueda, K. Watanabe, K. Yutani, and I. Kumagai. 1995. Role of Tyr residues in the contact region of anti-lysozyme monoclonal antibody HyHEL10 for antigen binding. *J Biol Chem* 270: 18551–18557.

110. Starovasnik, M.A., M.P. O'Connell, W.J. Fairbrother, and R.F. Kelley. 1999. Antibody variable region binding by Staphylococcal protein A: thermodynamic analysis and location of the Fv binding site on E-domain. *Protein Sci* 8: 1423–1431.

111. Yarema, K.J. 2001. New directions in carbohydrate engineering: a metabolic substrate-based approach to modify the cell surface display of sialic acids. *Biotechniques* 31: 384–393.

112. Kuschert, G.S., F. Coulin, C.A. Power, A.E. Proudfoot, R.E. Hubbard, A.J. Hoogewerf, and T.N. Wells. 1999. Glycosaminoglycans interact selectively with chemokines and modulate receptor binding and cellular responses. *Biochemistry* 38: 12959–12968.

113. Fromm, J.R., R.E. Hileman, E.E. Caldwell, J.M. Weiler, and R.J. Linhardt. 1995. Differences in the interaction of heparin with arginine and lysine and the importance of these basic amino acids in the binding of heparin to acidic fibroblast growth factor. *Arch Biochem Biophys* 323: 279–287.

114. Fromm, J.R., R.E. Hileman, E.E. Caldwell, J.M. Weiler, and R.J. Linhardt. 1997. Pattern and spacing of basic amino acids in heparin binding sites. *Arch Biochem Biophys* 343: 92–100.

115. Tyler-Cross, R., M. Sobel, L.E. McAdory, and R.B. Harris. 1996. Structure-function relations of antithrombin III-heparin interactions as assessed by biophysical and biological assays and molecular modeling of peptide-pentasaccharide-docked complexes. *Arch Biochem Biophys* 334: 206–213.

116. Quesenberry, M.S., R.T. Lee, and Y.C. Lee. 1997. Difference in the binding mode of two mannose-binding proteins: demonstration of a selective minicluster effect. *Biochemistry* 36: 2724–2732.

117. Novokhatny, V. and K. Ingham. 1997. Thermodynamics of maltose binding protein unfolding. *Protein Sci* 6: 141–146.

118. Seelig, A., T. Alt, S. Lotz, and G. Holzemann. 1996. Binding of substance P agonists to lipid membranes and to the neurokinin-1 receptor. *Biochemistry* 35: 4365–4374.

119. Wieprecht, T., M. Beyermann, and J. Seelig. 1999. Binding of antibacterial magainin peptides to electrically neutral membranes: thermodynamics and structure. *Biochemistry* 38: 10377–10387.

120. Srivastava, D.K., S. Wang, and K.L. Peterson. 1997. Isothermal titration microcalorimetric studies for the binding of octenoyl-CoA to medium chain acyl-CoA dehydrogenase. *Biochemistry* 36: 6359–6366.

121. Zhang, F. and E.S. Rowe. 1994. Calorimetric studies of the interactions of cytochrome c with dioleoylphosphatidylglycerol extruded vesicles: ionic strength effects. *Biochim Biophys Acta* 1193: 219–225.

122. Simon, S.A. and T.J. McIntosh. 1984. Interdigitated hydrocarbon chain packing causes the biphasic transition behavior in lipid/alcohol suspensions. *Biochim Biophys Acta* 773: 169–172.

123. Anchordoquy, T.J. 1999. Nonviral gene delivery systems, Part 1: Physical stability. *Biopharm* 12: 42–48.

124. Lo, Y.L. and Y.E. Rahman. 1995. Protein location in liposomes, a drug carrier: a prediction by differential scanning calorimetry. *J Pharm Sci* 84: 805–814.

125. Papahajopoulos, D., M. Moscarello, E.H. Eylar, and T. Isac. 1975. Effects of proteins on thermotropic phase transitions of phospholipid membranes. *Biochim Biophys Acta* 401: 317–335.

126. Collins, D. and Y. Cha. 1994. Interaction of recombinant granulocyte colony stimulating factor with lipid membranes: enhanced stability of a water-soluble protein after membrane insertion. *Biochemistry* 33: 4521–4526.

127. Rourke, A.M., Y. Cha, and D. Collins. 1996. Stabilization of granulocyte colony-stimulating factor and structurally analogous growth factors by anionic phospholipids. *Biochemistry* 35: 11913–11917.

128. Maneri, L.R. and P.S. Low. 1989. Fatty acid composition of lipids which copurify with band 3. *Biochem Biophys Res Commun* 159: 1012–1019.

129. Lung-Nan, L., J.F. Brandts, M.J. Brandts, and V.V. Plotnikov. 2002. Determination of the volumetric properties of proteins and other solutes using pressure perturbation calorimetry. *Anal Biochem* 302: 144–160.

130. Helper, L.G. 1969. Thermal expansion and structure in water and aqueous solutions. *Can J Chem* 47: 4613.

131. Palma, R. and P.M. Curmi. 1999. Computational studies on mutant protein stability: the correlation between surface thermal expansion and protein stability. *Protein Sci* 8: 913–920.

132. Joly, M. 1965. *A Physico-Chemical Approach to the Denaturation of Proteins.* Academic Press, London.

133. Lumry, R. and H. Eyring. 1954. Conformation changes of proteins. *J Phys Chem* 58: 110–120.

134. Kauzmann, W. and R.B. Simpson. 1953. Kinetics of protein denaturation. III. Optical rotations of serum albumin, β-lactoglobulin, and pepsin in urea solutions. *J Am Chem* 75: 5154–5157.

135. Tanford, C. 1961. *Physical Chemistry of Macromolecules.* John Wiley & Sons, New York.

136. Remmele, R.L., Jr., J. Zhang, V. Dharmavaram, D. Balaban, M. Durst, A. Shoshitaiskvili, and H. Rand. 2004. Scan-rate dependent melting transitions of interleukin-1 receptor (type II): elucidation of meaningful thermodynamic and kinetic parameters of aggregation acquired from DSC simulations (submitted for publication).

137. Cooper, A., M.A. Nutley, and A. Wadood. 2000. Differential scanning microcalorimetry. In *Protein-Ligand Interactions: Hydrodynamics and Calorimetry.* S.E. Harding and B.Z. Chowdhry, editors. Oxford University Press, Oxford, 287–318.

138. Sanchez-Ruiz, J.M., J.L. Lopez-Lacomba, M. Cortijo, and P.L. Mateo. 1988. Differential scanning calorimetry of the irreversible thermal denaturation of thermolysin. *Biochemistry* 27: 1648–1652.

139. Conejero-Lara, F., J.M. Sanchez-Ruiz, P.L. Mateo, F.J. Burgos, J. Vendrell, and F.X. Aviles. 1991. Differential scanning calorimetric study of carboxypeptidase B, procarboxypeptidase B and its globular activation domain. *Eur J Biochem* 200: 663–670.

140. Privalov, G., V. Kavina, E. Freire, and P.L. Privalov. 1995. Precise scanning calorimeter for studying thermal properties of biological macromolecules in dilute solution. *Anal Biochem* 232: 79–85.

141. Kendrick, B.S., J.F. Carpenter, J.L. Cleland, and T.W. Randolph. 1998. A transient expansion of the native state precedes aggregation of recombinant human interferon-gamma. *Proc Natl Acad Sci U.S.A.* 95: 14142–14146.

142. Su, X.Y., A. Li Wan Po, and S. Yoshioka. 1994. A bayesian approach to Arrhenius prediction of shelf-life. *Pharm Res* 11: 1462–1466.

143. Matsuura, J.E., A.E. Morris, R.R. Ketchem, E.H. Braswell, R. Klinke, W.R. Gombotz, and R.L. Remmele, Jr. 2001. Biophysical characterization of a soluble CD40 ligand (CD154) coiled-coil trimer: evidence of a reversible acid-denatured molten globule. *Arch Biochem Biophys* 392: 208–218.

14

Vibrational Spectroscopy in Bioprocess Monitoring

Emil W. Ciurczak

INTRODUCTION

Bioprocesses are often quite sensitive to small changes in physical (temperature, particle size) and chemical (viability of initiator, purity of nutrients) influences. Because of this, seldom are two batches of any bioengineered material exactly the same (different starting materials, length of reaction, percent yield). The end point of the reaction and yield may vary from batch to batch for no easily discernable reason. Traditional methods of analysis are often too slow to allow the operator to make small batch corrections in real time. Spectrometric monitoring is an ideal solution to correct minute variations that could greatly affect the yield and purity of the final product. Modern spectrometers, equipped with powerful, high-speed computers and using advanced algorithms (chemometrics) are capable of analyzing and controlling the most complex reaction mixtures.

HARDWARE AND THEORY

Chromatographic systems have one thing in common: most depend on spectrophotometric detection devices, i.e., ultraviolet (UV), visible, fluorescent, and midrange infrared (MIR) spectrometers. High-performance liquid chromatography (HPLC) has been used to, in essence, purify (separate) the constituents from the matrix, then introduce them to a spectrometer for identification or quantification. One reason that spectrometers were not placed in a production setting was, in part, their fragility. In recent years, vibrational spectrometers have been developed that are robust enough to be placed in a production environment. The first instruments developed for process control were near-IR (NIR) spectrometers, usually filter-based. Recently, traditional MIR spectrometers have become available for industry and there are now a number of Raman instruments on the market.

Advances in instrumentation, such as diode arrays and ruggedized interferometers, have made IR and Raman instruments readily available for process work. NIR hardware has always been used more for production and quality control than laboratory and research work. They, too, have become smaller, faster, more rugged and, in 1980s dollars, less expensive. Explosion-proof enclosures allow close proximity to reactors containing solvents and can be operated in dusty locations (raw material handling situations).

Inexpensive, rugged fiber optics, equipped with every type of probes, make all these instruments amenable to processing measurements in real time. Recently, several manufacturers have been introducing wireless units that are smaller, more rugged, and faster than conventional spectrometers. Many "process" instruments from a decade ago were simply lab instruments fitted with explosion-proof enclosures and placed in factories. They are now being built specifically for the process environment.

Add to these new instruments the breakthroughs in computers (more power, speed, and the ability to go wireless) and software (e.g., chemometrics packages), and the capacity to monitor and affect bioprocesses in real time is more than possible; it is almost commonplace.

APPLICATIONS

Raman

While not the most common tool found in production, Raman is one of the more interesting. Its ability to "not see" water or glass makes it quite useful for most bioprocesses because they are almost exclusively run in water. Raman is based on the scattering of light. Most light shone upon a sample is either transmitted or elastically (without change) scattered. A very small portion of the light is scattered inelastically. This "Raman effect" causes some of the photons to gain energy and some to lose energy. When detected and analyzed, the spectrum resembles that of an IR instrument. The difference is that Raman requires a center of symmetry for a band to be seen, while IR requires a dipole (natural or induced). As a consequence, the spectra are complementary to IR spectra. Because water has a strong, permanent dipole, it is "invisible" to Raman, thus making it ideal for following aqueous reactions.

Unlike IR and NIR, Raman is best when used with a full-scanning device. Because the transfer of energy is so inefficient, the larger throughput of a Fourier transform (FT) device makes it the most logical instrument for the job. There are simpler filter and grating-based instruments on the market, but they are not as versatile.

A trivial yet important application is following ethanol production via a bioprocess. Sivakesava et al.[1] simultaneously measured glucose, ethanol, and the optical cell density of *Saccharomyces cerevisiae* during ethanol fermentation, using an off-line approach. Samples were brought to an instrument located near the fermentation tanks and the measurements made in short order. While they eventually used MIR due to the interfering scatter of the media, they proved that Raman could be used for this application.

Sivakesava et al. also used Raman (as well as FT-IR and NIR) to perform a simultaneous on-line determination of biomass, glucose, and lactic acid in lactic acid fermentation by *Lactobacillus casei*.[2] Partial least squares (PLS) and principal components regression (PCR) equations were generated after suitable wavelength regions were determined. The best standard errors were found to be glucose, 2.5 g/l; lactic acid, 0.7 g/l; and optical cell density, 0.23. Best numbers were found for FT-IR with NIR and Raman being somewhat less accurate (in this experiment).

Similar work was performed by Shaw et al.[3] in 1999 when they used FT-Raman, equipped with a charge coupled device (CCD) detector (for rapid measurements) as an on-line monitor for the yeast biotransformation of glucose to ethanol. An ATR (attenuated total reflectance) cell was used to interface the instrument to the fermentation tank. An Nd:YAG laser (1064 nm) was used to lower fluorescence interference and a holographic notch filter was employed to reduce Rayleigh scatter interference. Various chemometric approaches were explored and are explained in detail in their paper. The solution was pumped continuously through a bypass, used as a "window" in which measurements were taken.

A narrow beam, attainable with Raman, was used by Schuster et al.[4] to characterize the population distribution in *Clostridium* cultures. The technique was applied to the acetone–butanol (ABE) fermentation process in which the solventogenic *Clostridia* go through a complex cell cycle. After drying the cells on calcium fluoride carriers, single-cell spectra were obtained. Cells of different morphology showed different spectra. A number of cell components could be detected and varied in quantity. The approach was seen to be far faster than conventional methods.

Weldon et al.[5] used Raman in the surface-enhanced mode (SERS) to monitor bacterial (*P. acnes*) hydrolysis of triglycerides in lipid mixtures that model sebaceous gland secretions. While technically not a "process" monitor, it paves the way for methods to monitor specific moieties in very complicated matrices.

Similar work was performed by Maquelin et al.[6] in 2000. They used confocal Raman microspectroscopy to obtain spectra directly from microbial microcolonies on solid culture medium. The spectra were obtained after 6 h of culturing and were of most commonly encountered organisms. While depth studies showed varying quantitative levels of success, the qualitative (identification) of various bacterial strains was deemed a success.

Infrared

This "old reliable" technique, (also called "midrange IR" and the "FTIR" of Fourier-Transform IR) is useful in that it supplies the greatest amount of structural and, therefore, qualitative information. A photon of specific energy (in the IR region of the spectrum) is absorbed by a chemical bond within the molecules whose energy corresponds to that photon. The bond is excited and, depending upon the energy imparted, increases its vibration, bending, or rotational speed. This energy is then dispersed to the overall sample via intramolecular "bumping" or contact.

Because the energies of covalent (single, double, or triple) bonds throughout a molecule are quite specific to each pair of atoms, the resultant spectrum may be used to give a quite specific qualitative "picture" of the species present. The amount of light absorbed is also proportional (Beer's law) to the amount of material present and may be used for quantitative analyses.

The vibrational spectrum of a typical organic molecule is quite "rich" (with many peaks that are specific to the chemistry and environment of the species); thus, numerous wavelengths may be used to monitor any individual species. Conversely, using the entire spectrum of a mixture and using chemometric algorithms are useful for the simultaneous determination of several components.

Well-designed reflectance cells have made this strong absorber of water useful in the process vat. Using long path-length gas cells also allows the analyst to monitor head-space gases in order to follow bioreactions.

IR instruments are available in filter-based, grating-based, and FT-based models. The usual approach is to use a full-spectrum model to ascertain the working wavelengths for a particular reaction, then to apply simpler filter instruments to the process. This works where one, two, or three discrete wavelengths may be used for the analysis. If complex, chemometric models are used, and full-spectrum instruments are needed.

Several papers have been published in which, instead of concentrating on specific reactions, the technology was highlighted. One, by Marose et al.,[7] discusses the various optics, fiber optics, and the probe designs that allow *in situ* monitoring. They describe the various optical density probes used for biomass determination: *in situ* microscopy, optical biosensors, and specific sensors such as NIR and fluorescence.

Ehnholt et al.[8] produced a broad paper covering raw materials, and in-process and final-product measurements. While the uses are primarily in the food industry, the rancidity was often caused by microorganisms. One case involved off-flavor materials being produced in drying and curing ovens. Marker compounds (concomitant) released during the breakdown process (of saturated and unsaturated compounds) were nonenal, decenal, and octenone for the unsaturated aldehydes and ketones, and nonanal, decanal, and octanone for the saturated molecules. A 10-m folded path gas cell was used with an FT-IR for measurements down to 1 $\mu g/m^3$.

An entire thesis on using FT-IR/ATR for on-line monitoring and control of bioprocesses was published in 2000,[9] as well as a paper on the preliminary research leading to the book.[10] Most models were based on fermentation of *E. coli*. They used a 15-l fermentor equipped with a diamond composite ATR probe. The values determined (using PLS equations) for glucose and acetate were ±0.98 g/l and ±3.66 g/l, respectively. This was statistically equivalent to the precision of the reference method. Using these equations, the on-line measurements were used to control the levels of glucose (auto feed loop). A set point of 1.0 g/l was maintained at 1.03 ± 0.28 g/l. A set-point of 0.50 g/l was maintained at 0.51 ± 0.14 g/l. The controlled feeding allowed an increase in final cell mass and a decrease in acetate formation when compared with "normal" uncontrolled fed-batch fermentations.

The majority of citations may be broken down into two broad categories: (1) fermentations where the end products (e.g., alcohol) and feed materials are the focus, and (2) reactions where the bacteria–fungi, themselves, are the focus. While, at first blush, these appear quite interwoven, the separation will be seen as helpful in specific research.

In a paper that addresses both these topics, Gordon et al.[11] explain how they followed a corn mixture fermented by *Fusarium moniliforme* spores. They followed the concentrations of starch, lipids, and protein throughout the reaction. The amounts of *Fusarium* and even corn were also measured. A multiple linear regression (MLR) method was satisfactory, with standard errors of prediction (SEP) for the constituents being 0.37% for starch, 4.57% for lipid, 4.62% for protein, 2.38% for *Fusarium*, and 0.16% for corn. It may be inferred from the data that PLS or PCA (principal components analysis) may have given more accurate results.

Fayolle et al.[12] described work done on alcoholic fermentation, wherein they studied the effects of temperature and various calibration methods. The samples were removed and submitted for HPLC and other conventional analyses. The samples were used as is for MIR spectra generation. PLS-1 was used for equation construction. The test RSDs for glucose, fructose, glycerol, and ethanol were, respectively, 12.5, 6.1, 0.6, and 2.9 g/l. The wavelengths assigned to various components were also listed.

Bellon[13] also described fermentation control using MIR. MLR was first used for calibration and measurement of glucose, fructose, and ethanol. When PCR and PLS were applied later, the SEP for glucose went from 6.69 to 5.98 to 4.29 g/l, respectively. For fructose, the MLR to PCA to PLS progression was 7.58 to 5.89 to 6.61 g/l. Depending on the component and level of accuracy needed, an analyst may have to use different algorithms for different components.

Another fermentation process is described by Fayolle et al.[14] In this work, processes producing lactic acid bacteria were studied. Samples were extracted from the reactor and assayed by conventional methods and scanned in the IR. Equations were generated, using PLS, for lactose, galactose, lactic acid, and biomass. The SEP for each of these constituents was, respectively, 3.4, 1.5, 0.9, and 0.9 g/l.

Fayolle et al. used a remote system to monitor on-line fermentations.[15] Both the substrates (glucose, fructose, lactose, and galactose) and the metabolites (ethanol and lactic acid) were monitored. The equations used were built with PLS. The reference method was HPLC. For the alcohol fermentation, glucose, fructose, and ethanol had SEPs (in g/l) of 3.5, 4.5, and 3.8, respectively. For the lactic acid fermentation, the SEPs for lactose, galactose, and lactic acid were 4.1, 1.4, and 2.0, respectively.

Alcoholic and lactic fermentations were both monitored in a paper by Piqué et al.[16] The resultant equations all had R^2 values of at least two nines (> 0.99) when PLS was used to generate them. The standard errors for the components (transmission/ATR) in g/l were lactose, 0.36/0.46; galactose, 0.14/0.57; lactic acid, 0.21/0.95; sugars, 1.81/2.48; glycerol, 0.07/0.09; and ethanol, 0.57/0.98.

Kansiz et al. has published a paper wherein they used MIR and sequential injections to monitor an acetone–butanol fermentation process.[17] In this work, acetone, acetate, n-butanol, butyrate, and glucose were analyzed automatically, using computer-controlled sampling techniques. In this case, gas chromatography was the reference method. The SEPs for the components were acetone, 0.077; acetate, 0.063; butyrate, 0.058; n-butanol, 0.301; and glucose, 0.493 g/l. The authors state that the precision and accuracy of the MIR methods were as good as the reference method.

In another fermentation process, Mosheky et al.[18] reacted *Saccharomyces cerevisiae* with sugars and followed the progress of the fermentation with MIR-ATR. Two PLS models were used: one for sucrose, fructose, and glucose; and one for the ethanol. The authors did not specify SEPs for the experiment, but showed correlation coefficients of better than 0.998 for all analytes.

Crowley et al. performed some interesting work with *Pichia pastoris* in a fed-batch process.[19] The complex mixture was measured using a multibounce attenuated total reflectance (HATR) cell. The authors developed models for glycerol, methanol, and the product, a heterologous protein. The results are reported somewhat differently from "normal" chemometric results. The authors used root-mean square error (RMSE) for the product as a performance index and measured a percent difference for the methanol and glycerol.

The closer the RSME is to zero, the better the performance of the equation; the RMSE for the product was 0.021, an excellent value. The percent difference, where the value is less than or equal to 100, is better the higher it is; it was 90.1 for glycerol and 84.8 for methanol. With no experience in this measurement value, the present author will just report that Crowley and colleagues' claim that it is very good. All the correlation coefficients were 0.91 or better.

Karen Skinner determined the yeast glycogen content for beer production using MIR.[20] While not widely accepted as an important parameter for beer production, glycogen content is most often determined. Because all process applications, even raw material measurements, speed up the production of the final product, this application qualifies as a process measurement. Peak ratios were used to quantify the glycogen content of samples made into potassium bromide (KBr) pellets. Compared with the enzymatic reference method, the IR method had slightly higher coefficients of variance. The correlation coefficients were in the 0.98 range for most of the work.

Some very nice engineering and experimental work was performed by Pollard et al. in measuring a fungal fermentation in real time.[21] The work was performed on a 75-l fermentor and showed detection limits for fructose, glutamate, and proline in production matrices of 0.1, 0.5, and 0.5 g/l, respectively. Glucose and phosphate were measured in a 280-l pilot scale batch with detection limits for both of 0.1 g/l.

The SEP for glucose and phosphate were 0.16 and 1.8 g/l, respectively. SEP values for the other constituents were fructose, 0.44; glutamate, 0.6; and proline, 0.5 g/l. Much detail is covered in the engineering and operation of the monitoring hardware. Calculations are given for determining sensitivity vs. depth penetration of the probe. The paper also lists 44 references and may serve as a good reference source for others engaged in similar work.

Schuster et al. reported work on monitoring a complex acetone–butanol–ethanol (ABE) fermentation system.[22] They looked at the qualitative nature of the biomass as well as the solvents present in the liquid phase. A hierarchical cluster analysis was performed on samples from various times of the fermentation. The clusters were then classified using classical markers and analyses. The resultant table, combining qualitative interpretation and quantitative results, shows an interesting mosaic of the system over time. Total solvents, optical density, and butyric acid are given as numeric values in either absorbance units of g/l.

The qualitative status contained four headings: growth, solvent production, physiological status, and differentiation. The terms used for growth were growing, slowed or ceased, ceasing, and stopped. Solvent production, as might be expected, was just starting, running, and nearly finished. The physiological status used terms such as early solventogenic and late solventogenic. Differentiation was "just starting," "at maximum," "recently started," or "far." This is an ambitious paper.

The subject of a study by Desgranges et al.[23] was biomass estimation by MIR. The IR was supplemented by generated carbon dioxide measurement. Measurements were made on glucosamine, ergosterol, sucrose, and nitrogen. The trouble of correlating with CO_2 production was that fungal respiration continued for 72 h after the other parameter did not change, except for the sporulation rate.

As an interesting adjunct for this last paper, a note by Morrow and Crabb[24] actually follows biomass production by using MIR to measure the production of CO_2 by the reaction. A long path-length gas cell was used to follow the production of the gas and calculate the production of increasing biomass.

Some less-than-rapid methods may also be considered "in process" if they are used to modify or enhance the process. Cheung et al. did some work in examining alterations in extracellular substances during biofilm development of *Pseudomonas aeruginosa* on aluminum plates.[25] The FT-IR spectra, taken over days, showed changes by the fifth day. They believe that structural changes or modifications of the cell envelope had taken place. Structural interpretations were made through band assignments in the IR. Changes in spectra from the bottom, middle, and top of the plates and at the air–water interface suggested particular changes. The mechanisms of attachment differed by a $–COO^-$ interaction in the air–water interface, and by both $–COO^-$ and NH_3^+ groups beneath the water surface.

McGovern et al.[26] analyzed the expression of heterologous proteins in *E. coli* via pyrolysis mass spectrometry and FT-IR. The application was to α2-interferon production. To analyze the data, artificial neural networks (ANN) and PLS were utilized. Because cell pastes contain more mass than the supernatant, these were used for quantitative analyses. Both the MS and IR data were difficult to interpret, but the chemometrics used allowed researchers to gain some knowledge of the process. The authors show graphics indicating the ability to follow production via either technique.

The measurement technique was the crux of a paper by Acha et al.[27] discussing the process of the dechlorination of aliphatic hydrocarbons. An ATR–FTIR sensor was developed to monitor parts per million (ppm) of trichloroethylene (TCE), tetrachloroethylene (PCE), and carbon tetrachloride (CT) in the aqueous effluent of a fixed-bed dechlorinating bioreactor. It was found that the best extracting polymer was polyisobutylene (PIB) as a 5.8 μm film. This afforded detection limits of 2, 3, and 2.5 mg/l for TCE, PCE, and CT, respectively. The construction and operation of the measurement system are detailed in the paper.

In a review paper devoted to the chemometrics discussed in previous citations, Shaw et al.[28] go into details of the many applications of advanced chemometrics to bioprocess monitoring. Multivariate methods and numerous algorithms are explored in this work.

Similarly, in a paper by Sonnleitner,[29] many of the instruments available for monitoring bioprocesses are described. He discusses various conventional and nonconventional monitoring instruments and evaluates them for usefulness and benefits, also discussing their pitfalls.

Near-Infrared

The bands in NIR arise from the same vibrations, bends, and rotations as does the MIR spectrum. The difference is that, whereas MIR represents the first increase in energy (from ground state to first "excited" state), NIR comes from all the higher energy states: ground to second, third, etc., and excited states. Because each successive level is less likely to occur, the energy or absorption is ten times less than the level before it. This means low absorbances and long path lengths are common to NIR.

Instead of glowbars, as used in MIR, tungsten halogen lamps are the sources of light. The detectors are solid-state semiconductors such as lead sulfide (PbS) or indium gallium arsenide (InGaAs). These are orders of magnitude quieter than typical MIR detectors and often more sensitive.

The lower absorptivities and stronger light sources allow for deeper penetration into the process broth: centimeters vs. millimeters. NIR was also the first spectroscopic technique used for process control by large numbers of chemical manufacturers. As a consequence, more equipment and expertise are available for NIR applications than for MIR or Raman.

Because NIR was initially used for food and agriculture products, it has evolved as a technique for complex matrices. Many types of hardware have become available for NIR work: interference filters, gratings, interferometers, diode arrays, and acousto–optic tunable filters. And, as it was originally developed for complex mixtures, chemometrics has been an integral part of any NIR analysis for the last few decades. NIR practitioners are quite comfortable with multivariate equations and development of equations for complex matrices.

Because there are so many similar applications, only representative papers will be discussed here. A thorough search in Chemical Abstracts, Ovid, or any number of databases should find many more good references on this topic. These papers will be loosely broken into categories such as "biomass," although there is inevitable crossover.

Constituents/Metabolites

Some of the earliest work on NIR of bioprocesses was performed on the nutrients and metabolites in a fermentation broth. A "classic" paper (if 1996 is antiquity) was written by Hall et al.[30] on the determination of acetate, ammonia, biomass, and glycerol in *E. coli* fermentations. This early paper used NIR to simultaneously monitor all the above-mentioned parameters. The correlation coefficients were all better than 0.985 with variable SEPs: acetate, 0.7 g/l; ammonia, 7 mM; glycerol, 0.7 g/l; and biomass, 1.4 g/l. While later work with more modern equipment has attained better results, this remains as one of the first. The work was performed "at line" in a cuvette, but rapidly enough to be considered a process measurement.

The fermentation of 1,3-dehydroxyacetone (DHA) was studied with NIR by Varadi et al.[31] The purpose of the study was to determine optimum addition time during a batch-fed fermentation. DHA is used as a raw material for some cosmetics and drugs, and is an additive in food, tobacco, and rubber. The bioconversion of glycerol to DHA was followed with NIR and a series of different algorithms used to evaluate the data. A fast Fourier transform was performed on each spectrum, and several algorithms were applied. The SEPs for glycerol ranged from 1.82 to 0.82%, with PLS giving the best values. For DHA, the values ranged from 0.6 to 0.82% SEP, with PLS again giving the most accurate values. Using discriminant analysis, cluster plots were constructed to identify the conditions of the fermentation (e.g., pH).

A monensin fermentation broth was analyzed by NIR by Forbes et al.[32] Large numbers of statistical evaluations were performed, and a number of spectral interpretations accompany the test. The procedure for estimation of monensin and oil concentrations was not as important as the validation scheme used by the authors. The entire assay procedure was geared to satisfying the FDA and ICH requirements for a validation.

Selectivity, linearity, accuracy, precision, and robustness, as well as long-term drift, were addressed. The logic of the building of the equation used for prediction is detailed. The entire paper transcends a mere research paper and may serve as a blueprint for validation of NIR methods for bioprocessing.

Another paper, this one authored by Chung et al.,[33] discusses multiple measurements via NIR. This work describes simultaneous measurement of glucose, glutamine, ammonia, lactate, and glutamate at levels expected in typical bioreactor levels. While merely a feasibility study, the spectral interpretation and mathematical approach serve as excellent starting points for similar work with actual production processes.

Another variation on the fermentation of alcohol was the work done in a rice vinegar broth by Yano et al.[34] Using second derivative spectra from 1600 to 1760 nm, the ethanol and acetic acid concentrations were easily determined in the culture broth. Gas chromatography was used as the reference method.

Lactic acid fermentation was the topic of a paper by Vaccari et al.[35] In this work, lactic acid, glucose, and biomass were determined over the course of the reaction. The measurements were made in real time, using a bypass pump and flow-through cell for the NIR measurements. Instead of using normal chemometric statistics, the authors used correlation coefficients, mean of differences, standard deviation, student's t-test, and the student test parameter of significant difference to evaluate the results. Under these restrictions, the results appeared fairly good, with the biomass having the best set of statistics.

The simultaneous measurement of multiple components in a high-density recombinant *E. coli* production process was reported by Macaloney et al.[36] This well-written paper completely describes the methodology by which glycerol, acetate, ammonium, and dry weight were predicted. The manner in which these equations may be used as real-time control and analysis mechanisms is demonstrated. For comparison, the SEP values for glycerol, acetate, ammonium, and dry weight were, respectively, 0.6 g/l, 1.9 g/l, 13 mM, and 1.3 g/l. From a simplicity perspective, the equations were basic MLR in nature.

One interesting paper addresses amino acid production.[37] The authors describe a fed-batch process for production of amino acids, such as L-lysine (from *Corynebacterium glutamicium*) and L-threonine (from *Escherichia coli*). For the fermentation broth of the L-lysine, the optical density, ammonium, and L-lys were measured. For L-threonine, OD, ammonium, and L-thr were measured. For all materials, the values were deemed acceptable and comparable with the reference methods.

The last paper in this category describes monitoring of a submerged filamentous bacterial cultivation.[38] Arnold et al. studied a 12-l stirred tank where a strain of *Streptomyces fradiae* was used for fermentation. They followed ammonium, methyl oleate, glucose, and glutamate concentrations over time in the production of tylosin.

Both simple linear regression (SLR) and PLS equations were tested for accuracy and precision. The SEP values for methyl oleate, glucose, glutamate, and ammonium were, respectively, 0.65 (and 0.68, 0.57; three models), 1.35, 0.61, and 0.016. For an immensely complex system, these values are outstanding.

Biomass

In many cases, the amount of final product is the only thing that interests the production manager. In cases where the ratio of starting materials or the process conditions are not easily changed, knowing when the process is complete is critical. Only a few papers specifically mentioned biomass in their titles.

One, by Macaloney et al.,[39] discusses biomass and glycerol in an *E. coli K12* fermentation process. MLR was used wherein discrete wavelengths were identified to have relationships with the constituents in the broth. The correlation and SEP for the glycerol were 0.98 and 0.3 g/l, and for the biomass they were 0.99 and 0.2 g/l.

Vaidyanathan et al. studied the spectra of biomass for various microorganisms.[40] They studied *E. coli* (a recombinant strain from Eli Lilly and Co.), *Streptomyces fradiae C373.5* (Lilly), *Penicillium chrysogenum* (SKB), *Aspergillus niger B1-D,* and *Aureobasidiium pullulans IMI 145194.* All the peak wavelengths for all the organisms are listed in tables. The correlation coefficients, standard errors, etc., for all five experiments are listed but are too numerous to outline here. Suffice it to say that reasonable correlation and accuracy were achieved in this work.

One interesting paper by Suehara et al. used NIR to measure the cell mass in solid cultures of mushroom.[41] Because mushrooms grow in solid matrices, spectroscopic analyses are sometimes difficult to perform. With coffee grounds as the main medium, reflection NIR was used for the determinations. The correlation between the glucosamine (analyte determined) predicted and that found by the conventional method was 0.992 with an SEP of 0.346 mg/g.

A paper by Kasprow et al.[42] is important because it shows the realization that starting materials need to be analyzed on a routine basis just as with reaction products. Kasprow et al. discuss the correlation of fermentation yield with the yeast extract composition as seen by NIR. Using PLS for the correlations, models were constructed with a correlation of 0.996 and a standard error of 1.16 WSW. The authors used the models to predict yields, using different lots of yeast, and were quite satisfied with the results.

Hardware and Software

A number of authors wrote primarily to emphasize either the equipment or the algorithms used in their work. Hammond and Brookes wrote a paper focusing on the technique, not necessarily the results, of any single fermentation.[43] In this work, the authors discuss, of course, the experimental work performed and the results, but emphasized the definitions of terms used throughout NIR and the equations by which they are determined. The hardware used is diagrammed and the concept of "locally weighted" regression models is introduced. The actual biological activities followed included antibiotic production and enzyme activity in protease reactions. Detailed illustrations accompany the normal graphic data representations and tables. This decade-old paper holds up well.

While the results of the work performed by Vaidyanathan et al.[44] are scientifically useful, the thorough treatment the authors give to chemometrics is excellent. There is a detailed description of principal components (PCA) with a number of pictures of loadings to help explain the process. Using PC scores, three-dimensional representations of the samples are shown. This is a good paper for someone just beginning to use chemometrics.

In a series of controlled experiments, Dosi et al. used the conversion of glucose to lactic acid as a model for the potential of controlling (automatically by computer) the concentrations of the constituents.[45] Nice schematics for the reactor setup and connections to the computer/NIR spectrometer/microfiltration unit, etc., are shown. Six cases are described, each using a conventional batch process. Transition from batch mode to automatic was triggered by predefined criteria such as degree of substrate conversion or biomass concentration. Control charts and comparisons of NIR data with conventional assays are given for all six cases.

Another paper focusing on instrumentation was published by Cavinato et al.[46] The difference between this paper and the previous publications is that the authors utilized the short-wavelength NIR region from 700 to 1100 nm with a diode-array instrument in lieu of the standard holographic grating-based scanning equipment applied in the others. The ethanol fermentation that was followed allowed the shorter wavelength region (Vis/NIR) between 400 and 1100 nm to be evaluated. The authors concluded that there was less scattering and better penetration (due to lower absorptivities), allowing longer path lengths. These longer path lengths led to more representative samples that, in turn, gave better statistics. Yeast cytochromes were also followed during fermentation with equal success.

Yeung et al. addressed the question of whether or not pull samples gave as good equations as samples measured on-line.[47] The so-called add-back and process-stream samples were compared for statistical robustness in the equations they produced and their performance in predicting values in an on-line reaction. Materials such as biomass (optical density), protein, and RNA were measured by both techniques.

It was found that fewer PLS factors were needed for process stream samples to yield a decent model. The correlation coefficients were stronger and the SEPs lower in every case. One example is protein; add-back $r^2 = 0.891$ using four PLS factors, whereas the process-stream samples enabled a three-factor PLS model to be generated with an $r^2 = 0.915$. The respective SEP for each model was add-back = 4.351; process = 2.343. This paper seems to confirm the idea that samples taken during a process are more representative of that process.

Similar results were reported by Lennox et al.,[48] only using PCA instead of PLS. They give a very good explanation of PCA for any reader not previously acquainted with the technique. They then explain how PCA may be used in process analyses. The authors also mention, in passing, PLS and ANN. Some of the considerations addressed in the paper are the application of PCA and on-line considerations (e.g., data processing and missing data). Considerably less space is devoted to PLS and ANN applications, but they are addressed. An appendix shows the calculations used for confidence limits of T^2, using F-distribution confidence and limits (for SPE chart) based on chi-squared distribution.

An overview-type paper by Crowley et al.[49] addresses some differences between microbial and mammalian cultivations. The paper suggests different analytical approaches based on the length of reactions and concentration and number of products and/or reactants in the vessel. Diagrams for reactor-monitoring systems are shown, and methods for making control loops are discussed.

Two papers by Arnold et al.[50–51] were published in 2002 and 2003. These were complementary and described the method development and implementation strategies for fermentation monitoring and control. The first part, method development, goes into the spectral interpretation and actual equation-building for *E. coli* fermentation. Ammonium, glucose, glutamate, and methyl oleate are the constituents determined. The rationale for the wavelengths chosen is given, and the models built using MLR and PLS techniques are highlighted.

The second paper goes more into the hardware and programming needed for control loops in a production setting. Four fermentation models are used for this portion: *E. coli* (unicellular), *S. fradiae* (filamentous), *Pichia pastoris* (unicellular), and CHO-K1 (animal cell line). Fermentation profiles are shown and strategies are discussed.

Hagman and Sivertsson discussed the work performed at Pharmacia and Upjohn[52] on monitoring and controlling bioprocesses. They followed the protein production-derived form Chinese hamster cells (CHO-cells) in a 500-l reactor over a 3-month period. The diagrams of the flow cell and pumping/NIR system are displayed, and the logic behind the work outlined. External and on-line calibrations were performed.

One conclusion stated was that the more complex the matrix, the more samples were needed for calibration. PLS was used to calibrate for chemical and biological activities, and PCA was used to identify the qualitative occurrences within the reactor. Thus, numerical values and qualitative information could be gleaned simultaneously.

A number of authors emphasized the algorithms or experimental schemes used for their work over the actual numerical values calculated. These papers were meant to demonstrate the concepts in a global manner rather than to give a series of data for a single reaction.

Riley et al.[53] wrote about an "adaptive calibration scheme" for nutrients and by-products in insect cell bioreactors. In their work, the authors used both actual and synthetic samples to generate equations. In order not to be misled by unrelated correlations (combinations of analytes that coincidentally parallel one another), the synthetic samples contained ratios so that only the analytes in question would be correlated upon by the wavelengths chosen. Ranges and compositions were extended by the synthetic mixtures in order to cover all combinations possible in a production reactor. Spiked "spent" media were also used to build equations.

Using large- and small-volume samples, mixed with synthetically produced samples, excellent values were developed. Samples were analyzed for alanine, glucose, glutamine, and leucine in samples removed from an Sf-9 insect cell culture bioreactor. It was seen that purely synthetic standards led to poor predictions of actual runs. The best results were achieved from mixed — actual and synthetic — samples.

A related paper by Riley et al.[54] considers the work as a matrix-enhanced calibration procedure. Using glucose and glutamine in insect cell culture medium as a model, a complex mixture of over 20 components was analyzed. Levels of analytes, not in the expected concentrations, were introduced in the complex matrix. Using fresh and "spent" media as background, analytes were spiked at varying levels to produce orthogonal (nonparallel or noninterdependent) sample levels. These were used to produce equations for process evaluation. The best number and combination of spent and fresh samples were determined by evaluating the SEP and MPE (mean percent error) for each combination and resulting equation.

A relatively short but interesting paper by Arnold et al.[55] suggests different calibrations for different segments of a bioreaction. Using the cultivation of *S. fradiae* and monitoring the oils and tylosin, the reaction was followed for 150 h. It was determined that the extensive matrix changes over the entire reaction time would make calibration difficult. The process was broken into (1) 0 to 50 h, (2) 50 to 100 h, and (3) 100+ h.

The rationale for this is seen in some of the trends throughout these times: for instance, oil in (1) has a high concentration, in (2) it is decreasing, whereas in (3) it is depleted. A gas such as CO_2 is insignificant in (1), increasing in (2), and decreasing in (3). Thus, it makes sense to divide the reaction into three segments and develop an equation for constituents for each. Their results justified the decision.

Vaidyanathan et al. performed a critical evaluation of models developed for an industrial submerged bioprocess for antibiotic production.[56] Both transmission and reflection techniques were employed on a number of reactions. Oil and tylosin were the analytes followed throughout the reaction.

A large number of SLRs (Beer's law), MLR, and PLS equations were generated and evaluated. Various wavelength regions (omitting the major water peaks around 1420 and 1940 nm) were investigated as were the number of wavelengths or factors used in each equation. (Obviously, SLR, by definition, uses only one wavelength.) A large number of graphs are included to show the number of validation trials run by the authors. The authors conclude that careful model building does lead to equations that may be used in a production setting.

ANNs were discussed in detail in a paper by Li and Brown.[57] This group from the University of Rhode Island used both the more conventional PLS algorithm and ANN to compare which could be construed as better. While not going into detail as to how ANN actually works, the comparison is a nice piece of experimental work. The glucose-to-ethanol fermentation (by *Saccharomyses cerevisiae*) used here was somewhat trivial, but served well for this work.

The PLS equations were based on spectral features of the glucose and ethanol. The PLS equations (seven factors) produced an average SEP of 0.19% for glucose and 0.11% for ethanol. The ANN equations tended to have a bit lower SEP for all the experiments. The ANN algorithm used is briefly described in the paper.

The deconvolution of spectra is the topic of a paper by Vaidyanathan et al.[58] The authors use the somewhat complex matrix of mycelial bioprocesses for a model. Throughout the reactions of five different unicellular microorganisms, biomass, external proteins, penicillin, T-sugars, and ammonium were measured vs. time. Each analyte was justified from spectral interpretation. The spectral range used was from 700 to 2500 nm, with specific regions used for each experiment.

Many spectra are shown to support the authors' conclusions, as are tables. The rationale for using particular peaks is accompanied by R^2, SEC, and SEP for the equation developed from that wavelength or region. The sheer number of data prohibits even a summary in this chapter. The reader is directed to the paper for actual values. The soundness of the developed equations does seem to justify the authors' assignments of wavelengths to various components.

CONCLUSIONS

Having seen the number of papers devoted to bioprocess analyses utilizing vibrational spectroscopy, it cannot be considered an experimental tool any longer. Manufacturers are responding to pressure to make their instruments smaller, faster, explosion-proof, lighter, less expensive, and, in many cases, wireless. Processes may be followed in-line, at-line, or near-line by a variety of instruments, ranging from inexpensive filter-based to robust FT instruments. Raman, IR, and NIR are no longer just subjects of feasibility studies: they are ready to be used in full-scale production.

References

1. Sivakesava, S., Irudayaraj, J., and Demirci, A., *J. Ind. Microbiol. Biotech.*, **26**, 185–190 (2001).
2. Sivakesava, S., Irudayaraj, J., and Demirci, A., *Process Biochem.*, **37**, 371–378 (2001).
3. Shaw, A.D., Kaderbhal, N., Jones, A., Woodward, A.M., Goodcare, R., Rowland, J.J., and Kell, D.B., *Appl. Spectrosc.*, **53**(11), 1419–1428 (1999).
4. Schuster, K.C., Mertens, F., and Gapes, J.R., *Vib. Spectrosc.*, **19**(2), 467–477 (1999).
5. Weldon, M.K., M.M., Harris, A.B., and Stoll, J.K., *J. Lipid Res.*, **39**, 1896–1899 (1998).

6. Maquelin, K., Choo-Smith, L.P., van Vreeswijk, T., Endtz, H.P., Bennett, R., Bruining, H.A., and Puppels, G.J., *Anal. Chem.*, **72**, 12–19 (2000).

7. Marose, S., Lindemann, C., Ulber, R., and Scheper, T., *Trends Biotechnol.*, **17**(1), 30–34 (1999).

8. Ehntholt, D.J., Tayor, R.F., and Miseo, E.V., *ISA Trans.*, **31**(4), 67–73 (1992).

9. Doak, D.L., The use of FTIR-ATR spectroscopy for on-line monitoring and control of bioprocesses, 218 pp. Avail: UMI, Order No. DA9982862 From: *Diss. Abstr. Int.*, B 2001, **61**(8), 4277 (2000).

10. Doak, D.L. and Phillips, J.A., *Biotechnol. Prog.*, **15**(3), 529–539 (1999).

11. Gordon, S.H., Green, R.V., Wheeler, B.C., and James, C., *Biotechnol. Adv.*, **11**(3), 665–675 (1993).

12. Fayolle, P., Picque, D., Perret, B., Latrille, E., and Corrieu, G., *Appl. Spectrosc.*, **50**(10), 1325–1330 (1996).

13. Bellon, V., *Sensor. Actuat., B,* **12**, 57–64 (1993).

14. Fayolle, P., Picque, D., and Corrieu, G., *Vib. Spectrosc.*, **14**(2), 247–252 (1997).

15. Fayolle, P., Picque, D., and Corrieu, G., *Food Control,* **11**(4), 291–296 (2000).

16. Piqué, D., Lefier, D., Grappin, R., and Corrieu, G., *Anal. Chim. Acta,* **279**(1), 67–72 (1993).

17. Kansiz, M., Gapes, J.R., McNaughton, D., Lendl, B., and Schuster, K.C., *Anal. Chim. Acta*, **438**(1–2), 175–186 (2001).

18. Mosheky, Z.A., Melling, P.J., and Thomson, M.A., *Spectroscopy* **16**(6), 15–18, 20 (2001).

19. Crowley, J., McCarthy, B., Nunn, N.S., Harvey, L.M., and McNeil, B., *Biotechnol. Lett.*, **22**(24), 1907–1912 (2000).

20. Skinner, K.E., *J. Am. Soc. Brew. Chem.*, **54**(2), 71–75 (1996).

21. Pollard, D.J., Buccino, R., Conners, N.C., Kirschner, T.F., Olewinski, R.C., Saini, K., and Salmon, P.M., *Bioprocess Biosyst. Eng.*, **24**(1), 13–14 (2001).

22. Schuster, K.C., Mertens, F., and Gapes, J.R., *Vib. Spectrosc.*, **19**, 467–477 (1999).

23. Desgranges, C., Georges, M., Vergoignan, C., and Durand, A., *Appl. Microbiol. Biotechnol.*, **35**(2), 206–209 (1991).

24. Morrow, R.C. and Crabb, T.M., *Adv. Space Res.*, **26**, 289–298 (2000).

25. Cheung, H.Y., Sun, S.Q., Sreedhar, B., Ching, W.M., and Tanner, P.A., *J. Appl. Microbiol.*, **89**(1), 100–106 (2000).

26. McGovern, A.C., Ernill, R., Kara, B.V., Kell, D.B., and Goodacre, R., *J. Biotechnol.*, **72**(3), 157–167 (1999).

27. Acha, V., Meurens, M., Naveau, H., and Agathos, S.N., *Biotechnol. Bioeng.*, **68**(5), 473–487 (2000).

28. Shaw, A.D., Winson, M.K., Woodward, A.M., McGovern, A.C., Davey-Kaderbhai, N., Broadhurst, D., Gilbert, R.J., Taylor, E.M., Goodacre, D.B., Alsberg, B.K., and Rowland, J.J., *Adv. Biochem. Eng. Biotechnol.*, **66**, 83–113 (2000).

29. Sonnleitner, B., *Adv. Biochem. Eng. Biotechnol.*, **66**, 1–64 (2000).

30. Hall, J.W., M.B., McNeil, B., Rollins, M.J., Draper, I., Thompson, B.G., and Macaloney, G., *Appl. Spectrosc.*, **50** 102–108 (1996).

31. Varadi, M., Toth, A., and Rezessy, J., in *Making Light Work: Advanced Near-Infrared Spectroscopy International Conference for Near-Infrared Spectroscopy*, Eds: Murray, I. and Cowe, I.A., Meeting Date 1991, 382–386. VCH, Weinheim, Germany (1992).

32. Forbes, R.A., Luo, M.Z., and Smith, D.R., *J. Pharm. Biomed. Anal.*, **25**, 239–256 (2001).
33. Chung, H., Arnold, M.A., Rhiel, M., and Murhammer, D.W., *Appl. Spectrosc.*, **50**(2), 270–276 (1996).
34. Yano, T., Aimi, T., Nakano, Y., and Tamai, M., *J. Ferment. Bioeng.*, **84**(5), 461–465 (1997).
35. Vaccari, G., Dosi, E., Campi, A.L., Gonzolez-Vera, A., Matteuzzi, D., and Manto-vani, G., *Biotechnol. Bioeng.*, **43**(10), 913–917 (1994).
36. Macaloney, G., Hall, J.W., Rollins, M.J., Draper, I., Anderson, K.B., Preston, J., Thompson, B.G., and McNeil, B., *Bioprocess Eng.*, **17**, 157–167 (1997).
37. Schmidt, S., Kircher, M., Kasala, J., and Locaj, J., *Bioprocess. Eng.*, **19**(1), 67–70 (1998).
38. Arnold, S.A., Crowley, J., Vaidyanathan, S., Matheson, L., Mohan, P., Hall, J., Harvey, L.M., and Mcneil, B., *Enzyme Microbial Technol.*, **27**(9), 691–697 (2000).
39. Macaloney, G., Hall, J.W., Rollins, M.J., Draper, I., Thompson, B.G., and McNeil, B., *Biotechnol. Tech.*, **8**(4), 281–286 (1994).
40. Vaidyanathan, S., McNeil, B., and Macaloney, G., *Analyst*, **124**(2), 157–162 (1999).
41. Suehara, K.I., Nakano, Y., and Yano, T., *J. Near-Infrared Spectrosc.*, **6**(1–4), 273–277 (1998).
42. Kasprow, R.P., Lange, A.J., and Kirwan, D.J., *Biotechnol. Prog.*, **14**(2), 318–325 (1998).
43. Hammond, S.V. and Brookes, I.K., Near infrared spectroscopy — a powerful technique for at-line and on-line analysis of fermentations, *Harnessing Biotechnology for the 21st Century: Proceedings of the 9th International Biotechnology Symposium Exposition,* Eds: Ladisch, M.R. and Bose, A., pp. 325–333. American Chemical Society, Washington, D.C. (1992).
44. Vaidyanathan, S., Arnold, S., Matheson, L., Mohan, P., McNeil, B., and Harvey, L.M., *Biotechnol. Bioeng.*, **74**(5), 376–388 (2001).
45. Dosi, E., Vaccari, G., Campi, A.L., and Mantovani, G., *Near-Infrared Spectroscopy Future Waves: Proceedings of the 7th International Conference for Near-Infrared Spectroscopy,* Eds: Davies, Anthony, M.C. and Williams, P.C. (1996), Meeting Date 1995, pp. 249–254. NIR Publications, Chichester, UK. (1997).
46. Cavinato, A.G., Mayes, D.M., Ge, Z., and Callis, J.B., *Proceedings of Frontiers in Bioprocessing 2*, Meeting Date 1990, pp. 90–98. Eds: Todd, P., Sikdar, S.K., and Bier, M. American Chemical Society, Washington, D.C. (1992).
47. Yeung, K.S.Y., Hoare, M., Thornhill, N.F., Williams, T., and Vaghjiani, J.D., *Biotechnol. Bioeng.*, **(63)** 6, 684–693 (1999).
48. Lennox, B., Montague, G.A., Hiden, H.G., Kornfeld, G., and Goulding, P.R., *Biotechnol. Bioeng.*, **74**: 125–135 (2001).
49. Crowley, J., A.S., Harvey, L.M., and McNeil B., *Eur. Pharm. Rev.*, **5**(3), 134–138 (2000).
50. Arnold, S.A., Harvey, L.M., McNeil, B., and Hall, J.W., *Biopharm. Int.*, **15**(11), 26–34 (2002).
51. Arnold, S.A., Harvey, L.M., McNeil, B., and Hall, J.W., *Biopharm. Int.*, **16**(1), 47–49 (2003).
52. Hagman, A. and Sivertsson, P., *Process Contr. Qual.*, **11**(2), 125–128 (1998).
53. Riley, M.R., Arnold, M.A., Murhammer, D.W., Walls, E.L., and DelaCruz, N., *Biotechnol. Prog.*, **14**(3), 527–533 (1998).

54. Riley, M.R., Arnold, M.A., and Murhammer, D.W., *Appl. Spectrosc.,* **52**(10), 1339–1347 (1998).
55. Arnold, S.A., Matheson, L., Harvey, L.M., and McNeil, B., *Biotechnol. Lett.,* **23**(2), 143–147 (2001).
56. Vaidyanathan, S., Arnold, A., Matheson, L., Mohan, P., Macaloney, G., McNeil, B., and Harvey, L.M., *Spectrosc. Biotechnol. Prog.,* **16**(6), 1099–1106 (2000).
57. Li, Y. and Brown, C., *J. Near-infrared Spectrosc.,* **6**, 3–17 (1998).
58. Vaidyanathan, S., Harvey, L.M., and McNeil, B., *Anal. Chim. Acta,* **428**(1), 41–59 (2001).

Index

A

AAA. *See* Amino acid analysis

Absorbance detection techniques, in reversed-phase liquid chromatography, 52–53

    fluorescence, 52

    mass spectrometry, 52–53

    multiangle light-scattering detectors, 52

Absorption pharmacokinetics studies, 8

Affinity chromatography, reversed-phase liquid chromatography, compared, 59–61

Aliphatic hydrocarbons, dechlorination of, vibrational spectroscopy, 390

Amb 1 immunostimulatory complexes, in size exclusion chromatography, 106

Amines, fluorescamine reaction of, 18

Amino acid analysis, fluorescence, 18

Anaphylaxis, peanut-allergic individual, size exclusion chromatography, 106

Animal toxicity, testing on, 8

Anomalous behavior of proteins in reversed-phase liquid chromatography, 51–52

Apothecary development, overview, 1

Applications for protein assays, 19–21

Arrhenius plots, 369–370

B

Base deactivated columns, 37

BCA method, colorimetric assays, 15–16

Beer's law, 14, 17, 396

Bioanalytical laboratories, function of, overview, 6

Biologic License Application, 9

Biopharmaceutical development, 1–12

    capillary electrophoresis, 161–226

    enzyme-linked immunosorbent assay, 279–304

    high-performance hydrophobic interaction chromatography, 81–94

    high-performance ion-exchange chromatography, 67–80

    mass spectrometry, 227–278

    microcalorimetric approaches, 327–382

    nuclear magnetic resonance spectroscopy, 305–326

    protein measurement assay, 13–26

    reversed-phase liquid chromatography, 27–66

    size exclusion chromatography, 95–112

    slab gel electrophoresis, 113–160

    vibrational spectroscopy, 383–400

Bio-Rad DC protein assay, 15

Bradford method, colorimetric assays, 16